Introduction to Quantum Control and Dynamics

Advances in Applied Mathematics

Series Editors:
Daniel Zwillinger, H. T. Banks

Handbook of Peridynamic Modeling
Edited by Florin Bobaru, John T. Foster, Philippe H Geubelle, Stewart A. Silling

Linear and Complex Analysis for Applications
John P. D'Angelo

Advanced Engineering Mathematics with MATLAB, 4th Edition
Dean G. Duffy

Quadratic Programming with Computer Programs
Michael J. Best

Introduction to Radar Analysis
Bassem R. Mahafza

CRC Standard Mathematical Tables and Formulas, 33rd Edition
Edited by Daniel Zwillinger

The Second-Order Adjoint Sensitivity Analysis Methodology
Dan Gabriel Cacuci

Operations Research
A Practical Introduction, 2nd Edition
Michael Carter, Camille C. Price, Ghaith Rabadi

Handbook of Mellin Transforms
Yu. A. Brychkov, O. I. Marichev, N. V. Savischenko

Advanced Mathematical Modeling with Technology
William P. Fox, Robert E. Burks

Handbook of Radar Signal Analysis
Bassem R. Mahafza, Scott C. Winton, Atef Z. Elsherbeni

The Geometry of Special Relativity, Second Edition
Tevian Dray

Introduction to Quantum Control and Dynamics, Second Edition
Domenico D'Alessandro

https://www.routledge.com/Advances-in-Applied-Mathematics/book-series/CRC
ADVAPPMTH?pd=published,forthcoming&pg=1&pp=12&so=pub&view=list

Introduction to Quantum Control and Dynamics

Second Edition

Domenico D'Alessandro

CRC Press
Taylor & Francis Group
Boca Raton London New York

CRC Press is an imprint of the
Taylor & Francis Group, an **Informa** business

A CHAPMAN & HALL BOOK

Second edition published 2022
by CRC Press
6000 Broken Sound Parkway NW, Suite 300, Boca Raton, FL 33487-2742

and by CRC Press
2 Park Square, Milton Park, Abingdon, Oxon, OX14 4RN

First edition published by Chapman and Hall/CRC Press 2007

CRC Press is an imprint of Taylor & Francis Group, LLC

Library of Congress Cataloging-in-Publication Data
Library of Congress Control Number: 2020951718

ISBN: 978-0-367-50790-9 (hbk)
ISBN: 978-1-003-05126-8 (ebk)
ISBN: 978-0-367-76764-8 (pbk)

Typeset in CMR10
by KnowledgeWorks Global Ltd.

DOI: 10.1201/9781003051268

Contents

Preface to the First Edition

In many experimental setups, an electromagnetic field interacts with a system, such as an atom, a nucleus or an electron, whose dynamics follows the laws of quantum mechanics. While it is appropriate to treat the latter system as a quantum mechanical one, the electromagnetic field can often be treated as a classical field, giving predictions that agree with macroscopic observation. This is the so-called *semiclassical approximation*. With technological advances of the last decades, it is nowadays possible to shape the interacting electromagnetic field almost at will. This leads to a point of view which considers these experiments as *control experiments* where the electromagnetic field plays the role of the control, and the atom, nucleus or electron, or any other quantum mechanical system, is the object of the control.

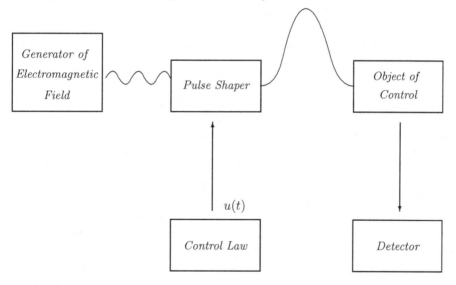

FIGURE 0.1: Scheme of a quantum control experiment.

Along with the above-mentioned technological opportunities, in the last two or three decades many more applications of quantum mechanical systems have emerged. These applications require the manipulation of quantum systems in a precise way. They include *control of chemical reactions, quantum computation* and *quantum metrology*, which is the science that studies how to measure physical quantities using quantum systems. The introduction of control theory in these sciences is important not only in the concrete implementation of experiments but also because it gives new insight into several theoretical questions at the heart of quantum physics. These include questions arising in the newly developed *quantum information theory*.

The state of a quantum system $\vec{\psi}$ is represented by a vector in a Hilbert space $\mathcal{H}$ whose dynamics follows a linear differential equation called the *Schrödinger equation*,

$$i\frac{d}{dt}\vec{\psi} = H(u(t))\vec{\psi}, \qquad \vec{\psi}(0) = \vec{\psi}_0. \qquad (1)$$

Here $H(u)$ is a linear Hermitian operator on the Hilbert space $\mathcal{H}$ which we have made a function of a control $u = u(t)$. This models the external action on the system and, in most cases we will treat, has the physical meaning of an interacting electromagnetic field. The state vector $\vec{\psi}$ is constrained to have norm equal to one and therefore it can be seen as a point on the complex sphere of radius one in $\mathbb{C}^n$, if the Hilbert space $\mathcal{H}$ has dimension n. The solution of (1) has the form

$$\vec{\psi}(t) = X(t)\vec{\psi}_0, \qquad (2)$$

where $X(t)$ is the solution of the *Schrödinger operator equation*

$$i\dot{X} = H(u(t))X, \qquad X(0) = 1. \qquad (3)$$

Here 1 denotes the identity operator. In the finite-dimensional case, where the dimension of the Hilbert space is n, X is an element of the Lie group of unitary matrices $U(n)$.

The first question one faces, when dealing with an experimental apparatus, concerns the type of experiments that can be performed. If the only available degree of freedom is the control, the corresponding mathematical question concerns what type of state transfer can be obtained, with (2) and (3), by opportunely varying the control, $u = u(t)$. This is a question of *controllability* for system (3). This question can be answered by applying results on the controllability of right invariant control systems on Lie groups. In particular, the *Lie Algebra Rank Condition* says that (under some weak assumptions on the set of the possible control functions) the set of values reachable from the identity for system (3) is the connected Lie group corresponding to the Lie algebra generated by the matrices $iH(u)$, as u varies in the set of possible values for the controls. This fundamental result is nothing but an application of general results of geometric control theory, which were developed by several authors in the seventies. Not only, it allows us to give an answer to a very natural question in quantum control experiments but also it opens up the possibility of applying results and ideas of Lie group and Lie algebra theory to the analysis and control of quantum systems.

This book is an introduction to the analysis and control of quantum dynamics which emphasizes the application of Lie algebra and Lie group theory. It was developed from lecture notes written during a course given at Iowa State University in the spring semester of 2004. This course was intended for advanced undergraduate and beginning graduate students in mathematics, physics and engineering. These lecture notes were expanded during a sabbatical visit at the Institute of Quantum Electronics at ETH Zurich in the fall 2004 and in the following two years. The book is written for an audience with a general mathematics and physics knowledge and introduces many concepts from the beginning. In particular, it presents introductory notions of quantum mechanics. The emphasis is on the mathematical concepts and the physics and applications are presented mostly at the level of fundamental physics rather than of precise implementation of given control experiments. However, a chapter at the end of the book presents some physical implementations of quantum control and dynamics from a more applied point of view. The subject is interdisciplinary in nature as it is at the intersection of several fields: quantum mechanics, control theory, Lie group theory, linear algebra, etc.

The book is organized as follows. Chapter 1 introduces the basics of quantum mechanics, with an emphasis on dynamics and a quantum information theory point of view. This chapter could be skipped by a reader already familiar with quantum mechanics. In Chapter 2, from fundamental physics, a class of models for quantum control systems is derived. This class describes many experimental situations and is the main class of models considered in the book. In Chapter 3, we study the controllability of quantum systems and, in the process, we introduce many concepts of Lie algebra and Lie group theory as well as Lie transformation groups. Chapter 4 deals with observability of quantum systems and the related problem of quantum state determination and measurement. Chapter 5 introduces Lie group decompositions as a tool to analyze dynamics and design control algorithms. In Chapters 6 and 7 we describe more methods for control. In particular, in Chapter 6 we deal with optimal control theory applied to quantum systems. Several optimal control design strategies are described there giving both analytic and numerical algorithms for optimal control. Chapter 7 presents several more methods for control, including Lyapunov control, adiabatic and STIRAP methods. All the control methods described in this book are *open loop*, i.e., the control law is designed before the experiment and then used during the experiment without modifications. Chapter 8 presents some topics in quantum information theory that are relevant to the study of the dynamics of quantum systems, in particular concerning *entanglement* and *entanglement dynamics*. Chapter 9 discusses physical set-ups of implementations and applications of quantum control and dynamics. Chapters 1, 2, 3, 5 and 8 should be read in the given order. The remaining chapters could be, for the most part, read independently after the first three chapters.

In writing this book I have benefited from the interaction with many people. In particular, I would like to remember the late Mohammed Dahleh, who was my Ph.D. advisor and first introduced me to the subject of control of quantum systems. His view of control theory as an interdisciplinary field and his curiosity for other areas of science and technology greatly stimulated my own interest in these areas. I was fortunate to work for many years with Francesca Albertini. She has been a very competent and effective collaborator in much of the research that led to this book. I learned a lot from the interaction with Raffaele Romano who has been an excellent research collaborator and gave me many useful suggestions on this book. I would like to thank Justin Peters, the chair of my department, who has always done his best to create the optimal conditions for me to work. Faculty and students at my department have been very kind and patient in helping me with several issues during the writing of this book. I am indebted to Wolfgang Kliemann for several useful discussions and his continuous encouragement in this work. I would like to thank Atac Imamoglu for his kind hospitality at ETH Zurich in the fall of 2004 and Geza Giedke for teaching me many interesting things during that visit. I had useful discussions on the subject of this book with many people. A partial list includes Claudio Altafini, Adriano Batista, Ferdinando Borsa, Ugo Boscain, Laura Cattaneo, Mehmet Dagli, Viatcheslav Dobrovitski, Richard Ng, Michele Pavon, Murti Salapaka, Johnathan Smith, Pablo Tamborenea and Lorenza Viola. My thanks also go to department of Physics at the University of Lecce (Italy) for its hospitality. In particular, I would like to thank Giulio Landolfi who invited me there, in the fall of 2006, to give a series of lectures, which helped to improve some parts of this book. I would like to thank James Fiedler for a careful reading of an early draft. Mehmet Dagli also read some parts of this manuscript and gave me useful suggestions. Finally, I would like to thank CRC Press for its assistance and for agreeing to publish this book and the National Science Foundation for financial support during this work.[1]

Credit goes to Ahmet Ozer for Figures 1.1, 2.6, 6.1, 6.2, 6.6, 7.1, 7.2, 7.3, 7.4, 8.1, 8.4, 9.2, 9.3, 9.4, 9.5, 9.6, 9.7, 9.8, 9.9, 9.10, and 9.11 while all the other figures were drawn by the author except Figures 1.3, 4.2, 6.4 and 6.5, which are by Mariano Di Maria at www.elabografica.it.

Domenico D'Alessandro
Ames, Iowa, U.S.A.

---

[1]This material is based upon work supported by the National Science Foundation under Grant No. ECS0237925. Any opinions, findings and conclusions or recommendations expressed in this material are those of the author and do not necessarily reflect the views of the National Science Foundation (NSF).

Preface to the Second Edition

It has been about thirteen years since the publication of the first edition and I can say that the book served the objective it was originally written for. When I wrote it, I wanted to write the book I would have liked to have as a graduate student. Starting to delve into the area of control of quantum systems with only a background in control theory and a certain level of mathematical maturity, I had to rely mostly on my adviser's guidance, some extra courses and help from fellow graduate students in the physics department. Since the field is highly interdisciplinary, I always thought it would have been very useful to have an introduction with all the basic notions in the same place.

During these years, I had the privilege to teach the material of this book in several institutions around the world. It was very rewarding to see students approaching this subject and to discuss the topics of the book with interested scholars. In this process, however, I discovered that the book contained errors. These were in some cases mathematical errors, in some cases typos and in some other cases things that I would have liked to express in a different way. For this reason, I continuously annotated my personal copy of the book with corrections and comments. This second edition is mostly the result of this note-taking activity during the past years. I hope that now the book is always mathematically and conceptually correct, more clear and ultimately more useful.

In these thirteen years the field has seen a large development which went in parallel with the related field of quantum information, computation and communication. I have made no attempt to give an updated comprehensive review of the latest results, keeping the book at an introductory level. The book contains, however, some new references which I thought were relevant. The only large addition is an extra chapter (Chapter 4) on the dynamical decomposition of quantum systems that are not controllable. I have decided to add this chapter, which is largely based on my own research, for several reasons. First of all, even though generically quantum systems are controllable, the presence of symmetries makes them often not controllable. Understanding their structure seems to me of fundamental value, just like the Kalman decomposition is of fundamental value in the study of linear systems of classical control theory. Moreover, this subject gave me the opportunity to introduce several notions concerning representation theory that were not presented in the first edition of the book. I believe that the basic ideas of this theory should be known to any serious student of quantum control and dynamics. Lastly, I found quite intriguing the fact that the quantum decomposition can be obtained with different techniques. This suggests that there may be more general principles to be discovered behind what is presented here. As a result of the introduction of a new Chapter 4, the numbering of the chapters has shifted by +1 with respect to the first edition. Former Chapter 4 is now Chapter 5, and so on.

Among the additions in this second edition is also a section on quantum control landscapes which is now part of the chapter on optimal control (section 7.6). The concept of 'landscapes' has become very popular in the quantum control community in the last few years and I thought an introductory book would not be complete without mentioning the basic result on this subject. This result explains why it is surprisingly easy to find numerically an optimal control for a quantum system. It has generated a large literature in the last years and the topic is still the subject of several investigations. Some other parts of the book have been expanded and updated. For example, a new section 3.2.4 discusses the fact that controllability is a generic property and the 'Notes and References' section of Chapter 10 now contains a brief discussion of the experiments that earned the 2012 Nobel Prize in Physics.

I was fortunate to have had many collaborations in the past thirteen years with colleagues that helped me improve my understanding of the material presented in this book. Acknowledging them all would be too long and any list would likely be incomplete. Among the people that directly helped with this second edition, I would like to thank the students that attended my classes based on this book. They have been very helpful and insightful on many occasions. I would like to thank Rebing Wu for his suggestions on the optimal control landscapes section. I would also like to thank Jonas Hartwig for teaching me several interesting things on representation theory which helped in the writing of Chapter 4. Francesco Ticozzi also gave me several suggestions on the book during a course I taught at the University of Padova.

Thanks go to CRC Press for agreeing to publish the second edition of this book and to the National Science Foundation for financial support during this work.[2]

Finally, at a personal level, I would like to remember my parents, Maria Pia and Leonardo, who always encouraged my passion for studying and discovery, and I would like to thank my wife, Phoebe, for her understanding of me spending long hours on this project.

Domenico D'Alessandro
Ames, Iowa, U.S.A.

[2]The second edition of this book is based upon work supported by the National Science Foundation under Grant No. ECCS 1710558. Any opinions, findings and conclusions or recommendations expressed in this material are those of the author and do not necessarily reflect the views of the National Science Foundation (NSF).

Chapter 1

Quantum Mechanics

We present some of the main notions of quantum mechanics and introduce some of the notation that will be used in the book. Some more concepts of interest in general quantum mechanics will also be introduced in other parts of the book as needed. There are several introductions to quantum mechanics at various levels. Some of the most popular are [66], [109], [115], [120], [193], [245].

1.1 States and Operators

1.1.1 State of a quantum system

1.1.1.1 States and Hilbert spaces

The **state** of a quantum mechanical system is represented by a *nonzero* vector in a separable Hilbert space. The word 'represented' means here that knowledge of the state vector gives complete information on the properties of the system. A **Hilbert space** $\mathcal{H}$ (see, e.g., [239] pg. 76 ff.) is a complex vector space with an *inner product*, that is, an operation $(\cdot, \cdot) : \mathcal{H} \times \mathcal{H} \to \mathbb{C}$ which satisfies the following properties:

1.
$$(\vec{y}, \vec{x}) = (\vec{x}, \vec{y})^*, \tag{1.1}$$

   where a^* denotes the complex conjugated of a;

2.
$$(\vec{x} + \vec{y}, \vec{z}) = (\vec{x}, \vec{z}) + (\vec{y}, \vec{z}),$$

3.
$$(\alpha \vec{x}, \vec{y}) = \alpha^*(\vec{x}, \vec{y}), \tag{1.2}$$

   for any complex scalar α, and,

4.
$$(\vec{x}, \vec{x}) \geq 0,$$

DOI: 10.1201/9781003051268-1

1

where equality holds if and only if $\vec{x} = 0$.

The *norm* of a vector $\vec{x}$, $||\vec{x}||$, is defined as $||\vec{x}|| := \sqrt{(\vec{x}, \vec{x})}$. The *metric* of $\mathcal{H}$ is defined using this norm, the distance between the vectors $\vec{x}$ and $\vec{y}$ being given by $||\vec{x} - \vec{y}||$. A Hilbert space is required to be *complete* with respect to this metric.[1]   A metric space is called *separable* if it contains a dense countable set. It can be proven that a Hilbert space is separable if and only if it admits a countable orthonormal basis $\{\vec{x}_j, j \in J\}$, J a countable set.   In this case $\{\vec{x}_j\}$ is such that

$$(\vec{x}_j, \vec{x}_k) = \delta_{jk}, \tag{1.3}$$

and every vector $\vec{x} \in \mathcal{H}$ can be written in a unique way as

$$\vec{x} = \sum_{j \in J} (\vec{x}_j, \vec{x}) \vec{x}_j. \tag{1.4}$$

Here we used the *Kronecker delta symbol* δ_{jk}

$$\delta_{jk} = 0, \qquad \text{if } j \neq k,$$

$$\delta_{jk} = 1, \qquad \text{if } j = k.$$

Remark 1.1.1 A *form* on a complex vector space $\mathcal{H}$ is a map $: \mathcal{H} \times \mathcal{H} \to \mathbb{C}$. The inner product as defined above is a *sesquilinear* form on $\mathcal{H}$ in that it is antilinear in the first argument, i.e., $(\alpha \vec{x} + \beta \vec{y}, \vec{z}) = \alpha^*(\vec{x}, \vec{z}) + \beta^*(\vec{y}, \vec{z})$ and linear in the second one, i.e., $(\vec{x}, \alpha \vec{y} + \beta \vec{z}) = \alpha(\vec{x}, \vec{y}) + \beta(\vec{x}, \vec{z})$. Some authors give the definition by imposing linearity in the first argument, replacing α^* with α in (1.2), which results in antilinearity in the second argument.

1.1.1.2   Dirac notation

In quantum mechanics, the state vectors are denoted using Dirac's notation, that is, the state ψ is denoted as $|\psi\rangle$ and called **ket**. The inner product between two kets $|\phi\rangle$ and $|\psi\rangle$ is denoted by $\langle \phi | \psi \rangle$. Notice that from (1.1)

$$\langle \phi | \psi \rangle = \langle \psi | \phi \rangle^*.$$

Often a basis of the underlying Hilbert space $\mathcal{H}$ is given by a set of orthonormal eigenvectors of a preferred linear operator on $\mathcal{H}$. In this case, an eigenvector corresponding to the eigenvalue λ is denoted by $|\lambda\rangle$.

---

[1]Recall that a Cauchy sequence $\{\vec{x}_n\}$ is such that, for every $\epsilon > 0$, there is an N such that $||\vec{x}_n - \vec{x}_m|| < \epsilon$ if $n > N$ and $m > N$. A metric space $\mathcal{H}$ is complete if every Cauchy sequence in $\mathcal{H}$ converges to an element of $\mathcal{H}$.

1.1.1.3 Physical states and normalization

Vectors $|\psi\rangle$ and $\alpha|\psi\rangle$, for any $\alpha \in \mathbb{C}$, $\alpha \neq 0$, represent in quantum mechanics the same **physical state**. For this reason, it is often more appropriate to speak about *rays* or *directions* in the Hilbert space $\mathcal{H}$ as representing the state of a quantum system. In choosing a representative among the vectors proportional to $|\psi\rangle$, one often picks a vector with unit norm. Therefore, it is often assumed that the vector representing the state of the system has norm equal to one, i.e., for every $|\psi\rangle$, $\langle\psi|\psi\rangle = |||\psi\rangle||^2 = 1$. Physical states can be viewed as points on a complex sphere with radius one in a Hilbert space, with points differing by a phase factor treated as the same state.

1.1.1.4 Dimension of a quantum system

In modeling a quantum system with an appropriate Hilbert space one chooses a set of states that 1) have a definite physical meaning, e.g., each state corresponds to a given energy or each state corresponds to a given position, etc. and 2) form an orthonormal basis of the Hilbert space which means that conditions (1.3) and (1.4) are verified. Conditions (1.3) and (1.4) are written for a countable set but they generalize for an uncountable set. In particular, in the latter case, the sum is replaced by an integral (cf. 1.1.2.4). According to the meaning of this preferred basis, we say that we work in a given *representation*. For example, if the chosen vectors represent states with given values of the energy, we say that we work in the energy representation. The **dimension** refers to the cardinality of the preferred set of states. It could be finite, countably infinite or noncountably infinite. In the energy representation, the dimension is called the **number of levels** (of energy). This terminology is justified by the measurement postulate which will be discussed in section 1.2, in that, the preferred states correspond to eigenvectors of linear operators on the Hilbert space.

The simplest, but very important, example of a quantum system is a two-dimensional (or two-level) system which has a pair of distinguished orthonormal states, say $|0\rangle$ and $|1\rangle$. Every physical state $|\psi\rangle$ of such a system can be written as a superposition of these two states as

$$|\psi\rangle = \alpha|0\rangle + \beta|1\rangle, \quad |\alpha|^2 + |\beta|^2 = 1,$$

where, as mentioned above, we imposed the normalization on the states.

A simple example of an uncountably infinite-dimensional quantum system is a particle which moves along a line whose only degree of freedom is the position x. The preferred set of states can be denoted by $\{|x\rangle\}$. The state $|x\rangle$ represents the state of a system which, when position is measured, is found with certainty in position x.

1.1.1.5   Example: Spin and spin-$\frac{1}{2}$ particles

Several quantum systems possess a property called **spin**. Spin is a purely quantum mechanical feature which has no direct classical counterpart. The spin j of a quantum mechanical system may be a positive integer or a half integer. It is a fixed quantity associated to a system, just like its mass. For a quantum system with spin, one can measure a **spin angular momentum**. Such a system, in a magnetic field, will be subjected to a force just like a classical system of a charged particle with orbital angular momentum (cf. Lorentz equation (2.10)). If the spin angular momentum of a system with spin j is measured, the result (in appropriate units) is in the set $\{-j, -(j-1), \ldots, (j-1), j\}$. The Hilbert space associated to a system with spin j and all the other degrees of freedom neglected has dimension $2j+1$. The simplest example is $j = \frac{1}{2}$. Particles that have spin $\frac{1}{2}$ include, for example, electrons. In this case there are two distinguished states which, upon measurement of the spin angular momentum, will give $+\frac{1}{2}$ and $-\frac{1}{2}$, with certainty. These two states span the underlying Hilbert space of the system.

One of the early experimental, demonstrations of the existence of spin is the Stern-Gerlach experiment, which is discussed in detail in several textbooks (see, e.g., [245]). The Stern-Gerlach experiment was also one of the first experimental demonstrations of physical phenomena that cannot be explained by classical mechanics alone. It will be discussed later in this chapter when we discuss measurement (see subsection 1.2.2.2).

The physical state of a spin $\frac{1}{2}$ particle (as well as the state of any two-level quantum system) is often represented by an arrow from the origin to a point on the unit sphere called the *Bloch sphere*. Let $|\frac{1}{2}\rangle$ ($|-\frac{1}{2}\rangle$) be the state which produces with certainty the result $\frac{1}{2}$ ($-\frac{1}{2}$) when the spin angular momentum is measured. Then the state

$$|\psi\rangle := \cos\left(\frac{\theta}{2}\right)|\frac{1}{2}\rangle + e^{i\phi}\sin\left(\frac{\theta}{2}\right)|-\frac{1}{2}\rangle$$

is represented by an arrow to a point with spherical coordinates θ and ϕ. See the 'pure state' in Figure 1.1.

1.1.2   Linear operators

Given two Hilbert spaces $\mathcal{H}_1$ and $\mathcal{H}_2$ a linear operator X, $X : \mathcal{H}_1 \to \mathcal{H}_2$, is a linear map from $\mathcal{H}_1$ to $\mathcal{H}_2$. It is called *bounded* if $\|X\| := \sup\{\|X\vec{x}\| : \|\vec{x}\| = 1\}$ is finite. If $\mathcal{H}_1$ and $\mathcal{H}_2$ are both finite-dimensional every linear operator X is bounded.[2]

---

[2]Cf. Example 1.3. in [68].

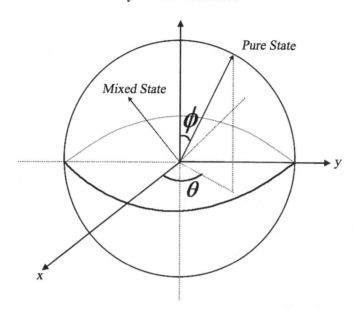

FIGURE 1.1: Pure and mixed states on the Bloch sphere. The mixed state vector has length strictly less than one; that is, its endpoint is inside the Bloch sphere. The endpoint of the pure state vector is on the surface of the Bloch sphere.

1.1.2.1 Bras

To any ket $|\psi\rangle$ is associated a linear operator $\langle\psi| : \mathcal{H} \to \mathbb{C}$ which, when applied to the vector $|\phi\rangle$, gives the number $\langle\psi|\phi\rangle$. This linear operator is called the **bra** associated with the ket $|\psi\rangle$. It is an element of the dual space of $\mathcal{H}$. If $\mathcal{H}$ has a countable basis, then we can find a countable orthonormal basis $\{|e_j\rangle\}$ and write $|\psi\rangle = \sum_j \alpha_j |e_j\rangle$. The bra corresponding to $|\psi\rangle$ is then given by (cf. Exercise 1.1)

$$\langle\psi| = \sum_j \alpha_j^* \langle e_j|. \tag{1.5}$$

1.1.2.2 Outer product

Given a ket $|\psi\rangle$ and a bra $\langle\phi|$, we define a linear operator $|\psi\rangle\langle\phi| : \mathcal{H} \to \mathcal{H}$, which is called the **outer product** of $\langle\phi|$ with $|\psi\rangle$. It maps the vector $|\chi\rangle$ to the vector $|\psi\rangle\langle\phi|\chi\rangle$, the vector $|\psi\rangle$ multiplied by the scalar $\langle\phi|\chi\rangle$. Notice this vector is in general not normalized.

1.1.2.3 Sums, linear combinations and compositions of operators

Given two (or more) linear operators X and Y, we can construct a new **sum** operator $X + Y$ which acts on a ket $|\psi\rangle$ as

$$(X + Y)|\psi\rangle = X|\psi\rangle + Y|\psi\rangle. \tag{1.6}$$

If α is a scalar, $\alpha \in \mathbb{C}$, and X a linear operator, the operator αX is defined as

$$(\alpha X)|\psi\rangle = \alpha(X|\psi\rangle). \tag{1.7}$$

Using the sum of two operators and product of an operator by a scalar, we can construct more linear operators by taking linear combinations. Consider, for instance, operators of the type $|\phi\rangle\langle\psi|$. The operator $|\phi_1\rangle\langle\psi_1| + \frac{1}{2}|\phi_2\rangle\langle\psi_2|$ is defined as

$$\left(|\phi_1\rangle\langle\psi_1| + \frac{1}{2}|\phi_2\rangle\langle\psi_2|\right)|\psi\rangle = |\phi_1\rangle\langle\psi_1|\psi\rangle + \frac{1}{2}|\phi_2\rangle\langle\psi_2|\psi\rangle.$$

Another way to construct new operators from old ones is by *composition* or *product*. If X and Y are two linear operators, the linear operator XY is defined as

$$XY|\psi\rangle = X(Y|\psi\rangle). \tag{1.8}$$

1.1.2.4 State expansion and completeness relations

Given a countable orthonormal basis in the Hilbert space $\mathcal{H}$, $\{|e_j\rangle\}$, $j = 1, 2, ...,$ we have

$$\sum_j |e_j\rangle\langle e_j| = \mathbf{1}, \tag{1.9}$$

where $\mathbf{1}$ is the identity operator. In fact, by writing $|\psi\rangle$ according to the **state expansion**,

$$|\psi\rangle = \sum_k \alpha_k |e_k\rangle, \tag{1.10}$$

and applying $\sum_j |e_j\rangle\langle e_j|$ to both sides, by using orthonormality of the $|e_j\rangle$'s, one obtains

$$\left(\sum_j |e_j\rangle\langle e_j|\right)|\psi\rangle = \sum_{j,k} \alpha_k |e_j\rangle\langle e_j|e_k\rangle = \sum_{j,k} \alpha_k |e_j\rangle\delta_{jk} = |\psi\rangle,$$

where we used $\langle e_j|e_k\rangle = \delta_{jk}$. By multiplying (1.10) on the left by $\langle e_j|$ and using again $\langle e_j|e_k\rangle = \delta_{jk}$, we have $\alpha_j = \langle e_j|\psi\rangle$. Relation (1.9) is called the **completeness relation**.

In the case of uncountably infinite basis of the Hilbert space, an orthonormal basis is $|x\rangle$, as x varies in an appropriate measurable set Ω with measure dx, such that

$$\langle x|x_1\rangle = \delta(x - x_1), \tag{1.11}$$

where $\delta(x - x_1)$ is the Dirac delta function centered at x_1.[3] The state $|\psi\rangle$ is expanded as

$$|\psi\rangle = \int_\Omega \psi(x)|x\rangle dx, \qquad (1.12)$$

for some function $\psi(x)$ on Ω. The function $\psi(x)$ is called the **wave function**. By multiplying both sides by $\langle x_1|$ and using (1.11) we obtain that $\psi(x_1) = \langle x_1|\psi\rangle$. Also, the completeness relation in this case becomes

$$\int_\Omega |x\rangle\langle x|dx = 1.$$

1.1.2.5 Hermitian, unitary operators and projections

Given a linear operator X and a ket $|\psi\rangle$, we can construct linear operators $\mathcal{H} \to \mathbb{C}$ in two different ways. We can consider the ket $X|\psi\rangle$ and take the corresponding bra. Denote this bra by $\langle X\psi|$. Recall that we have defined above in this section a bra as a linear operator $\mathcal{H} \to \mathbb{C}$. Alternatively, we can consider the linear operator $\langle\psi|X$, which multiplies an element of $\mathcal{H}$ by X first and then takes the inner product with $|\psi\rangle$, i.e., $\langle\psi|X : |\phi\rangle \to \langle\psi|(X|\phi\rangle)$. $\langle X\psi|$ and $\langle\psi|X$ are not always the same. However, if they are the same for every ket $|\psi\rangle$ in $\mathcal{H}$, then the operator X is said to be **Hermitian** or **self adjoint**. Given a bounded linear operator X, there exists a unique[4] linear operator, denoted by $X^\dagger$, acting on $\mathcal{H}$, such that for every $|\psi\rangle$ and $|\phi\rangle$,

$$\langle X\psi|\phi\rangle = \langle\psi|X^\dagger|\phi\rangle.$$

$X^\dagger$ is called the **adjoint** of X. An Hermitian operator X is such that $X = X^\dagger$.

Example 1.1.2 Consider the three-dimensional Hilbert space $\mathcal{H}$ spanned by an orthonormal basis $\{|0\rangle, |1\rangle, |2\rangle\}$. The linear operator X, $X : \mathcal{H} \to \mathcal{H}$, defined by

$$X|0\rangle := |0\rangle,$$

$$X|1\rangle := |0\rangle + |1\rangle,$$

$$X|2\rangle := |0\rangle,$$

---

[3]A rigorous definition of the Dirac delta function centered at x_1, $\delta(x - x_1)$, requires the introduction of the theory of distributions (see, e.g., [141]). Here, we shall only need the basic property, for every continuous function f,

$$\int_a^b f(x)\delta(x - x_1)dx = \begin{cases} f(x_1) & \text{if } x \in [a, b], \\ 0 & \text{if } x \notin [a, b]. \end{cases}$$

Whenever we shall use this property, we shall make the tacit assumption that the relevant function f is continuous.

[4]See, e.g., [68] Theorem 2.2.

is not self-adjoint. To see this consider the ket $|\psi\rangle := |1\rangle$. We have, from (1.5), $\langle X\psi| = \langle 0| + \langle 1|$. Given that state $|0\rangle$, we calculate $\langle X\psi|0\rangle = 1$, while $\langle\psi|X|0\rangle = \langle 1|0\rangle = 0$.

A special case of a Hermitian operator is a **projection**, which is a linear operator P having the property

$$P = P^2 = P^{\dagger}.$$

A **unitary operator** X is defined by the relation $XX^{\dagger} = X^{\dagger}X = 1$. This implies that, for states $|\psi\rangle$, $|\phi\rangle$,

$$\langle X\psi|X|\phi\rangle := \langle\psi|X^{\dagger}X|\phi\rangle = \langle\psi|\phi\rangle,$$

that is, unitary operators preserve the inner product.

1.1.3   State of composite systems and tensor product spaces

Consider two systems Σ_1 and Σ_2 with states represented by vectors in the Hilbert spaces $\mathcal{H}_1$ and $\mathcal{H}_2$, respectively. The state of the composite system (Σ_1 together with Σ_2) is represented by a vector in a Hilbert space which is the **tensor product** of $\mathcal{H}_1$ and $\mathcal{H}_2$, denoted by $\mathcal{H}_1 \otimes \mathcal{H}_2$. The tensor product of $\mathcal{H}_1$ and $\mathcal{H}_2$ can be constructed starting from orthonormal bases of $\mathcal{H}_1$ and $\mathcal{H}_2$: Consider an orthonormal basis $\{|e_j\rangle\}$ of $\mathcal{H}_1$ and an orthonormal basis $\{|f_k\rangle\}$ of $\mathcal{H}_2$. Then a basis of $\mathcal{H}_1 \otimes \mathcal{H}_2$ is constructed by putting together, in all the possible ways, one element of the basis of $\mathcal{H}_1$ and one element of the basis of $\mathcal{H}_2$. In the special case where $\mathcal{H}_1$ and $\mathcal{H}_2$ are finite-dimensional, with dimensions n and m respectively, a basis of $\mathcal{H}_1 \otimes \mathcal{H}_2$ is given by all the possible pairs $|e_j\rangle|f_k\rangle$, $j = 1, ..., n$, $k = 1, ..., m$. A generic element of the basis is denoted by $|e_j f_k\rangle$ or $|e_j\rangle \otimes |f_k\rangle$. When appearing between two vectors, the symbol $\otimes$ is called the *tensor product* between vectors. If $|e_j\rangle$ and $|f_k\rangle$ belong to bases of the Hilbert spaces $\mathcal{H}_1$ and $\mathcal{H}_2$, respectively, then $|e_j\rangle \otimes |f_k\rangle$ is the corresponding element of the basis of $\mathcal{H}_1 \otimes \mathcal{H}_2$. By requiring the tensor product operation between vectors to be linear with respect to both arguments, the definition extends to arbitrary vectors in $\mathcal{H}_1$ and $\mathcal{H}_2$. In particular, if system Σ_1 is in the state $|\psi_1\rangle$ and system Σ_2 is in the state $|\psi_2\rangle$, we say that the overall system is in the state $|\psi_1\rangle \otimes |\psi_2\rangle$. Expand $|\psi_1\rangle$ and $|\psi_2\rangle$ in the corresponding bases as $|\psi_1\rangle := \sum_j \alpha_j |e_j\rangle$ and $|\psi_2\rangle = \sum_k \beta_k |f_k\rangle$, respectively. Using linearity, we have that

$$|\psi_1\rangle \otimes |\psi_2\rangle := \sum_{j,k} \alpha_j \beta_k |e_j\rangle \otimes |f_k\rangle,$$

which shows that $|\psi_1\rangle \otimes |\psi_2\rangle$ can be expressed in the basis of $\mathcal{H}_1 \otimes \mathcal{H}_2$, i.e., an element of $\mathcal{H}_1 \otimes \mathcal{H}_2$.

1.1.3.1   Entanglement

If the state of a composite system (Σ_1 and Σ_2) can be written as the tensor product of two states, $|\psi_1\rangle$ and $|\psi_2\rangle$, i.e., as $|\psi_1\rangle \otimes |\psi_2\rangle$, then we may say that system Σ_1 is in the state $|\psi_1\rangle$ and system Σ_2 is in the state $|\psi_2\rangle$. However, there are vectors in the space $\mathcal{H}_1 \otimes \mathcal{H}_2$ that cannot be written as tensor products of two vectors (see Example 9.1.1 in Chapter 9). If the composite system is in one of these states then it is not possible to separate the states of systems Σ_1 and Σ_2. In these cases, the two systems are said to be **entangled**. Entanglement is one of the main features of quantum mechanical systems which is used in quantum computation. There are several theoretical questions related to it including criteria for the characterization and measure of the entanglement of two systems. We shall treat entanglement and entanglement dynamics in Chapter 9. Independent surveys can be found in [125] and [207].

1.1.3.2   Inner product in tensor spaces

The inner product $(\cdot, \cdot)$ in $\mathcal{H}_1 \otimes \mathcal{H}_2$ is constructed from the inner products in the Hilbert spaces $\mathcal{H}_1$ and $\mathcal{H}_2$. One first defines the inner product for tensor product states $|\psi_1\rangle \otimes |\psi_2\rangle$, $|\phi_1\rangle \otimes |\phi_2\rangle$, as

$$(|\psi_1\rangle \otimes |\psi_2\rangle, |\phi_1\rangle \otimes |\phi_2\rangle) := \langle \psi_1|\phi_1\rangle\langle \psi_2|\phi_2\rangle, \tag{1.13}$$

and then extends by (sesqui)linearity to linear combinations of tensor products (cf. Remark 1.1.1). In particular, if

$$|\psi_1\rangle := \sum_{j,k} \alpha_{j,k}|e_j\rangle \otimes |f_k\rangle,$$

and

$$|\psi_2\rangle := \sum_{j,k} \beta_{j,k}|e_j\rangle \otimes |f_k\rangle,$$

we calculate

$$(|\psi_1\rangle, |\psi_2\rangle) = \sum_{j,k} \alpha_{j,k}^* \left(|e_j\rangle \otimes |f_k\rangle, \sum_{l,m} \beta_{l,m}|e_l\rangle \otimes |f_m\rangle \right)$$

$$= \sum_{j,k,l,m} \alpha_{j,k}^* \beta_{l,m} \left(|e_j\rangle \otimes |f_k\rangle, |e_l\rangle \otimes |f_m\rangle \right),$$

where in the first equality we used antilinearity with respect to the first argument and in the second equality we used linearity with respect to the second argument. Since

$$(|e_j\rangle \otimes |f_k\rangle, |e_l\rangle \otimes |f_m\rangle) := \langle e_j|e_l\rangle\langle f_k|f_m\rangle = \delta_{jl}\delta_{km},$$

we obtain

$$(|\psi_1\rangle, |\psi_2\rangle) = \sum_{j,k,l,m} \alpha^*_{j,k}\beta_{l,m}\delta_{jl}\delta_{km} = \sum_{j,k} \alpha^*_{j,k}\beta_{j,k}.$$

1.1.3.3   Linear operators in tensor spaces

Consider two vector spaces $\mathcal{V}_1$ and $\mathcal{V}_2$ along with a linear operator $A : \mathcal{H}_1 \to \mathcal{V}_1$ and a linear operator $B : \mathcal{H}_2 \to \mathcal{V}_2$. The **linear operator** $A \otimes B$ maps elements of $\mathcal{H}_1 \otimes \mathcal{H}_2$ to elements of $\mathcal{V}_1 \otimes \mathcal{V}_2$ as follows. For any product state, $|\psi_1\rangle \otimes |\psi_2\rangle$, we have

$$(A \otimes B)|\psi_1\rangle \otimes |\psi_2\rangle := (A|\psi_1\rangle) \otimes (B|\psi_2\rangle) \qquad (1.14)$$

and the definition extends to general states by linearity.

This definition includes some important special cases. If A is a linear operator $A : \mathcal{H}_1 \to \mathbb{C}$ (a bra) and B a linear operator $B : \mathcal{H}_2 \to \mathbb{C}$, then $A \otimes B$ is defined as

$$(A \otimes B)|\psi_1\rangle \otimes |\psi_2\rangle := (A|\psi_1\rangle)(B|\psi_2\rangle), \qquad (1.15)$$

and then extended by linearity to states that are not tensor products. In the right-hand side of (1.15), the tensor product coincides with the product of two numbers since both spaces have dimension 1. A special case of this type of operator is the bra corresponding to $|\psi_1\rangle \otimes |\psi_2\rangle := |\psi_1, \psi_2\rangle$ since from (1.13) it follows that $\langle\psi_1, \psi_2| = \langle\psi_1| \otimes \langle\psi_2|$, where the symbol $\otimes$ has the meaning of tensor product of two operators.

Another special case of (1.14) is the outer product of $|\psi_1\rangle \otimes |\psi_2\rangle$ and $|\phi_1\rangle \otimes |\phi_2\rangle$, for states $|\psi_1\rangle$, $|\phi_1\rangle$ in $\mathcal{H}_1$ and $|\psi_2\rangle$, $|\phi_2\rangle$ in $\mathcal{H}_2$. This can be written as

$$|\psi_1, \psi_2\rangle\langle\phi_1, \phi_2| := |\psi_1\rangle\langle\phi_1| \otimes |\psi_2\rangle\langle\phi_2|. \qquad (1.16)$$

The operator on the right-hand side transforms $|v\rangle \otimes |w\rangle$ into $|\psi_1\rangle\langle\phi_1|v\rangle \otimes |\psi_2\rangle\langle\phi_2|w\rangle$ and it is linear.

From tensor product operators one can construct more operators by using the linear combination and composition rules in subsection 1.1.2.3 (cf. (1.6), (1.7), (1.8)). In particular, notice that (assuming $\mathcal{V}_1 = \mathcal{H}_1$ and $\mathcal{V}_2 = \mathcal{H}_2$)

$$(A_1 \otimes B_1)(A_2 \otimes B_2) = A_1 A_2 \otimes B_1 B_2.$$

Example 1.1.3 We illustrate some of the calculations introduced in this subsection and concerning composite systems. Assume systems Σ_1 and Σ_2 are two spin-$\frac{1}{2}$ particles with Hilbert spaces $\mathcal{H}_1$ and $\mathcal{H}_2$, both two-dimensional and spanned by the orthonormal basis $\{|0\rangle, |1\rangle\}$. Consider the vector $|\psi_1\rangle$ in $\mathcal{H}_1$, given by $|\psi_1\rangle := \frac{1}{\sqrt{2}}(|0\rangle + |1\rangle)$, and the vector $|\psi_2\rangle$ in $\mathcal{H}_2$, given by $|\psi_2\rangle := i|0\rangle$. We have

$$|\psi_1, \psi_2\rangle := |\psi_1\rangle \otimes |\psi_2\rangle = \frac{i}{\sqrt{2}}\left(|0\rangle \otimes |0\rangle + |1\rangle \otimes |0\rangle\right).$$

The state

$$|\phi\rangle := \frac{1}{\sqrt{2}} (|0\rangle \otimes |0\rangle + |1\rangle \otimes |1\rangle)$$

is an entangled state as it cannot be expressed as the tensor product of two states (cf. Chapter 8). We calculate

$$(|\psi_1\rangle \otimes |\psi_2\rangle, |\phi\rangle) = \frac{-i}{2},$$

and

$$|\psi_1, \psi_2\rangle\langle\phi|$$

$$= \frac{i}{2} (|0\rangle\langle 0| \otimes |0\rangle\langle 0| + |1\rangle\langle 0| \otimes |0\rangle\langle 0| + |0\rangle\langle 1| \otimes |0\rangle\langle 1| + |1\rangle\langle 1| \otimes |0\rangle\langle 1|).$$

The previous definitions and example concern a system composed of *two* subsystems, that is, a *bipartite system*. The definitions extend naturally to the case of *multipartite systems*, i.e., systems composed by more than two subsystems.

1.1.4   State of an ensemble; density operator

It is often the case that the object of study consists of an **ensemble** of a large number of identical systems in different states. In these cases, we need to have information on the fraction of systems in a given state. This information is recorded using an operator rather than a vector. This operator, called the **density operator** or **density matrix**, is defined in the following.

If there is a fraction $0 < w_j \leq 1$ of systems in the ensemble with state $|\psi_j\rangle$, j in a set $\mathcal{I}$, $\sum_{j \in \mathcal{I}} w_j = 1$, then the density operator describing the state of the ensemble is given by the linear combination of outer products

$$\rho := \sum_{j \in I} w_j |\psi_j\rangle\langle\psi_j|, \tag{1.17}$$

which is a linear operator $\mathcal{H} \to \mathcal{H}$. The density operator completely describes the state of the ensemble. Special cases are **pure ensembles** or **pure states** which are such that $w_j = 1$, for some j. They are described by density operators of the form,

$$\rho := |\psi_j\rangle\langle\psi_j|.$$

On the contrary, **mixed ensembles** (or **mixed states**) are described by density operators (1.17) with more than one w_j different from zero. The density operator has the following fundamental properties that can be proved directly using the definition.[5]

---

[5] Recall that the trace of an operator X on a Hilbert space with countable orthonormal basis $|e_k\rangle$ is given by $\mathrm{Tr}(X) = \sum_k \langle e_k|X|e_k\rangle$ and is independent of the choice of the orthonormal basis. In the uncountable case, the sum is replaced by an appropriate integral $\mathrm{Tr}(X) = \int_\Omega \langle x|X|x\rangle dx$, where $|x\rangle$, $x \in \Omega$, is an orthonormal basis (cf. (1.11)).

1. ρ is Hermitian and positive semidefinite.[6]

2.

$$\mathrm{Tr}(\rho) = 1. \tag{1.18}$$

3.

$$\rho^2 = \rho$$

   if and only if ρ represents a pure ensemble.

4.

$$0 < \mathrm{Tr}(\rho^2) < 1$$

   if ρ represents a mixed ensemble.

The set of density matrices is a convex subset of the space of Hermitian operators whose 'extreme' points represent pure states. By 'extreme points' we mean operators that cannot be written as convex combinations of two distinct operators in the set. This is a consequence of properties 3 and 4 above.[7]

Entanglement can be defined for mixed states as well. If a state for a composite system $\Sigma_1 + \Sigma_2$ can be written as a density matrix of the form $\rho_1 \otimes \rho_2$, then the state is called a *product state*. States of the form $\sum_j w_j \rho_j$, where $\sum_j w_j = 1$ and $w_j > 0, \forall j$, are called *separable states*. States that are not separable are called *entangled*. Entanglement will be treated in Chapter 9.

1.1.4.1   Example: Ensembles of two-level systems

The density matrix representing an ensemble of two-level systems can be written as

$$\rho = \frac{1}{2}(1 + x\sigma_x + y\sigma_y + z\sigma_z), \tag{1.19}$$

---

[6]A Hermitian operator $X : \mathcal{H} \to \mathcal{H}$ is called positive semidefinite if $\langle \psi | X | \psi \rangle \geq 0$, for every $|\psi\rangle \in \mathcal{H}$.

[7]Write a density operator ρ as

$$\rho := \lambda \rho_1 + (1 - \lambda)\rho_2, \quad 0 \leq \lambda \leq 1,$$

and calculate ρ^2. We have

$$\mathrm{Tr}(\rho^2) = \lambda^2 \mathrm{Tr}(\rho_1^2) + (1 - \lambda)^2 \mathrm{Tr}(\rho_2^2) + 2\lambda(1 - \lambda)\mathrm{Tr}(\rho_1 \rho_2).$$

If ρ is a pure state then $\mathrm{Tr}(\rho^2) = 1$, but to have this we need $\mathrm{Tr}(\rho_1^2) = \mathrm{Tr}(\rho_2^2) = 1$ and ρ_1 proportional to ρ_2 which implies by (1.18) $\rho_1 = \rho_2$. We also need $\lambda = 0$ or $\lambda = 1$. We have used here the Cauchy-Schwartz inequality for the Frobenius inner product of Hilbert-Schmidt operators, $|\mathrm{Tr}(\rho_1 \rho_2)| \leq \sqrt{\mathrm{Tr}(\rho_1^2)\mathrm{Tr}(\rho_2^2)}$, with equality if and only if ρ_1 and ρ_2 are linearly dependent.

where the matrices $\sigma_{x,y,z}$ are the *Pauli matrices*

$$\sigma_x := \begin{pmatrix} 0 & 1 \\ 1 & 0 \end{pmatrix}, \qquad \sigma_y := \begin{pmatrix} 0 & i \\ -i & 0 \end{pmatrix}, \qquad \sigma_z := \begin{pmatrix} 1 & 0 \\ 0 & -1 \end{pmatrix}. \qquad (1.20)$$

The state is represented by a vector (called *Bloch vector*) as in Figure 1.1 whose endpoint has coordinates x, y and z. In the case of a pure state the representative point on the Bloch sphere is the same as the one obtained with polar coordinates in Figure 1.1. If the state is a mixed state, $\text{Tr}(\rho^2) < 1$, i.e., $x^2 + y^2 + z^2 < 1$, the endpoint is *inside* the Bloch sphere.

1.1.4.2 Ensembles of composite systems

Consider two subsystems with associated orthogonal bases $\{|e_j\rangle\}$ and $\{|f_k\rangle\}$, respectively. A pure state of the overall system $|\psi\rangle$ is described by

$$|\psi\rangle = \sum_{j,k} \alpha_{j,k} |e_j\rangle \otimes |f_k\rangle.$$

If we have an ensemble of systems, with states

$$|\psi_l\rangle := \sum_{j,k} \alpha_{j,k,l} |e_j\rangle \otimes |f_k\rangle,$$

each with probability w_l, using linearity and the definition of outer product for tensor product kets (1.16), we derive the density operator ρ for the ensemble,

$$\rho = \sum_l \sum_{j,k} \sum_{r,s} w_l \alpha_{j,k,l} \alpha^*_{r,s,l} |e_j\rangle\langle e_r| \otimes |f_k\rangle\langle f_s|.$$

1.1.4.3 Density matrix describing a single system

Density matrices are also used to describe an unknown state of a single system. If a quantum system is in the state

$$\rho = \sum_k w_k |\psi_k\rangle\langle\psi_k|,$$

with $\sum_k w_k = 1$, $w_k > 0$, then there is probability w_k that the system is in the pure state $|\psi_k\rangle$.

1.1.5 Vector and matrix representation of states and operators

Much of the treatment in the following will focus on finite-dimensional systems. The Hilbert space is spanned by an orthonormal basis $\{|e_j\rangle\}$, $j = 1, ..., n$. Associating with $|e_j\rangle$ the column vector having all the entries equal to zero and a 1 in the j-th position, a ket $|\psi\rangle := \sum_{j=1}^{n} \alpha_j |e_j\rangle$ is represented by a column vector denoted by $\vec{\psi}$,

$$
\vec{\psi} := \begin{pmatrix} \alpha_1 \\ \cdot \\ \cdot \\ \cdot \\ \alpha_n \end{pmatrix}.
$$

The corresponding bra $\langle\psi|$ is represented by $\vec{\psi}^{\dagger}$, i.e., the conjugate transposed of $\vec{\psi}$, and we have

$$
\langle\psi|\phi\rangle = \vec{\psi}^{\dagger}\vec{\phi}.
$$

It is known from linear algebra that once a basis is fixed, linear operators are in one to one correspondence with $n \times n$ matrices. While $\vec{\psi}$ denotes the column vector associated with $|\psi\rangle$, we shall use the same notation for linear operators and the matrices that represent them. Therefore, if $|\phi\rangle = X|\psi\rangle$, with an operator X, then $\vec{\phi} = X\vec{\psi}$, with the matrix X in the corresponding vector-matrix notation.

1.1.5.1 Composite systems and Kronecker product of matrices

Consider two systems with Hilbert spaces $\mathcal{H}_1$ and $\mathcal{H}_2$ spanned respectively by orthonormal bases $\{|e_j\rangle\}$, $j = 1, ..., n$ and $\{|f_k\rangle\}$, $k = 1, ..., m$. Consider now the corresponding basis in $\mathcal{H}_1 \otimes \mathcal{H}_2$, $\{|e_j, f_k\rangle\}$. In Exercise 1.4, the reader is asked to verify that if $|\psi\rangle \in \mathcal{H}_1$ has column vector representation $\vec{\psi}$ and $|\phi\rangle \in \mathcal{H}_2$ has column vector representation $\vec{\phi}$, then $|\psi\rangle \otimes |\phi\rangle$ has column vector representation $\vec{\psi} \otimes \vec{\phi}$, where $\otimes$ denotes the Kronecker product of the two vectors. Here we have ordered the elements of the basis $\{|e_j, f_k\rangle\}$ so that $|e_{j_1}, f_{k_1}\rangle$ comes before $|e_{j_2}, f_{k_2}\rangle$ if $j_1 < j_2$ or, when $j_1 = j_2$, if $k_1 < k_2$. The **Kronecker product** ($\otimes$) of two general matrices A and B (with arbitrary dimensions) is defined as

$$
A \otimes B := \begin{pmatrix} a_{11}B & a_{12}B & \cdots & a_{1n}B \\ a_{21}B & a_{22}B & \cdots & a_{2n}B \\ \cdot & \cdot & \cdots & \cdot \\ \cdot & \cdot & \cdots & \cdot \\ \cdot & \cdot & \cdots & \cdot \\ a_{r1}B & a_{r2}B & \cdots & a_{rn}B \end{pmatrix}.
$$

A comprehensive discussion of the properties of the Kronecker product can be found in [136] pg. 242 ff. Some of the properties we shall use in the following are

1.

$$
(A \otimes B)(C \otimes D) = AC \otimes BD, \tag{1.21}
$$

2.

$$(A \otimes B)^\dagger = A^\dagger \otimes B^\dagger,$$

3.

$$\mathrm{Tr}(A \otimes B) = \mathrm{Tr}(A)\,\mathrm{Tr}(B),$$

4. If **1** denotes the identity matrix of appropriate dimensions

$$e^{A \otimes 1 + 1 \otimes B} = e^A \otimes e^B. \tag{1.22}$$

With the above ordering convention on the basis of the tensor product space, the matrix representative of the tensor product $A \otimes B$ of two operators is the Kronecker product of the two corresponding matrices.

1.1.5.2 Correspondence between abstract and linear algebra operations

Generalizing what said earlier, all the operations that we have defined in the abstract setting of vectors and operators have a counterpart in linear algebra. This is summarized in Table 1.1. Notice in particular from the table that Hermitian operators correspond to Hermitian matrices. The *abstract operation* with vectors and operators, and the corresponding *linear algebra operation*, with column or row vectors and matrices, are such that the following commutative diagram is satisfied, once a basis is chosen. The process of going from an abstract setting to a matrix-vector setting once a basis is chosen is called *coordinatization* in linear algebra.

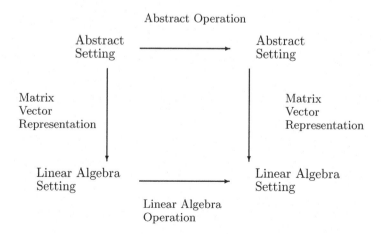

TABLE 1.1:   Abstract operations with vectors and operators and the corresponding linear algebra counterparts.

Abstract Operation	Linear Algebra Operation
$\langle\psi\|\phi\rangle$, inner product in $\mathcal{H}$	$\vec{\psi}^{\dagger}\vec{\phi}$, inner product in $\mathbb{C}^n$
$X\|\psi\rangle$, linear operator X acting on vector $\|\psi\rangle$	$X\vec{\psi}$, matrix X multiplying vector $\vec{\psi}$
$\|\psi\rangle\langle\phi\|$, outer product	$\vec{\psi}\vec{\phi}^{\dagger}$, column-row product
XY, composition of operators X, Y	XY, product of matrices X, Y
$\alpha X + \beta Y$, linear combination of operators X, Y	$\alpha X + \beta Y$ linear combination of matrices X, Y
$X^{\dagger}$, adjoint operator of X	$X^{\dagger}$ transposed, conjugate of matrix X
$\|\psi\rangle \otimes \|\phi\rangle$, tensor product of vectors $\|\psi\rangle$, $\|\phi\rangle$	$\vec{\psi} \otimes \vec{\phi}$, Kronecker product of vectors $\vec{\psi}$, $\vec{\phi}$
$A \otimes B$, tensor product of operators A, B	$A \otimes B$, Kronecker product of matrices A, B

1.2 Observables and Measurement

Measurement theory in quantum mechanics is the subject of much study. A summary and references can be found in [47]. There are in general several types of measurements that can be performed in a laboratory. In this section, we summarize the concepts concerning the standard Von Neumann-Lüders measurement. Extensions to more general types of measurements can be obtained using the formalism of operations and effects and this is treated in Appendix A.

1.2.1 Observables

In quantum mechanics, observed quantities are called **observables**. Each observable is associated with an Hermitian operator on the Hilbert space describing the system, $\mathcal{H}$. In several situations, specifically in several infinite-dimensional cases, the operator defined in this fashion does not map *norm bounded* elements of $\mathcal{H}$ (representing states) to norm bounded elements of $\mathcal{H}$. We illustrate this with an example.

Example 1.2.1 Consider a particle with a one-dimensional motion along the x axis. Assume the motion is not constrained to a bounded region so that the position x can assume all the values in $(-\infty, +\infty)$. In the position representation $\{|x\rangle\}$, the state $|\psi\rangle$ is written as (cf. (1.12))

$$|\psi\rangle = \int_{-\infty}^{+\infty} \psi(x)|x\rangle dx. \tag{1.23}$$

Since $|||\psi\rangle||$ is finite, the wave function $\psi(x)$ must be in $L^2(\mathbf{R})$.[8] The position observable is defined as the operator X which multiplies the wave function ψ by x. However, it is easy to construct examples where ψ is in $L^2(\mathbf{R})$ while $x\psi(x)$ is not.[9] Therefore, this operator does not map any element with finite norm to an element with finite norm.

---

[8]Given a measure space Ω, with associated measure dx, the vector space $L^2(\Omega)$ is the space of square integrable functions ψ, on Ω, i.e., functions satisfying

$$\left(\int_\Omega \psi^*(x)\psi(x)dx\right)^{\frac{1}{2}} < +\infty.$$

$L^2(\Omega)$ is a Hilbert space when equipped with the inner product

$$(\psi,\phi) := \int_\Omega \psi^*(x)\phi(x)dx.$$

An introduction to general L^p spaces, for general p, and general measure spaces, can be found in [239].
[9]Consider, for example, $\psi(x) := \frac{1}{x}$, for $x \geq 1$ and $\psi(x) \equiv 0$, for $x < 1$.

To circumvent this problem one defines a *domain* of a linear operator A, $\mathcal{D}(A)$ as the set of states in $\mathcal{H}$ where the linear operator A is bounded, i.e., $\||A|\psi\rangle\| < \infty$. A Hermitian operator is such that $\mathcal{D}(A) = \mathcal{D}(A^\dagger)$, and A and $A^\dagger$ coincide on the common domain. It is assumed that $\mathcal{D}(A)$ is dense in $\mathcal{H}$.

1.2.1.1   Spectrum and spectral theorem

Consider $\mathbf{R}$ as a measurable space.[10] A *projection valued measure*, dP, on $\mathbf{R}$ is defined as a map which associates with every measurable set $\Omega \subseteq \mathbf{R}$, a projection operator $P(\Omega)$, and satisfies the following two axioms

1. $P(\mathbf{R}) = \mathbf{1}$,

2. If $\{\Omega_n\}$ is a countable family of disjoint measurable sets $P(\cup_n \Omega_n) = \sum_n P(\Omega_n)$.

Associated with a projection valued measure is a *spectral family*, that is, a family of commuting projections P_λ, with $\lambda \in \mathbf{R}$, given by the Lebesgue integral

$$P_\lambda := \int_{-\infty}^\lambda dP.$$

The spectral family has several properties. In particular $\lambda_1 \le \lambda_2$ implies range$(P_{\lambda_1}) \subseteq$ range(P_{λ_2}). Moreover, P_λ is continuous from the right, i.e., $\lim_{\epsilon \to 0^+} P_{\lambda+\epsilon} = P_\lambda$, and $\lim_{\lambda \to \infty} P_\lambda = \mathbf{1}$, $\lim_{\lambda \to -\infty} P_\lambda = \mathbf{0}$, with $\mathbf{0}$ the zero operator. A point $\lambda \in \mathbf{R}$ is said to be *stationary* for the spectral family $\{P_\lambda\}$ if there exists $\bar\epsilon > 0$, such that for every ϵ, $\bar\epsilon > \epsilon > 0$,

$$P_{\lambda+\epsilon} - P_{\lambda-\epsilon} = \int_{\lambda-\epsilon}^{\lambda+\epsilon} dP = \mathbf{0}.$$

The **spectral theorem** (see, e.g., [5]) gives a resolution of every self-adjoint operator in terms of a projection valued measure and a spectral family. It is stated as follows.

Theorem 1.2.2 *To every self-adjoint operator A, there corresponds a unique projection-valued measure dP_A and a spectral family $\{P_{A\lambda}\}$, such that*

$$A = \int_{\mathbf{R}} \lambda dP_A, \tag{1.24}$$

on every $|\psi\rangle$ in $\mathcal{D}(A)$, the domain of A.

The **spectrum** of a self-adjoint operator A is defined as the set of non-stationary points of the corresponding spectral family. Points in the spectrum are called *eigenvalues*.

---

[10]In most of the discussion here we follow [47].

1.2.1.2 Special cases of the spectral decomposition

A self-adjoint operator A is said to have *discrete spectrum* if all the points in the spectrum of A are isolated points. An isolated point λ_0 in the spectrum is, by definition, such that there exists an $\epsilon > 0$ with $P_{A\lambda} - P_{A\lambda_0} = \mathbf{0}$ for all $\lambda_0 \leq \lambda < \lambda_0 + \epsilon$, and $P_{A\lambda_0} - P_{A\lambda} \neq \mathbf{0}$, for all $\lambda_0 - \epsilon < \lambda < \lambda_0$ (cf. Figure 1.2). In this case, the spectrum is a countable set, $\{\lambda_j\}$.

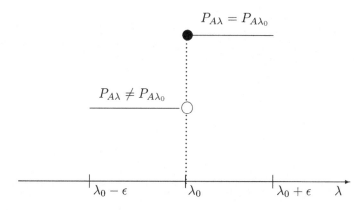

FIGURE 1.2: The spectral family has a *jump* at an isolated point of the spectrum.

Define[11] (cf. Exercise 1.9)

$$P_j := P_{A\lambda_j} - P_{A\lambda_{j-1}}. \tag{1.25}$$

The spectral theorem takes the form

$$A = \sum_j \lambda_j P_j. \tag{1.26}$$

A vector $|e_{jk}\rangle$ in the range of P_j is called an *eigenvector* corresponding to the eigenvalue λ_j. It has the property that $A|e_{jk}\rangle = \lambda_j|e_{jk}\rangle$. Choosing an orthonormal basis of range(P_j), $\{|e_{jk}\rangle\}$, we can write P_j as

$$P_j = \sum_k |e_{jk}\rangle\langle e_{jk}|.$$

If rank $P_j = 1$ (>)1 then λ_j is said to be a *nondegenerate (degenerate) eigenvalue*.

---

[11] Choose $P_{A\lambda_{j-1}} = \mathbf{0}$ if λ_j is the minimum value in the spectrum.

A continuous spectrum is such that it contains no isolated points. Generalizing the definitions given above for the discrete case, an eigenvector $|e\rangle$ corresponding to the eigenvalue λ_0 is such that, for every $\lambda > \lambda_0$, $|e\rangle$ is in the range of

$$\int_{\lambda_0}^{\lambda} dP_A,$$

and we have, using (1.24),

$$A|e\rangle = \lambda_0|e\rangle.$$

As an example of an operator with continuous spectrum, consider the position operator X of Example 1.2.1. The spectral family in this case (cf. [47] section 2.1) is such that, with $|\psi\rangle$ in (1.23),

$$P_{X\lambda}|\psi\rangle = \int_{-\infty}^{\lambda} \psi(x)|x\rangle dx. \tag{1.27}$$

The corresponding spectral measure can be written as

$$dP_X = |x\rangle\langle x|dx, \tag{1.28}$$

which if we use (1.11) gives (1.27). Accordingly, the position operator X is written as

$$X = \int_{-\infty}^{+\infty} x|x\rangle\langle x|dx. \tag{1.29}$$

There are also observables with mixed discrete and continuous spectrum.

1.2.2   The measurement postulate

1.2.2.1   Results of measurement and modification of the state

The Von Neumann-Lüders **measurement postulate** states that, when an observable A is measured, the result is an eigenvalue of A. Let J be a measurable subset of the spectrum of A, and let P_J be the projection

$$P_J := \int_J dP_A,$$

where dP_A is the projection measure associated to A. If the state at the moment of the measurement is $|\psi\rangle$, the probability of obtaining a value λ in J, as the result of the measurement, is given by

$$\Pr(\lambda \in J) = \langle\psi|P_J|\psi\rangle = \|P_j|\psi\rangle\|^2. \tag{1.30}$$

If the measurement gives a value in J, the state is modified according to the rule

$$|\psi\rangle \rightarrow \frac{P_J|\psi\rangle}{\sqrt{\langle\psi|P_J|\psi\rangle}}.$$

The result of a measurement is certain if and only if the system is already in an eigenvector of the observable being measured, in which case we obtain the corresponding eigenvalue and the state of the system is left unchanged.

In quantum mechanics therefore a measurement has the effect of modifying the state. This has several consequences for control theory and practice and, in particular, it complicates the use of *feedback*, namely the possibility of choosing the control after a reading of the state. The modification of the state has to be taken into account in feedback control schemes. This problem has given rise to a large literature. Using different and more sophisticated notions of measurement (see, e.g., [295], [308] and references therein), one can still design feedback control laws for quantum systems by finding the right balance between the information obtained through measurement and the level of modification of the state. This book only deals with open-loop (i.e., non feedback) control.

Denote by $\Pr(\lambda_j)$ the probability that the result of the measurement is λ_j. The *expectation value* for the measurement of A, when the system is in the state $|\psi\rangle$, is given by, in the discrete spectrum case,

$$\langle A \rangle_\psi := \sum_j \lambda_j \Pr(\lambda_j) = \sum_j \lambda_j \langle \psi | P_j | \psi \rangle = \langle \psi | \sum_j \lambda_j P_j | \psi \rangle = \langle \psi | A | \psi \rangle,$$

(1.31)

where we used the spectral decomposition (1.26). The same formula holds for the continuous spectrum case (cf. Exercise 1.10).

As we have already mentioned, it is common practice to denote by $|\lambda\rangle$ the eigenvector corresponding to the eigenvalue λ. With this notation, for a discrete spectrum $\{\lambda_j\}$ the formula of the spectral theorem can be written (in terms of one-dimensional projectors) as

$$A = \sum_j \lambda_j |\lambda_j\rangle\langle\lambda_j|,$$

where the λ_j, are possibly repeated. For a continuous spectrum Ω, the operator A can be written as

$$A = \int_\Omega x|x\rangle\langle x|dx,$$

where the projection-valued measure dP_A is written in terms of the eigenvectors $|x\rangle$, i.e., $dP_A := |x\rangle\langle x|dx$. This is the generalization of what was said in (1.28) and (1.29), for the position operator.

Consider an observable with discrete nondegenerate spectrum $\{\lambda_j\}$, namely the rank of each projection P_j associated with λ_j is one.[12] In terms of the

---

[12]The generalization of what we are going to say to the general possibly degenerate case is straightforward involving only some heavier notation.

eigenvectors $|\lambda_j\rangle$, a state $|\psi\rangle$ is expanded as

$$|\psi\rangle = \sum_j \alpha_j |\lambda_j\rangle,$$

and the projection associated to λ_j is $P_j = |\lambda_j\rangle\langle\lambda_j|$. According to (1.30), the probability of a result λ_j when A is measured and the state is $|\psi\rangle$ is

$$\langle\psi|P_j|\psi\rangle = |\alpha_j|^2.$$

Generalizing this to a continuous time spectrum, if $|\psi\rangle$ is expanded as in (1.12) in terms of the eigenvectors, $|x\rangle$, of A and writing the projection corresponding to a result in the set J as

$$P_J = \int_J |x\rangle\langle x|dx,$$

using (1.11), we obtain that the probability of having a result in the set J is given by (cf. (1.30))

$$\Pr(\lambda \in J) = \langle\psi|P_J|\psi\rangle = \int_J \psi^*(x)\psi(x)dx. \tag{1.32}$$

1.2.2.2   Example of measurement: The Stern-Gerlach experiment

In the *Stern-Gerlach experiment*, an oven contains an ensemble of particles with spin $\frac{1}{2}$ (silver atoms in the original version of the experiment). These particles exit the oven, are aligned through a narrow slit and pass through an inhomogeneous magnetic field.[13] In the magnetic field they experience a force which deflects their paths. In particular, the z-component of the force F_z is approximately proportional to the (spin) angular momentum in the z direction S_z, according to the relation

$$F_z \approx kS_z\frac{\partial B_z}{\partial z},$$

for some constant k and where B_z is the z-component of the magnetic field. In the Stern-Gerlach experiment, the particles are then collected on a plate. Therefore, the position of a particle collected on the plate indicates the force acting on the particle and therefore its angular momentum. As the angular

---

[13]If we think of this experiment with classical electrodynamics, we have spinning charges in a magnetic field. The spinning charges behave as magnets which tend to align themselves with the field. In fact, the torque experienced is proportional to their angular momentum. Moreover, the particles experience a magnetic force depending on their orientation when considered as magnets due to the fact that the external field is not homogeneous. The South pole of the magnet-particle will be repelled from the South pole of the external magnet a little less (or more) than the North pole does from the North pole of the external magnet.

momenta of the particles in the oven are completely random, in principle, classically one would expect them to be distributed continuously on the wall. However, the Stern-Gerlach experiment has shown that only two locations on the wall are possible for the particles. These correspond to the values of the spin angular momentum $+\frac{1}{2}$ and $-\frac{1}{2}$.

The Stern-Gerlach experiment was one of the first experimental demonstrations of the quantum mechanical measurement postulate. A scheme of the Stern-Gerlach experiment is shown in Figure 1.3. Further discussion can be found, e.g., in [245].

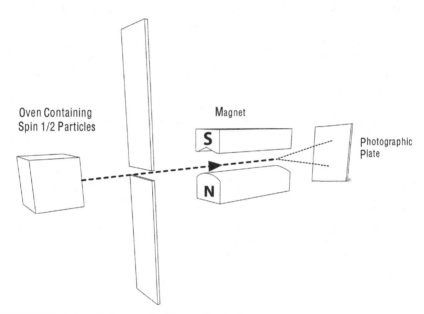

FIGURE 1.3:   Scheme of a Stern-Gerlach experiment.

1.2.2.3   Simultaneous measurement of observables; compatible and incompatible observable and uncertainty relations

In the rest of this section, we shall refer to the case of measurement of observables with discrete spectrum. The generalization to continuous spectrum can be obtained without much difficulty using the formalism of the spectral theorem.

Consider observables with discrete spectrum. A set of r observables,

$$\{S_1, ..., S_r\},$$

which are pairwise commuting, i.e., $[S_j, S_k] := S_j S_k - S_k S_j = 0;\ j, k = 1, ..., r,$

are said to be **compatible**. It follows from results in linear algebra and operator theory (see, e.g., [68], [135]) that a set of observables are compatible if and only if they have the same eigenvectors. These eigenvectors form a countable basis of $\mathcal{H}$, $\{|e_k\rangle\}$. In this case, each observable S_j can be expanded, with possibly repeated eigenvalues, as

$$S_j = \sum_k \lambda_{kj}|e_k\rangle\langle e_k|. \tag{1.33}$$

The projections $|e_k\rangle\langle e_k|$ are common to all the observables, while the eigenvalues λ_{kj} may be different for different j.

Consider two observables A and B and assume, for simplicity, that they are nondegenerate (i.e., every eigenspace has dimension 1). The measurement of A (while the system is in a *pure* state) gives a result λ and the state collapses into the corresponding eigenstate $|\lambda\rangle$. This is an eigenstate of B as well. Therefore, a measurement of B does not change the state. After the measurement of B a measurement of A gives again the value λ, with certainty. In other words, the measurement of B does not change the result for the measurement of A. For this reason commuting observables are called *compatible*.

The measurement postulate for the measurement of compatible observables, $\{S_1, S_2, ..., S_l\}$ is a generalization of the one for a single observable discussed above. It says that if the state immediately before the measurement is $|\psi\rangle$ and the results of the measurements are the eigenvalues $\lambda_1, ..., \lambda_l$ of the observables $S_1, S_2, ..., S_l$, with projectors (as in (1.25)) $P_1, P_2, ..., P_l$, respectively, then the state after the measurement will be

$$|\psi_1\rangle := \frac{P_l P_{l-1} ... P_1|\psi\rangle}{||P_l P_{l-1} ... P_1|\psi\rangle||}. \tag{1.34}$$

The probability of obtaining the l-ple of results, $\lambda_1, ..., \lambda_l$, is given by

$$\Pr(\lambda_1, ..., \lambda_l) = \langle\psi|P_l P_{l-1} ... P_1|\psi\rangle.$$

Notice that if $P_k P_j = 0$, where P_k and P_j correspond to two eigenvalues λ_k and λ_j for observables S_k and S_j, then the two values λ_k and λ_j cannot appear simultaneously as result of the measurement. The probability of obtaining them at the same time is zero.

Remark 1.2.3 The maximum number of linearly independent compatible observables for a system of dimension n is n, i.e., the maximum number of linearly independent pairwise commuting Hermitian matrices.

Incompatible observables are such that

$$AB - BA := [A, B] \neq 0.$$

The **Heisenberg uncertainty relation** quantitatively describes the uncertainty in the measurement of two incompatible observables. Consider an observable A and its expectation value in some state $|\psi\rangle$, $\langle A\rangle_\psi$. Define $\Delta A := A - \langle A\rangle_\psi \mathbf{1}$. The expectation value of $(\Delta A)^2$, $\langle(\Delta A)^2\rangle_\psi$, measures how much A is expected to deviate from its mean value at the state $|\psi\rangle$ and therefore the uncertainty in the measurement of A at the state $|\psi\rangle$. Defining the same quantity for the observable B, $\langle(\Delta B)^2\rangle_\psi$, and using the definition of the commutator $[A, B] := AB - BA$, the uncertainty relation is[14]

$$\langle(\Delta A)^2\rangle_\psi \langle(\Delta B)^2\rangle_\psi \geq \frac{1}{4}|\langle[A, B]\rangle_\psi|^2. \tag{1.35}$$

Therefore, a decrease in the uncertainty in the measurement of A implies an increase in the uncertainty in the measurement of B and vice versa.

Formula (1.35) follows from the Cauchy-Schwartz inequality

$$|\langle\phi_1|\phi_2\rangle|^2 \leq \langle\phi_1|\phi_1\rangle^2 \langle\phi_2|\phi_2\rangle^2,$$

applied for $|\phi_1\rangle = \Delta A|\psi\rangle$, $|\phi_2\rangle = \Delta B|\psi\rangle$, which leads to

$$|\langle\psi|\Delta A\Delta B|\psi\rangle|^2 \leq \langle\Delta A\rangle_\psi^2 \langle\Delta B\rangle_\psi^2. \tag{1.36}$$

Write $\Delta A\Delta B := \frac{1}{2}[\Delta A, \Delta B] + \frac{1}{2}\{\Delta A, \Delta B\}$ where we have introduced the anti-commutator, $\{A, B\} := AB + BA$ which is Hermitian if A and B are Hermitian. Since $[\Delta A, \Delta B]$ ($\{\Delta A, \Delta B\}$) is skew-Hermitian (Hermitian), $\langle\psi|[\Delta A, \Delta B]|\psi\rangle$ ($\langle\psi|\{\Delta A, \Delta B\}|\psi\rangle$) is purely imaginary (purely real). Therefore, the left-hand side of (1.36) is equal to $\frac{1}{4}|\langle\psi|[\Delta A, \Delta B]|\psi\rangle|^2 + \frac{1}{4}|\langle\psi|\{\Delta A, \Delta B\}|\psi\rangle|^2 \geq \frac{1}{4}|\langle\psi|[\Delta A, \Delta B]|\psi\rangle|^2$. Noticing that $[\Delta A, \Delta B] = [A, B]$, we obtain formula (1.35) (cf. [245], section 1.4).

1.2.3 Measurements on ensembles

Assume that we perform the measurement of an observable A on an ensemble described by the density operator

$$\rho := \sum_k w_k |\psi_k\rangle\langle\psi_k|. \tag{1.37}$$

We want to write, in terms of density operator, the state ρ_j corresponding to the ensemble of systems giving the result λ_j. After the measurement, the systems in the state $|\psi_k\rangle$ giving the result λ_j will be in the state $\frac{P_j|\psi_k\rangle}{\sqrt{\langle\psi_k|P_j|\psi_k\rangle}}$. The fraction of such systems is proportional both to w_k and to $\Pr(\lambda_j)$, that

---

[14]Note that the operator $[A, B]$ is in general not Hermitian, in particular if A and B are Hermitian, $[A, B]$ is skew-Hermitian, i.e., $[A, B]^\dagger = -[A, B]$. However, we can still define the 'expectation value' as $\langle[A, B]\rangle_\psi := |\langle\psi|[A, B]|\psi\rangle|$.

is, it is proportional to $w_k Pr(\lambda_j) = w_k \langle \psi_k | P_j | \psi_k \rangle$. Therefore, the density matrix representing the sub-ensemble of systems which have given result λ_j is

$$\rho_j := \sum_k f w_k \langle \psi_k | P_j | \psi_k \rangle \frac{P_j | \psi_k \rangle \langle \psi_k | P_j}{||P_j | \psi_k \rangle||^2},$$

for some factor f. Since $||P_j | \psi_k \rangle||^2 = \langle \psi_k | P_j | \psi_k \rangle$, this simplifies as

$$\rho_j := f P_j \left(\sum_k w_k | \psi_k \rangle \langle \psi_k | \right) P_j = f P_j \rho P_j,$$

and $f = [\mathrm{Tr}(P_j \rho P_j)]^{-1}$ so that $\mathrm{Tr}(\rho_j) = 1$. From (1.30) and the definition of the density matrix (1.37), it follows (cf. Exercise 1.6) that $\mathrm{Tr}(P_j \rho P_j)$ is the probability of finding the result λ_j for a measurement on an ensemble ρ. We have

$$\rho_j = [\mathrm{Tr}(P_j \rho P_j)]^{-1} P_j \rho P_j. \tag{1.38}$$

A measurement where systems that give the same result are collected in sub-ensemble is called **selective**. In a **nonselective** measurement, we assume to observe the expectation value of the observable A given by

$$\langle A \rangle_\rho := \sum_k w_k \langle \psi_k | A | \psi_k \rangle = \sum_k w_k \, \mathrm{Tr}(| \psi_k \rangle \langle \psi_k | A) = \mathrm{Tr}(\rho A). \tag{1.39}$$

After the measurement, the state ρ is transformed into

$$\rho' = \sum_j P_j \rho P_j, \tag{1.40}$$

as the sub-ensembles ρ_j corresponding to the different outcomes are combined together with coefficients $\mathrm{Tr}(P_j \rho P_j)$ to form the new ensemble.

Selective and nonselective measurements can also be defined on single systems whose state is represented by a density matrix (cf. 1.1.4.3). In this situation, a selective measurement which gives a result λ_j will project the state onto the state ρ_j as in (1.38). If the measurement is performed but the result is not read then the state after the measurement will be of the form ρ' in (1.40).

Because of the property

$$\mathrm{Tr}(\rho A) = \mathrm{Tr}(A\rho),$$

it is often stated that observables and states are 'dual' concepts. Both are represented by Hermitian operators and the expectation value of the observable A if the system is in a state ρ is equal to the expectation of the observable ρ in a state A.

1.3 Dynamics of Quantum Systems

The energy of a quantum system is a very important observable that plays a fundamental role in the description of the dynamics of the system. The associated operator is called the **Hamiltonian operator** and it is usually denoted by H. H might be time dependent and in fact this is typically the case for quantum control systems as studied here. A time-dependent Hamiltonian models an energy exchange with an external system which plays the role of a controller.

1.3.1 Schrödinger picture

1.3.1.1 Schrödinger equation

Let $|\psi(t)\rangle$ be the state of the system at time t. In quantum mechanics, it is postulated that $|\psi(t)\rangle$ evolves according to the **Schrödinger equation**

$$i\hbar \frac{d}{dt}|\psi(t)\rangle = H(t)|\psi(t)\rangle, \tag{1.41}$$

where i is the imaginary unit and $\hbar$ is the Planck constant. If $|\psi(0)\rangle$ is the initial state we have

$$|\psi(t)\rangle = X(t)|\psi(0)\rangle, \tag{1.42}$$

with the linear operator $X(t)$ satisfying the differential equation

$$i\hbar \frac{d}{dt}X(t) = H(t)X(t), \tag{1.43}$$

with initial condition equal to the identity operator. The operator $X(t)$ is called the **evolution operator** or **propagator** and equation (1.43) will be referred to as **Schrödinger operator equation**. It follows from the fact that H is Hermitian that, at every time t, $X(t)$ is a unitary operator. In particular, this implies that the evolution does not modify the norm of the state vector, which we have set equal to one. In the finite-dimensional case, in the vector matrix representation, X can be represented as a unitary matrix of dimension $n \times n$ (n being the dimension of the underlying Hilbert space $\mathcal{H}$). In this case, X varies in the Lie group of unitary matrices $U(n)$.

1.3.1.2 Schrödinger wave equation

Consider the Schrödinger equation (1.41) for the state $|\psi\rangle$ in an infinite-dimensional Hilbert space. The state $|\psi\rangle$ has the expansion (1.12). In particular applying $\langle x|$ to both sides of (1.41), and using $\psi(x,t) := \langle x|\psi(t)\rangle$, we obtain

$$i\hbar \frac{\partial \psi(x,t)}{\partial t} = \langle x|H(t)|\psi(t)\rangle. \tag{1.44}$$

Since $|\psi(t)\rangle$ has the linear expansion (1.12) and H is a linear operator acting on $|\psi(t)\rangle$, H is uniquely determined by a linear operator on the coefficient $\psi(x,t)$ which we still denote by H and still call Hamiltonian operator. Therefore, the right-hand side of (1.44) can be written, using (1.12), as

$$\langle x|H(t)|\psi(t)\rangle = \int_\Omega (H(t)\psi(x_1,t))\langle x|x_1\rangle dx_1.$$

Using (1.11) and the fundamental property of the Dirac delta function

$$\int_\Omega f(x)\delta(x - x_1)dx = f(x_1),$$

for every continuous function $f = f(x)$, if $x_1 \in \Omega$ we obtain the **Schrödinger wave equation**

$$i\hbar\frac{\partial\psi(x,t)}{\partial t} = H(t)\psi(x,t). \tag{1.45}$$

Notice that H here represents a linear operator on a space of functions.

In practice, the Hamiltonian operator $H = H(t)$ in (1.45) is derived from an expression of the energy of the system in the classical setting and a process called *canonical quantization* which associates quantum mechanical operators to classical quantities. We shall talk about the quantization procedure in Chapter 2.

Example 1.3.1 (Particle in a potential well) Consider a particle constrained to moving along the direction of the x-axis. Assume the (classical) total energy given by

$$H_{cl} = \frac{1}{2m}p^2 + V(x), \tag{1.46}$$

where p is the kinematical momentum, in the x direction, i.e., $p = m\dot{x}$, with m the mass of the particle, so that the first term represents the kinetic energy. V is the potential energy. Let us assume that $V(x)$ has a square shape, i.e., it is zero in an interval $(0, L)$ and equal to V_0 at the locations 0 and L. Assuming V_0 very large, we obtain the potential well in Figure 1.4. This scheme may represent a particle with negative charge such as an electron confined between two negatively charged plates. The system is also referred to as a *particle in a box*. Classically, in the absence of friction and viscous forces, the particle moves back and forth between the two walls of the box. To describe the system in quantum mechanics, we choose a Hilbert space spanned by the position eigenvectors $\{|x\rangle\}$ so that the state $|\psi\rangle$ can be written as $|\psi(t)\rangle := \int_{\mathbf{R}} \psi(x,t)|x\rangle dx$. As we shall see in Chapter 2, the rules of canonical quantization, applied to this situation, associate with p the operator $-i\hbar\frac{d}{dx}$, i.e., differentiation of the wave function by x and multiplication by $-i\hbar$, and with $V(x)$ the operator multiplying the wave function $\psi(x,t)$ by $V(x)$. Therefore, the quantum mechanical Hamiltonian is given by

$$H = -\frac{\hbar^2}{2m}\frac{\partial^2}{\partial x^2} + V(x), \tag{1.47}$$

so that Schrödinger wave equation (1.45) reads as

$$i\hbar\frac{\partial\psi(x,t)}{\partial t} = -\frac{\hbar^2}{2m}\frac{\partial^2\psi(x,t)}{\partial x^2} + V(x)\psi(x,t). \tag{1.48}$$

States corresponding to a given value E of the energy observable H are eigenstates of H. Since we assume that energy is conserved, we assume that when we measure energy we obtain a value E with certainty. The eigenstates are found from the eigenvalue-eigenvector equation for H, which is called the *time-independent Schrödinger wave equation*. That is,

$$-\frac{\hbar^2}{2m}\frac{d^2}{dx^2}\psi(x) + V(x)\psi(x) = E\psi(x). \tag{1.49}$$

If $0 < x < L$, we set $V(x) = 0$ and we impose boundary conditions at $x = 0$ and $x = L$. The general solution of (1.49) is of the form

$$\psi(x) = A\sin(kx) + B\cos(kx),$$

with $k^2 = \frac{2mE}{\hbar^2}$, with free coefficients A and B. Since $|\psi(x)|^2$ is the probability density of finding the particle in position x, and the particle is very unlikely to be found very close to the walls, the boundary conditions are chosen as $\psi(0) = \psi(L) = 0$, which gives $B = 0$. Moreover, since A cannot be zero (it would mean that the particle could not be found anywhere in the well), we must have $kL = \sqrt{\frac{2mE}{\hbar^2}}L = n\pi$, for $n = 1, 2, \dots$. This gives that the possible values for the energy E are in a discrete set and are given by

$$E_n = \frac{n^2\pi^2\hbar^2}{2mL^2}.$$

The solution ψ, corresponding to energy E_n, is given by

$$\psi_n(x) = A\sin(\frac{n\pi}{L}x),$$

where A is any number with absolute value $\sqrt{\frac{2}{L}}$ obtained from the requirement that $\int_0^L \psi^*(x)\psi(x)dx = 1$. Plots of ψ_n for the first three energy levels are shown in Figure 1.4. Notice in particular, for $n \geq 2$, the existence of 'nodes', i.e., points in the interval $(0, L)$ where ψ_n is equal to zero, i.e., the particle has probability to be found equal to zero.

The solution of the Schrödinger wave equation (1.48) for a system with energy E_n has the form $\psi(x,t) = e^{\frac{-iE_nt}{\hbar}}\psi_n(x)$.

The system of a particle in a quantum well of potential and various generalizations of it serve as models for several physical systems and in particular for quantum dots (see, e.g., [91]). It is possible to introduce a control in the model by shaping the potential energy with time (see, e.g., [267]) in order to obtain a desired behavior for the wave function.

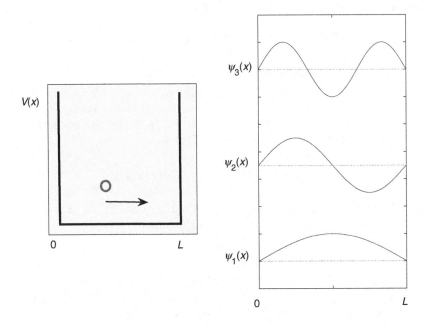

FIGURE 1.4: One-dimensional motion of a single particle in a quantum well and resulting wave functions for the first three values of the energy.

1.3.1.3   Liouville's equation and evolution of ensembles

Assume the initial state of the system is represented by a density matrix

$$\rho(0) := \sum_k w_k |\psi_k(0)\rangle\langle\psi_k(0)|.$$

Then, according to (1.42), ρ varies with time as

$$\rho(t) = \sum_k w_k X(t)|\psi_k(0)\rangle\langle\psi_k(0)|X^\dagger(t) = X(t)\rho(0)X^\dagger(t), \qquad (1.50)$$

where $X(t)$ is the solution of (1.43). By differentiating (1.50) taking into account (1.43), we obtain the differential equation for the density matrix known as **Liouville's equation** (also called **Von Neumann equation**)

$$i\hbar\frac{d}{dt}\rho = [H(t), \rho] := H(t)\rho - \rho H(t). \qquad (1.51)$$

1.3.2   Heisenberg and interaction pictures

In classical mechanics, the description of a system is done through observables (e.g., momentum, position, energy) rather than states. Therefore, it seems reasonable that a better parallel with classical mechanics could be obtained if we study the dynamics of observables rather than that of states. This approach is called the **Heisenberg picture**. Given an observable A, we define an operator function of time, $A^H = A^H(t)$, as

$$A^H(t) := X^\dagger(t)AX(t), \qquad (1.52)$$

where X is the propagator in (1.43). From (1.52) and (1.43), it follows that $A^H(t)$ satisfies

$$i\hbar \frac{dA^H}{dt} = [A^H, H], \qquad (1.53)$$

where H is the Hamiltonian of the system in (1.43), possibly function of time. This is known as the **Heisenberg equation**. Notice that the expectation value of the observable A at time t can be calculated both as $\mathrm{Tr}(A^H(t)\rho(0))$ and as $\mathrm{Tr}(A\rho(t))$ since from (1.50) we have

$$\mathrm{Tr}(A\rho(t)) = \mathrm{Tr}(AX(t)\rho(0)X^\dagger(t)) = \mathrm{Tr}(X^\dagger(t)AX(t)\rho(0)) = \mathrm{Tr}(A^H(t)\rho(0)),$$

using the property $\mathrm{Tr}(AB) = \mathrm{Tr}(BA)$. More generally, in the Heisenberg picture, one considers the state of the system constant and the observables varying with time and studies the dynamics of observables using Heisenberg equation (1.53).

In many cases the Hamiltonian H has the form

$$H := H_0 + H_I(t),$$

with H_0 a constant Hamiltonian (sometimes called *internal Hamiltonian*) and H_I a (possibly) time-varying Hamiltonian (sometimes called *interaction Hamiltonian*). Given the Schrödinger equation

$$i\hbar \frac{d}{dt}|\psi(t)\rangle = (H_0 + H_I(t))|\psi(t)\rangle,$$

it is possible to eliminate the term H_0 by making a change of coordinates

$$|\psi\rangle \to e^{\frac{iH_0 t}{\hbar}}|\psi\rangle. \qquad (1.54)$$

Defining $|\tilde{\psi}\rangle := e^{\frac{iH_0 t}{\hbar}}|\psi\rangle$, we obtain from (1.41)

$$i\hbar \frac{d}{dt}|\tilde{\psi}\rangle = H_I'(t)|\tilde{\psi}\rangle, \qquad (1.55)$$

with

$$H_I'(t) := e^{\frac{iH_0 t}{\hbar}}H_I(t)e^{-\frac{iH_0 t}{\hbar}}. \qquad (1.56)$$

This description of the dynamics is called **interaction picture**. In many cases, one works in a basis of eigenvectors of the Hamiltonian H_0 which is therefore represented by a diagonal matrix. Each eigenvector corresponds to a state having energy given by the corresponding eigenvalue of H_0. In the given basis, the matrix representation of H_0 is diagonal. Therefore, the coordinate transformation (1.54) does not affect the relative magnitude of the components of the state vector which are the same for $|\psi\rangle$ and $|\tilde{\psi}\rangle$.

1.4   Notes and References

It goes without saying that we have just scratched the surface of a treatment of quantum mechanics. The emphasis of this book will be on finite-dimensional systems but we introduced concepts in general when it was not too much more complicated to do so. We also followed a somehow 'axiomatic' presentation without discussing why certain things happen. A scientist interested in experiments would point out that quantum mechanics has given optimal predictions for experimental outcomes since its invention at the beginning of the twentieth century. However, several scientists have studied the *foundations* of quantum mechanics and developed alternative mathematical models which would explain some of the axioms. This study aims at answering some of the basic foundational questions. For instance, what is the meaning of a system being in the superposition of two states $|\psi\rangle := \alpha|0\rangle + \beta|1\rangle$? and what actually happens to the system when we do a measurement which makes it collapse to one state? Our 'axiomatic' presentation follows the standard textbook idea that a system is not really in a definite state before the measurement and only the act of measuring it forces it to reveal its state. This is the so-called *Copenhagen interpretation*. Other interpretations and theoretical treatments giving equivalent experimental predictions are possible (e.g., *hidden variables theories*, etc). An introduction to foundational questions in quantum mechanics can be found in [208].

A related issue is that of the connection between classical and quantum mechanics. On one hand, the quantum mechanical Hamiltonian to be used in the Schrödinger equation is obtained from the classical Hamiltonian in Hamiltonian mechanics via a process called *quantization* (see next chapter). On the other hand, one would like to obtain classical mechanics as a limit of quantum mechanics. One approach to achieve the latter task is to describe quantum dynamics via the *Feynman path integral formalism*. This approach is based on Lagrangian rather than Hamiltonian classical mechanics (cf. Appendix B). In this method, one calculates, as a path integral, the inner product $\langle x_1 | \psi(T) \rangle$ whose absolute value gives the probability of finding the state in $|x_1\rangle$ after an evolution for time T (cf. (1.30) with $P_J = |x_1\rangle\langle x_1|$), starting from the initial condition $|x_0\rangle$. A path integral is an integral over all the trajectories from x_0 to x_1 of the form

$$\int \mathcal{D}x(t) e^{i\frac{S(x(t))}{\hbar}}, \tag{1.57}$$

where $\mathcal{D}x(t)$ is a volume element in the (infinite-dimensional) space of trajectories from x_0 to x_1, $\hbar$ is the Planck constant and $S(x(t))$ is the *action* associated to a trajectory. This is (cf. (B.1) in Appendix B)

$$S(x) := \int_0^T L(x(t), \dot{x}(t))dt,$$

where L is the Lagrangian associated with the dynamics of the system. As $\hbar \to 0$ (classical limit), the term $e^{i\frac{S(x(t))}{\hbar}}$ in the integral (1.57) will result in high frequency oscillations which would give a nearly zero contribution to the integral for any trajectory except for the trajectory which satisfies the principle of least action, i.e., the classical trajectory predicted by classical Lagrangian mechanics. We refer to [187] for an enlightening introduction to the path integral formalism and some of its applications.

Finally, we mention the attempts in theoretical physics to conciliate quantum mechanics with general relativity. Such a field goes under the name of *quantum gravity* which is a general term for several approaches to the problem including *string theory, loop quantum gravity* etc. (see, e.g., [210] for an overview).

1.4.1 Interpretation of quantum dynamics as information processing

In quantum information theory, the state $|\psi\rangle$ of a (finite-dimensional) quantum system encodes information. In particular, in typical implementations, the information is encoded in a number of two-level systems called *qubits*. If a particular system performs a given unitary evolution X_T during an interval of time $[0, T]$, the information in $|\psi(0)\rangle$ is manipulated to obtain

$$|\psi(T)\rangle = X_T|\psi(0)\rangle.$$

X_T is also called a *quantum logic gate* as it may be seen as performing a *logic operation* on $|\psi(0)\rangle$. This way of performing quantum information processing is often referred to as the *circuit model* for quantum computation [207].

As an example, consider a two-level system with basis given by $|0\rangle$ and $|1\rangle$, and with constant Hamiltonian H defined by $H|0\rangle = -i\hbar|1\rangle$, $H|1\rangle = i\hbar|0\rangle$. If we write $|\psi(t)\rangle = \alpha_1(t)|0\rangle + \alpha_2(t)|1\rangle$, the differential equation corresponding to the Schrödinger equation for $\vec{\psi}(t) = \begin{pmatrix} \alpha_1(t) \\ \alpha_2(t) \end{pmatrix}$ is

$$\frac{d}{dt}\vec{\psi} = \begin{pmatrix} 0 & 1 \\ -1 & 0 \end{pmatrix}\vec{\psi}.$$

This equation can be solved explicitly. The solution is given by

$$\vec{\psi}(t) = \begin{pmatrix} \cos(t) & \sin(t) \\ -\sin(t) & \cos(t) \end{pmatrix}\vec{\psi}(0).$$

For $T = \frac{\pi}{2}$ we obtain

$$\vec{\psi}(T) = \begin{pmatrix} 0 & 1 \\ -1 & 0 \end{pmatrix}\vec{\psi}(0).$$

This dynamics performs the logic operation NOT in that it *inverts* the state which is transferred from $|1\rangle$ to $|0\rangle$ and from $|0\rangle$ to $-|1\rangle$ (notice the minus sign plays the role of an overall phase factor. The state $-|1\rangle$ is physically the same as the state $|1\rangle$). In general, and this is one of the main features of *quantum computation* as compared to *classical* computation, the state $|\psi\rangle$ may be not in one of the two basis states but in a *superposition* of the two before and/or after the computation (evolution). This fact is useful in several algorithms. Algorithms are, in this context, controlled evolution of a number of qubits. Such *quantum algorithms* were shown in some cases to be more powerful than classical algorithms (see, e.g., [207] for an introduction to quantum computation). A quantum algorithm is obtained with a *cascade* of r quantum logic gates $X_1, ..., X_r$ on a certain number of qubits. After r computations with an overall time T, $|\psi(T)\rangle = X_r X_{r-1} \cdots X_1 |\psi(0)\rangle$. The computations $X_1, ..., X_r$ can be performed by the same system by using different Hamiltonians H in different intervals. The Hamiltonians may be changed by changing the experimental setup (cf. section 10.3) and/or by interaction with an external field. The choice of an appropriate sequence of Hamiltonians is a *control theory* problem.

1.4.2   Direct sum versus tensor product for composite systems

We have said that the Hilbert space describing the total state of two systems with Hilbert spaces $\mathcal{H}_1$ and $\mathcal{H}_2$ is the tensor product space $\mathcal{H}_1 \otimes \mathcal{H}_2$. This is often accepted as a postulate of quantum mechanics. One might wonder why the direct sum of vector spaces, $\mathcal{H}_1 \oplus \mathcal{H}_2$, is not used instead.

Consider for simplicity two finite-dimensional systems of dimension n_1 and n_2, and two observables, for example energy of the single subsystems, represented by Hamiltonians H_1 and H_2. Consider the generic case where these Hamiltonians are not degenerate so that there are n_1 possible eigenvalues (values for the energy) for system 1 and n_2 possible values of the energy for system 2. Assume that we measure simultaneously the energy of system 1 and system 2. It is intuitive that the possible values for the the *total* energy for the total system are the sums of the two values of the energies. The number of possible values, in the generic case, will be $n_1 n_2$. So the total energy operator on the composite system has to have $n_1 n_2$ eigenvalues. This is the dimension of the tensor product of the two spaces. If we took the direct sum, its dimension would be $n_1 + n_2$. Such a choice therefore is not consistent with the measurement postulate.[15]

---

[15]The interpretation reported in this subsection was suggested to the author by Raffaele Romano.

1.5   Exercises

The goal of the following exercises is to acquaint the student with the notations of quantum mechanics and with translating them into the language of linear algebra.

Exercise 1.1 Prove formula (1.5) starting with the finite-dimensional case. In the infinite-dimensional (countable) case interpret $|\psi\rangle := \sum_{j=1}^{\infty} a_j |e_j\rangle$ as the $\lim_{b\to\infty} |\psi_b\rangle$ with $|\psi_b\rangle := \sum_{j=1}^{b} \alpha_j |e_j\rangle$ and the linear operator on the right-hand side of (1.5) as $|\phi\rangle \to \lim_{b\to\infty} \alpha_j^* \langle e_j |\phi\rangle$.

In the Exercises 1.2 through 1.5, we assume finite-dimensional quantum systems.

Exercise 1.2 Consider a Hilbert space $\mathcal{H}$, with basis $\{|e_j\rangle\}$, $j = 1, ..., n$. A vector $|\psi\rangle = \sum_{j=1}^{n} \alpha_j |e_j\rangle$ is represented by the column vector $\vec{\psi} := \begin{pmatrix} \alpha_1 \\ \vdots \\ \alpha_n \end{pmatrix}$.

If $|\phi\rangle = \sum_{j=1}^{n} \beta_j |e_j\rangle$ and $|\psi\rangle = \sum_{j=1}^{n} \alpha_j |e_j\rangle$, find the matrix representation of the outer product $|\psi\rangle\langle\phi|$. Verify that it is equal to $\vec{\psi}\vec{\phi}^{\dagger}$.

Exercise 1.3 An ensemble of two-level systems has $\frac{1}{3}$ of the systems in the state

$$|\psi_1\rangle = \frac{1}{\sqrt{2}}|0\rangle + \frac{i}{\sqrt{2}}|1\rangle,$$

and $\frac{2}{3}$ in the state

$$|\psi_2\rangle = |1\rangle.$$

Write an expression for the density matrix ρ representing the total system in terms of the states $|0\rangle$ and $|1\rangle$ and then use this along with the result of Exercise 1.2 to write a matrix for the operator ρ in the basis $|0\rangle$ and $|1\rangle$.

Exercise 1.4 (Tensor Product and Kronecker Product) Consider two vectors $|\psi\rangle$ and $|\phi\rangle$ in Hilbert spaces $\mathcal{H}$ and $\mathcal{H}_2$, respectively, with given bases $\{|e_j\rangle\}$ and $\{f_k\rangle\}$, respectively. Consider the column vector representations of $\psi\rangle$ and $|\phi\rangle$, i.e., $\vec{\psi}$ and $\vec{\phi}$. Consider the basis of the space $\mathcal{H}_1 \otimes \mathcal{H}_2$, defined as $\{|e_j\rangle \otimes |f_k\rangle\}$ with the ordering convention defined as follows: $|e_{j_1}, f_{k_1}\rangle$ comes before $|e_{j_2}, f_{k_2}\rangle$ if $j_1 < j_2$ or, when $j_1 = j_2$, if $k_1 < k_2$. Verify that the matrix vector representation of the tensor product $|\psi\rangle \otimes |\phi\rangle$, in the basis $\{|e_j\rangle \otimes |e_k\rangle\}$, is the Kronecker product of the two matrix vector representations. In other terms, verify that the following commutative diagram, which is a special case of the one in subsection 1.1.5, holds:

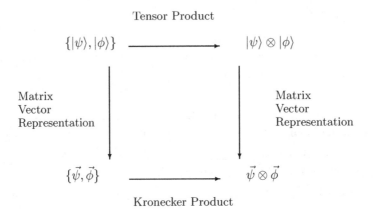

Tensor Product

$$\{|\psi\rangle, |\phi\rangle\} \xrightarrow{\hspace{3cm}} |\psi\rangle \otimes |\phi\rangle$$

Matrix Vector Representation   Matrix Vector Representation

$$\{\vec{\psi}, \vec{\phi}\} \xrightarrow{\hspace{3cm}} \vec{\psi} \otimes \vec{\phi}$$

Kronecker Product

Exercise 1.5 The matrix representation of the energy operator H of a two-level system is

$$H = \begin{pmatrix} \frac{3}{2} & \frac{i}{2} \\ \frac{-i}{2} & \frac{3}{2} \end{pmatrix}, \tag{1.58}$$

in a given orthonormal basis $|0\rangle$, $|1\rangle$.

- Find the two possible values E_1 and E_2 of the energy for this system.

- Using the spectral decomposition of the matrix (1.58), write the energy operator in Dirac notation, namely as the sum of appropriate outer products of states. What are the projectors P_j in this case?

- Assume the state of the system is $|0\rangle$. In what states may we find the system after a measurement of the energy, and with what probability?

- Write a differential equation which gives the evolution of the components of the column vector representing an arbitrary state $|\psi\rangle$ (in the basis $|0\rangle$, $|1\rangle$). Assume $\hbar = 1$.

Exercise 1.6 Use the measurement postulate (1.30) and the definition of density matrix (1.37), to show that $\mathrm{Tr}(P_j \rho P_j)$ is the probability of finding the result λ_j for a measurement on an ensemble ρ.

Exercise 1.7 The spectral theorem discussed in subsection 1.2.2 is a generalization of the spectral decomposition for matrices. Consider the matrix

$$A := \frac{1}{3} \begin{pmatrix} 4 & 2 & 1 \\ 2 & 4 & -1 \\ 1 & -1 & 1 \end{pmatrix},$$

and find its spectral decomposition (1.26) by identifying eigenvalues λ_j and corresponding projections P_j.

Exercise 1.8 Derive equation (1.55).

Exercise 1.9 Prove that the operators defined in (1.25) are such that the product of two of them is equal to **0**.

Exercise 1.10 Prove formula (1.31), in the continuous spectrum case, i.e., the expectation value of A at the state $|\psi\rangle$, $\langle A\rangle_\psi$, is equal to $\langle\psi|A|\psi\rangle$. Use the spectral theorem and the fact that the element of the spectral family corresponding to an interval $[\lambda_1, \lambda_2)$ of the spectrum is

$$\int_{\lambda_1}^{\lambda_2} dP_A.$$

Exercise 1.11 Prove that every final-dimensional complex inner product space is automatically complete. It is therefore a Hilbert space. Prove that, in addition, it is separable.

Exercise 1.12 Consider a couple of two-level systems. Give an example of an entangled pure state and a pure state that is not entangled. Explain why the states you propose are or are not entangled. Give an example of a separable mixed state that is not product state.

Exercise 1.13 Consider a two-level system with incompatible observables $S_x := \begin{pmatrix} 0 & 1 \\ 1 & 0 \end{pmatrix}$, $S_y := \begin{pmatrix} 0 & i \\ -i & 0 \end{pmatrix}$. Find a set of (normalized) states $|\psi\rangle$ so that equality holds in the inequality (1.35).

Chapter 2

Modeling of Quantum Control Systems; Examples

Quantum systems whose dynamics depends on one or more control functions will be called *quantum control systems*. In a typical situation the objects of control are charged particles while the control functions are appropriately shaped electromagnetic fields. To derive the appropriate models, in this chapter we describe elements of the theory of field-particle interaction. We start with the classical theory of electrodynamics and show how to obtain the quantum mechanical description through a process called *canonical quantization*. We present physical examples of (finite-dimensional) quantum control systems of interest in applications and we highlight the fact that such systems have a common mathematical structure. They are bilinear systems whose evolution is determined by a curve in the Lie group of unitary matrices. In a parallel to classical control theory, we identify a *state*, an *input* and an *output* in our description of quantum control systems.

2.1   Classical Theory of Interaction of Particles and Fields

Consider N charged particles in an electromagnetic field. The Hilbert space describing this system may be chosen to be spanned by the eigenvectors of the *position observable*. This is called the *position representation*. More precisely, if there are N particles involved, with coordinates $\{x_j, y_j, z_j\}$, $j = 1, 2, ..., N$, the Hilbert space is spanned by the eigenvectors $|\vec{r}\rangle := |x_1, y_1, z_1, ..., x_N, y_N, z_N\rangle$. These eigenvectors correspond to the eigenvalues $x_1, y_1, z_1, ..., x_N, y_N, z_N$ of the observables associated to the coordinates of particles $1, 2, ..., N$. The state of the system $|\psi(t)\rangle$ is expanded as (see (1.12) in Chapter 1)

$$|\psi(t)\rangle = \int_{\mathbf{R}^{3N}} \psi(\vec{r}, t)|\vec{r}\rangle d\vec{r}. \tag{2.1}$$

DOI: 10.1201/9781003051268-2

The function $\psi(\vec{r}, t) = \langle \vec{r} | \psi \rangle$, $\in L^2(\mathbf{R}^{3N})$, $\forall t$, is the wave function and the function $|\psi(\vec{r}, t)|^2$ is the (time-dependent) probability density for the particles to be in a particular region of space. More precisely, if there are N particles involved in the process, with coordinates $\{x_j, y_j, z_j\}$, $j = 1, 2, ..., N$, let $d\vec{r}$ be the infinitesimal volume element in $3N$-dimensional space. Then the probability of being in the $3N$-dimensional region I is

$$\mathrm{Pr}(I, t) := \int_I |\psi(\vec{r}, t)|^2 d\vec{r}.$$

This is a special case of (1.32).

The wave function $\psi(\vec{r}, t)$ satisfies the *Schrödinger wave equation*

$$i\hbar \frac{\partial \psi(\vec{r}, t)}{\partial t} = H(t)\psi(\vec{r}, t). \tag{2.2}$$

The Hamiltonian operator is obtained from the classical expression of the energy of the system[1] and the process of *canonical quantization* which associates quantum mechanical operators to physical variables. The quantum mechanical modeling of a system of charged particles interacting with an electromagnetic field involves the study of Lagrangian and Hamiltonian mechanics (cf. Appendix B) for systems with an uncountable number of degrees of freedom. This study is the subject of books on (nonrelativistic) quantum electrodynamics (see, e.g., [67] [129]). In the remainder of this section, we describe the main points.

Notation: In the remainder of this section and in the following two sections we shall often deal with classical variables and their corresponding quantum mechanical operators. We shall use the convention of denoting by $\hat{A}$ the quantum mechanical operator corresponding to the variable A. In particular, in these sections, the classical energy is denoted by H while the quantum mechanical Hamiltonian is denoted by $\hat{H}$.

2.1.1   Classical electrodynamics

2.1.1.1   Total energy and Maxwell equations

We start with the classical expression for the total energy H of a system of moving particles and the associated electromagnetic field. This is the sum of

---

[1]More precisely, it is obtained from the associated Hamiltonian of classical Hamiltonian mechanics. There are cases where this Hamiltonian does not coincide with the energy (cf. [67]) but these cases will not be treated here.

the kinetic energy of the particles and the energy of the field, i.e.,[2]

$$H = \sum_{j=1}^{N} \frac{1}{2} m_j \left(\frac{d\vec{r}_j}{dt} \right)^2 + \frac{\epsilon_0}{2} \int_{\mathbf{R}^3} \vec{E}^2(\vec{r}) + c^2 \vec{B}^2(\vec{r}) \, d\vec{r}, \qquad (2.3)$$

where $\vec{E}$ and $\vec{B}$ are the electric and magnetic fields respectively, c is the speed of light in vacuum, ϵ_0 is the free space *electric permittivity*. The quantity

$$\mu_0 := \frac{1}{\epsilon_0 c^2}$$

is called the *magnetic permeability*.[3] The potential energy of the particles is not explicitly present in (2.3) because it is contained in the field energy.

Fields and particles interact in that the motion of the particles is determined by the field and, at the same time, the particles establish the field. Moreover, the interaction among particles happens through the field. More specifically, once the motion of the particles is given, the electromagnetic field evolves

---

[2] A justification of this formula and the following formula (2.28) in terms of Lagrangian and Hamiltonian mechanics is presented in Appendix B. In particular see (B.31).

[3] In the following we shall often use the symbol ∇ of vector calculus. The symbol ∇ is defined as

$$\nabla := \frac{\partial}{\partial x} \vec{i} + \frac{\partial}{\partial y} \vec{j} + \frac{\partial}{\partial z} \vec{k}.$$

If ϕ is a function $\nabla \phi$ is the vector field

$$\nabla \phi = \frac{\partial \phi}{\partial x} \vec{i} + \frac{\partial \phi}{\partial y} \vec{j} + \frac{\partial \phi}{\partial z} \vec{k}.$$

If $\vec{A}$ is a vector field

$$\vec{A} = M\vec{i} + N\vec{j} + P\vec{k}$$

then $\nabla \cdot \vec{A}$ is a scalar function equal to the formal scalar product of ∇ and $\vec{A}$. It is defined by

$$\nabla \cdot \vec{A} = \frac{\partial M}{\partial x} + \frac{\partial N}{\partial y} + \frac{\partial P}{\partial z}.$$

Analogously, $\nabla \times \vec{A}$ is a vector field equal to the formal cross product of ∇ and $\vec{A}$, which is

$$\nabla \times \vec{A} = \left(\frac{\partial P}{\partial y} - \frac{\partial N}{\partial z} \right) \vec{i} + \left(\frac{\partial M}{\partial z} - \frac{\partial P}{\partial x} \right) \vec{j} + \left(\frac{\partial N}{\partial x} - \frac{\partial M}{\partial y} \right) \vec{k}.$$

The Laplacian operator ∇^2 acting on a vector field $\vec{A} = M\vec{i} + N\vec{j} + P\vec{k}$ gives $\nabla^2 \vec{A} = \nabla^2 M\vec{i} + \nabla^2 N\vec{j} + \nabla^2 P\vec{k}$ with, for a smooth function f, $\nabla^2 f := \frac{\partial^2 f}{\partial x^2} + \frac{\partial^2 f}{\partial y^2} + \frac{\partial^2 f}{\partial z^2}$.

according to Maxwell's equations:[4]

$$\nabla \cdot \vec{E} = \frac{\rho}{\epsilon_0}, \tag{2.4}$$

$$\frac{1}{\mu_0}(\nabla \times \vec{B}) - \epsilon_0 \frac{\partial \vec{E}}{\partial t} = \vec{J}, \tag{2.5}$$

$$\nabla \times \vec{E} + \frac{\partial \vec{B}}{\partial t} = 0, \tag{2.6}$$

$$\nabla \cdot \vec{B} = 0. \tag{2.7}$$

Here we have used the symbol $\rho = \rho(\vec{r}, t)$ (function of position and time) for the *charge density*[5] (charge per unit volume), and $\vec{J} = \vec{J}(\vec{r}, t)$ for the *current density* (current across unit area). The units used in (2.4) – (2.7) are *SI* units. In particular $\vec{J}$ is measured in $Coul/(m^2 \times s)$ while the charge density ρ is measured in $Coul/m^3$.

For a discrete distribution of N charges q_j at locations $\vec{r}_j$, $j = 1, ..., N$, the charge density is given by

$$\rho(\vec{r}) = \sum_{j=1}^{N} q_j \delta(\vec{r} - \vec{r}_j), \tag{2.8}$$

while $\vec{J}$ is given by

$$\vec{J}(\vec{r}) = \sum_{j=1}^{N} q_j \dot{\vec{r}}_j \delta(\vec{r} - \vec{r}_j). \tag{2.9}$$

---

[4]In the Maxwell equations, the first equation (2.4) is called *Gauss's law*. It expresses the fact that the flux integral across any closed surface is equal to the net charge enclosed inside the surface (multiplied by the factor $\frac{1}{\epsilon_0}$). In formulas we have

$$\int_S \vec{E} \ \vec{n} \, dS = \frac{1}{\epsilon_0} \int_V \rho \, dV,$$

which is obtained by performing a space integral of both sides of (2.4) and then applying the divergence theorem to the left-hand side. The third equation (2.6) expresses the so called Faraday's electromagnetic induction namely: a time-varying magnetic field creates an electric field. It is also called *Faraday's law*. The second equation (2.5), called *Ampere's law*, expresses the reciprocal fact, that a time-varying electric field, along with a moving charge $\vec{J}$, generates a magnetic field. The last equation, (2.7), expresses the fundamental property that, unlike the electric field, which has electric charges as point sources, the magnetic field has no point source. It can also be written applying the divergence theorem as

$$\int_S \vec{B} \cdot \vec{n} \, dS = 0,$$

where S is any closed surface.

[5]To follow standard notations, we denote the charge density function by ρ as there is no possibility of confusion with the density matrix introduced in the previous chapter.

Given an electromagnetic field, the dynamics of a particle of charge q is determined by *Lorentz equation*

$$\vec{F} = q(\vec{E} + \vec{v} \times \vec{B}), \tag{2.10}$$

where $\vec{F}$ denotes the force acting on the particles and the fields are evaluated at the point where the particle is located. The vector $\vec{v}$ is the velocity of the particle.

Summarizing the system of particles and field evolves according to the following diagram. The motion of the particles determines the field according to Maxwell's equations. The field has a back-action on the charges according to Lorentz equation. The variables associated with the charged particles are the positions $\vec{r}_j$ and the velocities $\dot{\vec{r}}_j$ of the (discrete) set of particles $j = 1, ..., N$. The variables associated with the field are the electric field $\vec{E} := \vec{E}(\vec{r})$, and the magnetic field $\vec{B} := \vec{B}(\vec{r})$, parametrized by the continuous space variable $\vec{r}$.

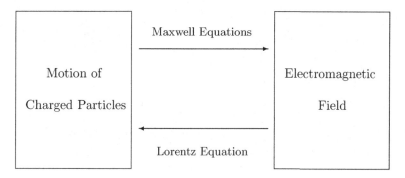

2.1.1.2   Vector and scalar potential

It is common practice, and it is instrumental for the transition from the classical to the quantum theory, to express the electric and magnetic field in terms of *vector and scalar potentials*. Doing this, the two *homogeneous* Maxwell equations, (2.6) and (2.7), are automatically satisfied. More specifically, equation (2.7) and the vector identity

$$\nabla \cdot (\nabla \times \vec{A}) \equiv 0, \qquad \forall \vec{A} := \vec{A}(\vec{r}), \tag{2.11}$$

suggest that (2.7) is automatically satisfied if we express $\vec{B}$ as the curl of another vector field, i.e., we set

$$\vec{B} := \nabla \times \vec{A}. \tag{2.12}$$

The vector field $\vec{A}$ is called **vector potential**. Replacing (2.12) into (2.6), we can write (2.6) as

$$\nabla \times (\vec{E} + \frac{\partial \vec{A}}{\partial t}) = 0. \tag{2.13}$$

The vector identity

$$\nabla \times \nabla \phi = 0, \tag{2.14}$$

valid for every (smooth) scalar function $\phi = \phi(\vec{r})$, suggests to write $\vec{E}$ as

$$\vec{E} = -\frac{\partial \vec{A}}{\partial t} - \nabla \phi, \tag{2.15}$$

for some scalar function ϕ. This way, (2.13) and therefore (2.6) is automatically satisfied. The function ϕ is called the **scalar potential**. The scalar and vector potentials have to satisfy equations that are obtained by placing (2.12) and (2.15) into the (inhomogeneous) first two Maxwell equations (2.4), (2.5). The equations obtained are

$$\nabla^2 \phi + \frac{\partial}{\partial t}(\nabla \cdot \vec{A}) = -\frac{\rho}{\epsilon_0}, \tag{2.16}$$

$$\nabla(\nabla \cdot \vec{A}) - \nabla^2 \vec{A} + \frac{1}{c^2}\frac{\partial}{\partial t}\nabla \phi + \frac{1}{c^2}\frac{\partial^2 \vec{A}}{\partial t^2} = \mu_0 \vec{J}, \tag{2.17}$$

where, in the last one, we used the vector identity

$$\nabla \times (\nabla \times \vec{A}) = \nabla(\nabla \cdot \vec{A}) - \nabla^2 \vec{A}. \tag{2.18}$$

The variables $\vec{A}$ and ϕ replace $\vec{E}$ and $\vec{B}$ as field variables and equations (2.16) and (2.17) replace the Maxwell's equations in determining how the motion of charged particles affects the field.

2.1.1.3 Gauge and gauge transformation

The observable quantities in the Maxwell equations are the electric and magnetic fields, $\vec{E}$ and $\vec{B}$. The vector and scalar potentials $\vec{A}$ and ϕ themselves are not observable but have been defined to express the Maxwell equations in a simpler form. There exists a certain freedom in the choice of the **gauge**, i.e., the pair $\{\vec{A}, \phi\}$. In particular, given a gauge, $\vec{A}$ and ϕ, one can define a new one, $\vec{A}'$ and ϕ', equivalent to $\vec{A}$ and ϕ in that it gives the same electric and magnetic fields. This is achieved through the **gauge transformation**

$$\vec{A}' := \vec{A} + \nabla \eta, \tag{2.19}$$

$$\phi' := \phi - \frac{\partial \eta}{\partial t}, \tag{2.20}$$

where η is a function of $\vec{r}$ and t. Different gauge transformations can be chosen which simplify equations (2.16) and (2.17) in different ways and are more or less appropriate depending on the physical situation.

The **Coulomb gauge** is the most used gauge in problems where the classical electromagnetic theory is translated into quantum mechanics. It is defined by choosing $\vec{A}$ so that[6]

$$\nabla \cdot \vec{A} = 0. \tag{2.21}$$

If we use (2.21) in (2.16), (2.17), we obtain the two equations

$$\nabla^2 \phi = -\frac{\rho}{\epsilon_0}, \tag{2.22}$$

$$-\nabla^2 \vec{A} + \frac{1}{c^2}\frac{\partial}{\partial t}\nabla\phi + \frac{1}{c^2}\frac{\partial^2 \vec{A}}{\partial t^2} = \mu_0 \vec{J}, \tag{2.23}$$

the first of which is *Poisson's equation*, which only involves the scalar potential ϕ.

2.1.1.4 Separation of the electrostatic and radiative energy term

Consider now the term describing the electromagnetic field energy in (2.3),

$$\frac{\epsilon_0}{2}\int_{\mathbf{R}^3} \vec{E}^2(\vec{r}) + c^2\vec{B}^2(\vec{r})d\vec{r}. \tag{2.24}$$

A portion of the energy in (2.24) is due to the Coulomb interaction between particles. This is called *electrostatic energy*. The remaining energy is called *radiative energy*.

To separate the electrostatic and radiative energy in (2.24), we recall (see, e.g., [24] pp. 92 – 93) that every sufficiently smooth vector field $\vec{V}$, going to zero sufficiently fast at infinity, can be written as[7] (cf. Exercise 2.4)

$$\vec{V} = \vec{V}_L + \vec{V}_T. \tag{2.25}$$

The *irrotational* (also called *longitudinal*) part, $\vec{V}_L$, is by definition a vector field such that

$$\nabla \times \vec{V}_L = 0,$$

while the *divergence free* part (also called *transversal* part), $\vec{V}_T$, is a vector field satisfying

$$\nabla \cdot \vec{V}_T = 0.$$

Condition (2.21) for the Coulomb gauge means that the vector potential is chosen purely transversal (cf. Figures 2.1 and 2.2).

---

[6] Given an arbitrary $\vec{A}$, the vector potential $\vec{A}'$ in the Coulomb gauge is obtained by choosing η solution of Poisson's equation (cf. (2.19))

$$-\nabla \cdot \vec{A} = \nabla^2 \eta.$$

[7] This result is known as *Helmholtz decomposition theorem*.

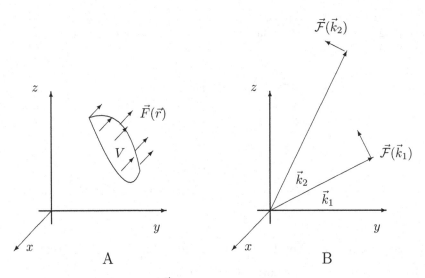

FIGURE 2.1:   Assume $\vec{F}$ represents the velocity of a fluid flow with constant density. For any region V in space the balance of material in and out of V is given by the surface integral $\int_{\partial V} \vec{F} \cdot \vec{n} dS$. According to Gauss's divergence theorem this integral is equal to the volume integral $\int_V \nabla \cdot \vec{F} d\vec{r}$. For a divergence-free vector field $\vec{F}$ this integral is always zero for every region V. So the balance of material in and out of V is zero for any region V (cf. Part A). The terminology 'transversal' derives from consideration of the spatial Fourier transform (cf. Exercise 2.4) of the vector field $\vec{F}$, $\mathcal{F}(\vec{k})$. This is perpendicular to $\vec{k}$ for every vector $\vec{k}$, as in Part B.

Under appropriate boundary conditions (the vector fields go to zero at infinity faster than $\frac{1}{|\vec{r}|^2}$), we have[8]

$$\int_{\mathbf{R}^3} \vec{V}_L^i(\vec{r}) \cdot \vec{V}_T^i(\vec{r}) \, d\vec{r} = 0. \tag{2.26}$$

Applying this to the energy of the field in (2.3) (recall that $\vec{B}$ is purely

---

[8]This fact is an immediate consequence of *Parseval-Plancherel theorem*

$$\int_{\mathbf{R}^3} \vec{F}^*(\vec{r}) \cdot \vec{G}(\vec{r}) d\vec{r} = \int_{\mathbf{R}^3} \vec{\mathcal{F}}^*(\vec{k}) \cdot \vec{\mathcal{G}}(\vec{k}) d\vec{k},$$

where $\vec{\mathcal{F}}$ and $\vec{\mathcal{G}}$ are the spatial Fourier transforms of the fields $\vec{F}$ and $\vec{G}$, respectively (cf. Exercise 2.4). It follows observing that the spatial Fourier transform of a longitudinal (transversal) vector field is parallel (orthogonal) to $\vec{k}$ (at every point $\vec{k}$) so that the Fourier transforms of longitudinal and transverse fields are orthogonal (see also, e.g., [67] section 1.B).

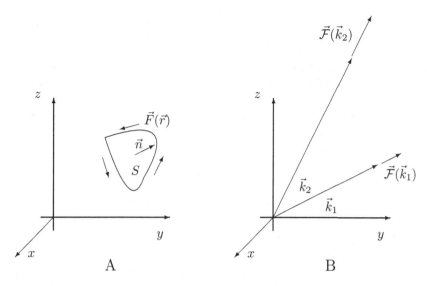

FIGURE 2.2: According to Stokes' theorem the *circulation* of a vector field $\vec{F}$ along the border of a surface S in $\mathbf{R}^3$, i.e., the line integral $\int_{\partial S} \vec{F} \cdot d\vec{l}$, is equal to the surface integral across S of the curl of $\vec{F}$, that is $\int_S \nabla \times \vec{F} \cdot \vec{n} dS$. For an irrotational vector field $\vec{F}$, $\nabla \times \vec{F}(\vec{r}) = 0$ for every $\vec{r}$. Therefore, the circulation is zero for every surface. The terminology 'transversal' derives from the consideration of the spatial Fourier transform (cf. Exercise 2.4) of the vector field $\vec{F}$, $\vec{\mathcal{F}}(\vec{k})$. This is parallel to $\vec{k}$ for every vector $\vec{k}$, as in Part B.

transversal because of (2.7)) we can write

$$\frac{\epsilon_0}{2} \int_{\mathbf{R}^3} \vec{E}^2(\vec{r}) + c^2 \vec{B}^2(\vec{r}) \, d\vec{r} = \frac{\epsilon_0}{2} \int_{\mathbf{R}^3} \vec{E}_L^2(\vec{r}) \, d\vec{r} + \frac{\epsilon_0}{2} \int_{\mathbf{R}^3} \vec{E}_T^2(\vec{r}) + c^2 \vec{B}_T^2(\vec{r}) \, d\vec{r}.$$

The first term is the electrostatic potential energy of the system due to the Coulomb interaction among the particles. In fact, for a discrete distribution of charges (2.8) we can prove that (see the following Remark 2.1.1)

$$\frac{\epsilon_0}{2} \int_{\mathbf{R}^3} \vec{E}_L^2(\vec{r}) \, d\vec{r} = \frac{1}{8\pi\epsilon_0} \sum_{k \neq j} \frac{q_k q_j}{|\vec{r}_k - \vec{r}_j|} := V_{coul}. \tag{2.27}$$

We write the total energy H of the system as

$$H = \sum_{j=1}^{N} \frac{1}{2} m_j \left(\frac{d\vec{r}_j}{dt} \right)^2 + V_{coul} + \frac{\epsilon_0}{2} \int_{\mathbf{R}^3} \vec{E}_T^2(\vec{r}) + c^2 \vec{B}^2(\vec{r}) d\vec{r}, \tag{2.28}$$

separating the kinetic energy of the charged particles (first term), the electrostatic potential energy due to the Coulomb interaction among charged particles (second term, defined in (2.27)) and the radiative energy of the field (third term).

Remark 2.1.1 (Proof of formula (2.27)) Formula (2.27) is valid modulo a constant term (theoretically infinity) which is independent of the mutual positions of the particles. This term is called *self-energies* and is typically not included in the expression of the energy. To derive (2.27), use (2.15) and the fact that in the Coulomb gauge $\vec{A}_L = 0$ to write

$$\frac{\epsilon_0}{2} \int_{\mathbf{R}^3} \vec{E}_L^2(\vec{r})\, d\vec{r} = \frac{\epsilon_0}{2} \int_{\mathbf{R}^3} (\nabla\phi)^2\, d\vec{r}, \tag{2.29}$$

since $\nabla\phi$ is purely longitudinal from (2.14). Recall now the integration by parts formula given by Green's first identity (see, e.g., [261]). For every volume V and smooth function ϕ,

$$\int_V (\nabla\phi)^2\, d\vec{r} = \int_{\partial V} \phi \vec{n} \cdot \nabla\phi\, dS - \int_V \phi \nabla^2\phi\, d\vec{r},$$

where the first integral on the right-hand side is a surface integral over ∂V, the boundary of V. Using this in (2.29) we can write

$$\frac{\epsilon_0}{2} \int_{\mathbf{R}^3} (\nabla\phi)^2 d\vec{r} = \frac{\epsilon_0}{2} \left(\lim_{V \to \infty} \int_{\partial V} \phi \vec{n} \cdot \nabla\phi\, dS - \int_V \phi \nabla^2\phi\, d\vec{r} \right). \tag{2.30}$$

We assume that the potential ϕ is chosen to be zero at infinity in a way so that the limit of the first integral of the right-hand side is zero. Using this and equation (2.22), we obtain

$$\frac{\epsilon_0}{2} \int_{\mathbf{R}^3} \vec{E}_L^2(\vec{r})\, d\vec{r} = \frac{1}{2} \int_{\mathbf{R}^3} \phi\rho\, d\vec{r}. \tag{2.31}$$

Poisson equation (2.22), assuming a discrete distribution of charges (2.8), has solution vanishing at infinity

$$\phi(\vec{r}) = \frac{1}{4\pi\epsilon_0} \sum_{j=1}^{N} \frac{q_j}{|\vec{r} - \vec{r}_j|}. \tag{2.32}$$

By plugging expressions (2.8) and (2.32) into (2.31) some terms become infinity. These terms are called *self-energies*. They are not considered as they do not depend on the mutual positions of the various charges. Neglecting these terms, we obtain (2.27).

2.1.1.5 System in the presence of an externally applied electro-magnetic field

The most interesting situation for our purposes is when the system of particles and field is interacting with an *external system of particles and field*.

In the external system, the motion of the charges, i.e., the functions $\rho(\vec{r}, t)$ and $\vec{J}(\vec{r}, t)$, are fixed and independent of the field. This means in practice that the back reaction of the field on the particles is very weak and-or it is compensated by other forces in the experiment. These functions determine the external field according to Maxwell's equations. Because of the linearity of Maxwell's equations, the total field will be the sum of the proper field of the particles and the external field. Also, from the linearity of (2.16) and (2.17) the vector and scalar potentials are the sum of the ones proper of the system and the ones of the external system. The situation is summarized in the following diagram which extends the one given in (2.1.1.1). The external field plays the role of the control $u(t)$.

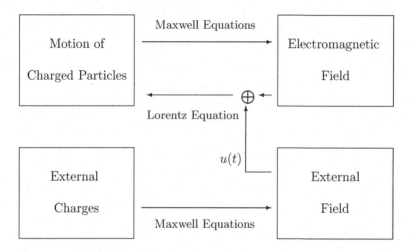

In the presence of an external field, the energy of the system of particles and field (excluding the external system) is not constant with time and it has to be modified (cf. [67] Complement C_{II} and pg. 174) by adding a potential energy term which, in the case of a discrete system of particles, takes the form

$$V_e := \sum_{j=1}^{N} q_j \phi_e(\vec{r}_j, t), \tag{2.33}$$

where $\phi_e(\vec{r}_j, t)$ is the scalar potential associated with the external system. This leads to the expression of the energy associated with a system subject to an external field (cf. (2.28) and (2.27))

$$H = \sum_{j=1}^{N} \frac{1}{2} m_j \left(\frac{d\vec{r}_j}{dt} \right)^2 + V_{coul} + \sum_{j=1}^{N} q_j \phi_e(\vec{r}_j, t) + \frac{\epsilon_0}{2} \int_{\mathbf{R}^3} \vec{E}_T^2(\vec{r}) + c^2 \vec{B}^2(\vec{r}) d\vec{r},$$

$$\tag{2.34}$$

with ϕ_e the external potential, while the external vector potential is hidden in the last integral in (2.34) which contains contributions from both the field generated by the particles and the external field.

Example 2.1.2 Let us reconsider the system of a particle in a potential well of Example 1.3.1 and its possible realization as a system of one charged particle in an external field. In particular, we may assume the particle to be an electron confined between two layers of negative charges at $x = 0$ and $x = L$, for example in a semiconductor. The charges of the two layers represent the *external charges* which set up the external field. In a one-dimensional problem, the x component of the electric external field $E_x := E_x(x)$ has large magnitude and it is negative close to $x = 0$ and it has large magnitude and it is positive close to $x = L$, while it is approximately equal to zero for $0 < x < L$. Consider equation (2.23). Since there are no external currents, i.e., $\vec{J} = 0$, we can take the external vector potential $\vec{A}_e$ equal to zero.[9] The external scalar potential ϕ_e satisfies (cf. (2.15))

$$-E_x(x) = \frac{d\phi_e}{dx}, \qquad (2.35)$$

and integrating this we obtain a shape of ϕ_e which is constant (and negative) equal to some $-\phi_{e0}$ until $x = 0$ increases until a value zero and then decreases at $x = L$ again to $-\phi_{e0}$. The potential energy for a particle at position x is obtained according to (2.33) as a function of the position of the particle x as $V_e = e\phi_e(x)$ where e is the charge of the electron. It has the shape of a quantum well as considered in Example 1.3.1 in the case $V_0 := -e\phi_{e0}$ very large. In fact the general expression of the energy (2.34) gives (1.46) in this case with $V = V_e$. The discussion is summarized in Figure 2.3. Since there is only one charge in the well $V_{coul} = 0$. If there is more than one charge the term V_{coul} is added to the energy. This scheme can be modified by adding a further external *control* electric field possibly time varying which modifies the shape of the well, for example by placing the system in a capacitor. In that case the model naturally extends. There will be an extra (control) potential energy obtained integrating (2.35) with the control field playing the role of E_x. This is a common scheme to control electron motion in semiconductors quantum dots (see, e.g., [267]). Some analysis on the effect of an external field on the motion of a single electron in a quantum well is given in [206].

---

[9]We can assume Coulomb gauge here so that $\vec{A}_e$ is purely transversal while $\frac{\partial}{\partial t}\nabla\phi_e$ is purely longitudinal from (2.14). Therefore, (2.23) becomes $\frac{1}{c^2}\frac{\partial^2 \vec{A}_e}{\partial t^2} - \nabla^2 \vec{A}_e = -\frac{1}{c^2}\frac{\partial}{\partial t}\nabla\phi_e$. On the left-hand side the partial derivative with respect to t is purely transversal because $\vec{A}_e$ is purely transversal. Moreover $\nabla^2 \vec{A}_e$ is also purely transversal, because from formula (2.18) $\nabla^2 \vec{A}_e = -\nabla \times (\nabla \times \vec{A}_e)$, and because of the vector identity (2.11). Equality of two vector fields, one purely longitudinal and one purely transversal implies that they are both zero.

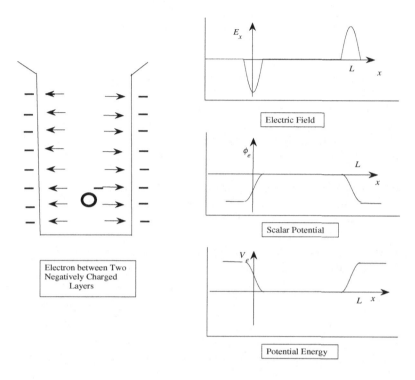

FIGURE 2.3: Electrostatic realization of the motion of a single charged particle in a potential well. An electron is confined between two negatively charged layers. A possible shape of the electric field E_x, the scalar potential ϕ_e and the potential energy V_e is indicated.

2.2 Quantum Theory of Interaction of Particles and Fields

2.2.1 Canonical quantization

Canonical quantization is the process which associates quantities in classical mechanics with operators in quantum mechanics. This process allows us to translate the classical description of a system to the quantum mechanical one and, in particular, to obtain the quantum mechanical Hamiltonian from the classical Hamiltonian. The general idea is that, given a system, one considers the canonical variables of Lagrangian mechanics and the associated canonical momenta (cf. Appendix B and [67]). With a canonical variable x and associated canonical momentum p, one associates operators $\hat{x}$ and $\hat{p}$ and the underlying Hilbert space can be chosen as spanned by the eigenvectors of $\hat{x}$ (or $\hat{p}$). The way $\hat{x}$ and $\hat{p}$ are defined has to be such that[10]

$$[\hat{x}, \hat{p}] = i\hbar \mathbf{1}. \tag{2.36}$$

The Hamiltonian operator, which determines the dynamics of the system, according to the Schrodinger equation, is obtained by transforming the expression of the Hamiltonian of classical mechanics (which is written only in terms of canonical variables and associated momenta) by replacing all the canonical variables and momenta by the corresponding operator. In this process *sums* are replaced by *sums of operators* and *products* are replaced by *compositions of operators* (cf. 1.1.2.3). Care is needed due to the fact that, even though the classical variable x commutes with its associated momentum p, the corresponding operators $\hat{x}$ and $\hat{p}$ do not commute (cf. (2.36)). A symmetrized form for xp can be used, that is $\frac{1}{2}(\hat{x}\hat{p} + \hat{p}\hat{x})$. More in general, any observable in classical mechanics which is a a function of canonical variables and momenta is replaced by the corresponding formula in terms of operators. When we make this association between certain classical variables and quantum mechanical operators we say that we '*quantize*' such variables.

Remark 2.2.1 In the passage from classical variables to quantum mechanical operators, one would like some of the structure of classical mechanics to have a clear counterpart in quantum mechanics. The impossibility of achieving this is the subject of several no-go type of theorems such as Groenewold's theorem [124]. Moreover, there are several ways to establish such a correspondence [19]. For the case of systems of particles and field quantization, the original work of the founders of quantum mechanics only referred to the variables concerning the particles, in particular position and momentum, leaving the fields as classical variables. This is also known as *first quantization*. *Second quantization* refers to the situation where both field and particles variables are quantized and in fact one can see particles as originated by fields (and vice versa) as in *quantum field theory* (see, e.g., [94]).

Consider now the classical system of charged particles and fields treated in the previous section. In order to give its quantum mechanical description, we should first identify the canonical variables and momenta and express the Hamiltonian (2.34) in terms of these. The system consists of two parts, the particles and the field, and there will be canonical variables and momenta associated with both particles and field. The corresponding operators will act on the Hilbert space associated with the particles motion, $\mathcal{H}_P$ or on the Hilbert space associated with the field $\mathcal{H}_F$, respectively. The total Hilbert space is $\mathcal{H}_P \otimes \mathcal{H}_F$ (cf. subsection 1.1.3), while $\mathcal{H}_P$ itself is the tensor product of Hilbert

---

[10]For a discussion and motivations for this requirement and relation with classical mechanics see, e.g., Chapter 15 of [192].

spaces each associated to one particle. The variables corresponding to the field in the Hamiltonian (2.34), i.e., $\vec{E}_T$, $\vec{B}$, $\vec{A}$ are replaced by the corresponding operators which are functions of operators associated to canonical variables and momenta of the field.

We shall not treat in detail the quantization of the electromagnetic field in this book as this is the object of study of *quantum optics*. We refer to [67] for a complete treatment. However, since the charged particles are the object of the control we shall focus on the particles part of the Hilbert space $\mathcal{H}_P$. We start by identifying the canonical variables and momenta for the particles, then we quantize them and, in the next subsection, we give the expression of the quantum mechanical Hamiltonian.

2.2.1.1   Canonical variables and momentum for the charged particles

The canonical variables for a system of N charged particles in an electromagnetic field are given by $\vec{r}_j$, $j = 1, ..., N$. The **canonical momentum** associated with $\vec{r}_j$, which we denote by $\vec{p}_j$, for particle j, is defined as the sum of the *kinematical momentum* $m_j\dot{\vec{r}}_j$ and a term proportional to the vector potential at the location of the particle, given by $q_j\vec{A}_{tot}(\vec{r}_j)$.[11] One has

$$\vec{p}_j := m_j\dot{\vec{r}}_j + q_j\vec{A}_{tot}(\vec{r}_j). \qquad (2.37)$$

Here $\vec{A}_{tot}$ is the total vector potential sum of the vector potential $\vec{A}$ of the system and the vector potential of the applied external field $\vec{A}_e$, i.e.,

$$\vec{A}_{tot}(\vec{r}, t) := \vec{A}(\vec{r}, t) + \vec{A}_e(\vec{r}, t).$$

With this definition, the energy (2.34) reads

$$H = \sum_{j=1}^{N} \frac{1}{2m_j}(\vec{p}_j - q_j\vec{A}_{tot}(\vec{r}_j))^2 + V_{coul} + \sum_{j=1}^{N} q_j\phi_e(\vec{r}_j, t) \qquad (2.38)$$

$$+ \frac{\epsilon_0}{2}\int_{\mathbf{R}^3} \vec{E}_T^2(\vec{r}) + c^2\vec{B}^2(\vec{r})d\vec{r}.$$

2.2.1.2   Rules of canonical quantization for variables describing the charged particles

We work in the position representation and therefore we describe the canonical quantization specifying how the quantum mechanical operators act on the

---

[11]In a description in terms of Lagrangian mechanics definition (2.37) follows from taking the derivative of the Lagrangian associated with the system with respect to $\dot{\vec{r}}_j$. This derivation is presented in Appendix B (cf. formula (B.25)) along with an overview of Lagrangian and Hamiltonian mechanics as applied to the system of interacting charged particles and fields. A complete treatment can be found in [67], [118].

wave function. In particular the full system of particles and field lives in the Hilbert space $\mathcal{H}_P \otimes \mathcal{H}_F$ and assume that field space has countable orthonormal basis $\{|k\rangle\}$, $k = 1, 2, ...$ The state of the full system is written in the position (and $\{|k\rangle\}$) representation as

$$|\psi_{full}\rangle := \sum_k \int_{\mathbf{R}^{3N}} \psi_{full}(\vec{r}, k)|\vec{r}\rangle \otimes |k\rangle d\vec{r}.$$

The state associated to the particles is obtained by calculating the *partial trace* with respect to the field degrees of freedom (cf. 9.1.1.3), which gives

$$|\psi\rangle = \int_{\mathbf{R}^{3N}} \psi(\vec{r})|\vec{r}\rangle d\vec{r},$$

where the *wave function* $\psi(\vec{r})$ is defined by

$$\psi(\vec{r}) := \sum_k \psi_{full}(\vec{r}, k).$$

The *rules of canonical quantization* associate with the canonical momentum $\vec{p}_j$ of the j-th particle in (2.37) the operator $\hat{p}_j$, which is a triple of operators $\hat{p}_j := \{\hat{p}_{jx}, \hat{p}_{jy}, \hat{p}_{jz}\}$, each corresponding to one component of $\vec{p}_j$. The operator $\hat{p}_j$ acts on the state of the particles $|\psi(t)\rangle$ in (2.1) according to

$$\hat{p}_{jl}\psi(\vec{r}, t) := -i\hbar\frac{\partial}{\partial l}\psi(\vec{r}, t), \qquad l = x_j, y_j, z_j, \qquad j = 1, ..., N,$$

or, in compact vector notation,

$$\hat{p}_j\psi(\vec{r}, t) = -i\hbar\nabla_j\psi(\vec{r}, t), \qquad j = 1, ..., N,$$

where ∇_j denotes the gradient with respect to coordinates of the j-th particle.

The rules of canonical quantization associate with the position

$$\vec{r}_j := \{x_j, y_j, z_j\}$$

of the j-th particle the operator $\hat{r}_j := \{\hat{x}_j, \hat{y}_j, \hat{z}_j\}$. $\hat{r}_j$ acts on the state of the particles $|\psi(t)\rangle$ as a multiplication by the corresponding component, i.e.,

$$\hat{m}_j\psi(\vec{r}, t) = m_j\psi(\vec{r}, t), \qquad m_j = x_j, y_j, z_j, \qquad j = 1, ..., N,$$

or, in compact vector notation,

$$\hat{r}_j\psi(\vec{r}, t) = \vec{r}_j\psi(\vec{r}), \qquad j = 1, ..., N.$$

From these definitions it follows that the following commutation relation (cf. (2.36)) holds (cf. Exercise 2.2)

$$[\hat{p}_{jl}, \hat{m}_k] = -i\hbar\delta_{lm}\delta_{jk}, \qquad l, m = x, y, z, \qquad j, k = 1, ..., N. \qquad (2.39)$$

Example 2.2.2 Going back to Example 2.1.2 and applying these quantization rules with the Hamiltonian (2.34) specialized to this case, we obtain the Hamiltonian of the form (1.47) of Example 1.3.1. This Hamiltonian can be modified by adding an extra (time-varying) control electric field which results in an extra term for the scalar potential as already discussed in Example 2.1.2.

2.2.2 Quantum mechanical Hamiltonian

Let us consider again the full classical Hamiltonian for a system for interacting particles and fields in the presence of an external driving field (2.34). Using the definition of the canonical momentum (2.37), we have

$$H = \sum_{j=1}^{N} \frac{1}{2m_j} \left(\vec{p}_j - q_j \vec{A}_{tot}(\vec{r}_j, t) \right)^2 + V_{coul} + \sum_{j=1}^{N} q_j \phi_e(\vec{r}_j, t) \qquad (2.40)$$

$$+ \frac{\epsilon_0}{2} \int_{\mathbf{R}^3} \vec{E}_T^2(\vec{r}) + c^2 \vec{B}^2(\vec{r}) \, d\vec{r}.$$

In anticipation of applying canonical quantization, we identify the variables associated with the particles, $\vec{p}_j$ and $\vec{r}_j$, and the ones associated with the field $\vec{E}$, $\vec{A}$, $\vec{B}$. We write the energy Hamiltonian (2.40) as

$$H := H_R + H_P + H_I,$$

with

$$H_R := \frac{\epsilon_0}{2} \int_{\mathbf{R}^3} \vec{E}_T^2(\vec{r}) + c^2 \vec{B}^2(\vec{r}) \, d\vec{r},$$

$$H_P := \sum_{j=1}^{N} \frac{1}{2m_j} (\vec{p}_j)^2 + V_{coul},$$

$$H_I := -\sum_{j=1}^{N} \frac{q_j}{m_j} \vec{p}_j \cdot \vec{A}_{tot}(\vec{r}_j, t) + \sum_{j=1}^{N} \frac{q_j^2}{2m_j} \vec{A}_{tot}^2(\vec{r}_j, t) + \sum_{j=1}^{N} q_j \phi_e(\vec{r}_j, t). \quad (2.41)$$

H_R is the energy of the field only. It is transformed into an Hamiltonian operator which acts only on the part of the Hilbert space $\mathcal{H}_P \otimes \mathcal{H}_F$ concerning the field, i.e., on $\mathcal{H}_F$. This term, once quantized, has the form $\hat{H}_R := 1 \otimes \hat{H}_{RF}$ for an operator $\hat{H}_{RF}$ on $\mathcal{H}_F$.

The term H_P gives a quantum mechanical operator $\hat{H}_P$ according to the rules of canonical quantization. The operator $\hat{H}_P$ is given by

$$\hat{H}_P = \sum_{j=1}^{N} -\frac{1}{2m_j} \hbar^2 \nabla_j^2 + \hat{V}_{coul},$$

where from (2.27) the operator $\hat{V}_{coul}$ multiplies the wave function by

$$V_{coul} := \frac{1}{8\pi\epsilon_0} \sum_{k \neq j} \frac{q_k q_j}{|\vec{r}_k - \vec{r}_j|}.$$

All the terms appearing in H_I in (2.41) contain variables concerning both the particles and the field, both the external field and the field created by the particles. This corresponds to the *interaction* part of the quantum mechanical Hamiltonian which contains operators acting simultaneously on the particles Hilbert space $\mathcal{H}_P$ and the field Hilbert space $\mathcal{H}_F$. These are possibly time-varying operators as a consequence of the presence of the external field. Since $\vec{A}_{tot} = \vec{A} + \vec{A}_e$, we write H_I as

$$H_I = H_{IED} + H_{Ie},$$

where

$$H_{IED} := \sum_{j=1}^{N} \frac{q_j^2}{2m_j} \left(\vec{A}^2(\vec{r}_j, t) + 2\vec{A} \cdot \vec{A}_e(\vec{r}_j, t) \right) - \sum_{j=1}^{N} \frac{q_j}{m_j} \vec{p}_j \cdot \vec{A}(\vec{r}_j), \quad (2.42)$$

and

$$H_{Ie} : \sum_{j=1}^{N} \frac{q_j^2}{2m_j} \vec{A}_e^2(\vec{r}_j, t) - \sum_{j=1}^{N} \frac{q_j}{m_j} \vec{p}_j \cdot \vec{A}_e(\vec{r}_j, t) + \sum_{j=1}^{N} q_j \phi_e(\vec{r}_j, t). \quad (2.43)$$

The term H_{IED} models the *electrodynamic* interaction between particles and fields, possibly time varying, because of the presence of the external field. This term is the interaction which happens through the vector potential $\vec{A}$ as opposed to the *electrostatic* interaction which is the Coulomb interaction V_{coul}. The term H_{Ie} models the interaction with the external field only. The quantum mechanical operator associated to this term is

$$\hat{H}_{Ie} = \sum_{j=1}^{N} \frac{q_j^2}{2m_j} \hat{A}_e^2(\vec{r}_j, t) + \sum_{j=1}^{N} \frac{q_j}{m_j} i\hbar \nabla_j \hat{A}_e(\vec{r}_j, t) + \sum_{j=1}^{N} q_j \hat{\phi}_e(\vec{r}_j, t).$$

According to the rules of (first) canonical quantization the operators associated with the external fields correspond to multiplication by functions. Therefore, in this expression, $\hat{\phi}_e(\vec{r}_j, t)\psi = \phi_e(\vec{r}_j, t)\psi$, where the second expression is multiplication of two functions. The operator $\hat{A}_e = \hat{A}_e(\vec{r}_j, t)$ applied to the wave function ψ gives a vector whose components are the components of A_e multiplied by ψ. We have $\hat{A}_e(\vec{r}_j, t)\psi = \psi \vec{A}_e(\vec{r}_j, t)$, where the second term represents multiplication of a vector function by a scalar. When applied to

$\hat{A}_e(\vec{r}_j, t)\psi$, ∇_j returns a $\nabla_j \cdot \vec{A}_e(\vec{r}_j, t)\psi$, which is a scalar function again.[12] The operator $\hat{A}_e^2(\vec{r}_j, t)$ is multiplication of ψ by the function $\vec{A}_e^2(\vec{r}_j, t)$.

We illustrate the mathematical definitions with a simple example without making reference to a particular physical situation.

Example 2.2.3 Consider a system of two particles in 3-D space. The coordinates are $\vec{r} := \{x_1, y_1, z_1, x_2, y_2, z_2\}$. Let the wave function be

$$\psi = \psi(\vec{r}) = \psi(x_1, y_1, z_1, x_2, y_2, z_2) = \frac{1}{\pi}e^{-(x_1^2 + x_2^2)}.$$

Consider an external scalar and vector potential ϕ_e and $\vec{A}_e$, independent of time and given by

$$\phi_e = 2 - x^2,$$

and

$$\vec{A}_e = y\vec{i} + x\vec{j},$$

which is in the Coulomb gauge since $\nabla \cdot A_e = 0$. We calculate

$$\hat{\phi}_e(\vec{r}_1)\psi = \frac{2 - x_1^2}{\pi}e^{-(x_1^2 + x_2^2)},$$

$$\hat{A}_e(\vec{r}_1)\psi = \frac{1}{\pi}e^{-(x_1^2 + x_2^2)}(y_1\vec{i} + x_1\vec{j}),$$

and

$$\nabla_1 \cdot \hat{A}_e(\vec{r}_1)\psi = \frac{-2x_1 y_1}{\pi}e^{-(x_1^2 + x_2^2)}.$$

2.3   Introduction of Approximations, Modeling and Applications to Molecular Systems

In modeling real systems of charged particles and fields the Hamiltonian (2.40) is almost never considered in full. Appropriate approximations are introduced depending on the particular situation at hand. Consider the *control*

---

[12]Notice that there is ambiguity in this quantization process. Since we are quantizing the product $\vec{p}_j \cdot \vec{A}_e(\vec{r}_j, t)$ we might as well have used the operator $i\hbar\vec{A}_e(\vec{r}_j, t) \cdot \nabla_j$ or the symmetrized form $\frac{1}{2}i\hbar(\vec{A}_e(\vec{r}_j, t) \cdot \nabla_j + \nabla_j \cdot \vec{A}_e(\vec{r}_j, t))$. However, as we assume $\vec{A}_e$ is chosen in the Coulomb gauge all these choices are equivalent. In fact, from the product rule, we have

$$\nabla_j \cdot (\vec{A}_e(\vec{r}_j)\psi) = (\nabla_j \cdot \vec{A}_e(r_j))\psi + \vec{A}_e(\vec{r}_j) \cdot \nabla_j\psi,$$

and using the fact that, in the Coulomb gauge, $\nabla_j \cdot \vec{A}_e(\vec{r}_j) = 0$, we have

$$\nabla_j \cdot (\vec{A}_e(\vec{r}_j)\psi) = \vec{A}_e(\vec{r}_j) \cdot \nabla_j\psi.$$

of molecular or atomic dynamics where the objectives of the control may include: maximizing the amount of a desired product in a chemical reaction, expediting a given reaction, or dissociating a molecule. The system of interest consists of the electrons and nuclei of the (reacting) atoms. These are treated as charged particles in an electromagnetic field, as in the previous section. The underlying Hilbert space for the particles is spanned by the position eigenvectors of such particles, and therefore it is uncountably infinite-dimensional.

2.3.1 Approximations for molecular and atomic systems

2.3.1.1 Neglecting electrodynamic interactions; the semiclassical approximation

In the control of molecular or atomic systems it can be assumed that the particles (electrons and nuclei) are sufficiently near so that the Coulomb interaction alone is the dominant term which takes into account the interaction among particles. The electrodynamic interaction terms corresponding to (2.42), and in particular the vector potential $\vec{A}$, are therefore considered small and neglected. This also allows us to decouple the part of the system referring to the own system field, corresponding to the part of the Hilbert space $\mathcal{H}_F$, from the other system variables. The only interaction of the particles with a field considered is the one with the *external field*. Therefore, we can consider only the part of the Hilbert space modeling the particles and the Hamiltonian contains no field other than the external field. Considering the external fields as stronger than the fields responsible for the electrodynamic interaction and therefore neglecting these ones is referred to the *semiclassical approximation*. In this situation the classical external fields are assumed not affected by the interaction with the particles and they are not quantized. They are treated as external classical inputs.

The classical Hamiltonian, after these approximations, is written as

$$II_P + II_{le} - \sum_{j=1}^{N} (\vec{p}_j)^2 + V_{coul} - \sum_{j=1}^{N} \frac{q_j}{m_j} \vec{p}_j \cdot \vec{A}_e(\vec{r}_j, t) \qquad (2.44)$$

$$+ \sum_{j=1}^{N} \frac{q_j^2}{2m_j} \vec{A}_e^2(\vec{r}_j, t) + \sum_{j=1}^{N} q_j \phi_e(\vec{r}_j, t).$$

2.3.1.2 The long-wavelength approximation and the electric dipole interaction

The last three terms in the Hamiltonian (2.44) model the interaction between the external field and the motion of the particles. In fact, they depend on both $\vec{A}_e$, ϕ_e and the positions of the j-th particle $\vec{r}_j$. It is common to rewrite them in terms of the dot product of the electric field $\vec{E}$ and the dipole

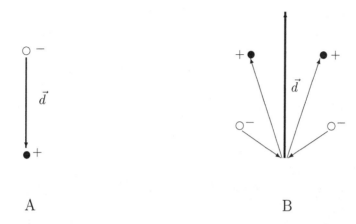

FIGURE 2.4: The electric dipole moment $\vec{d}$ of a system of one positive and one negative charge (Part A) such as an hydrogen atom and the electric dipole moment of a system of two positive and two negative charges (Part B).

moment(cf. examples in Figure 2.4)

$$\vec{d} = \sum_{j=1}^{N} q_j \vec{r}_j, \tag{2.45}$$

where, in the case of a molecule or atom, the q_j's correspond to the charge of the electrons and protons.[13]

This requires a gauge transformation and an approximation. First perform a gauge transformation of the form (2.19), (2.20), with η given by

$$\eta = \eta(\vec{r}, t) = -\vec{r} \cdot \vec{A}_e(\vec{r}_0, t),$$

where $\vec{r}_0$ is a fixed position that can be taken at the center of mass of the molecule. The new vector and scalar potentials are given by

$$\vec{A}'_e(\vec{r}, t) = \vec{A}_e(\vec{r}, t) - \vec{A}_e(\vec{r}_0, t),$$

$$\phi'_e(\vec{r}, t) = \phi_e(\vec{r}, t) + \vec{r} \cdot \frac{\partial \vec{A}_e(\vec{r}_0, t)}{\partial t}.$$

---

[13]For a single atom, the protons may be considered giving no contribution as they may be assumed concentrated in one point.

We notice that this gauge transformation keeps $\vec{A}$ in the Coulomb gauge, as $\nabla \cdot \vec{A}' = 0$ when $\nabla \cdot \vec{A} = 0$. Take $\vec{r}_0 = \vec{0}$ and assume, without loss of generality, that ϕ_e had been chosen so that $\phi_e(\vec{0}, t) \equiv 0$ for every t.[14] In the expression of the classical Hamiltonian (2.44) ϕ_e, $\vec{A}_e$ and $\vec{p}_j$ have to be replaced by the new ϕ'_e, $\vec{A}'_e$ and $\vec{p}'_j$. We now make the *long-wavelength approximation* assuming that the field does not change significantly near $\vec{0}$ and that the particles stay close to $\vec{0}$, i.e., $\vec{r}_j \approx \vec{0}$, for all $j = 1, ..., N$. Therefore, we have

$$\vec{A}'_e(\vec{r}_j, t) := \vec{A}_e(\vec{r}_j, t) - \vec{A}_e(\vec{0}, t) \approx 0, \tag{2.46}$$

and we expand to first order in $\vec{r}$

$$\phi'_e(\vec{r}, t) \approx \phi_e(\vec{0}, t) + \frac{\partial}{\partial t}\vec{A}_e(\vec{0}, t) \cdot \vec{r} + \nabla\phi_e(\vec{0}, t) \cdot \vec{r}.$$

Using (2.15) and the assumption on the value of ϕ_e at the origin, we have

$$\phi'_e(\vec{r}, t) \approx -\vec{E}_e(\vec{0}, t) \cdot \vec{r},$$

where $\vec{E}_e$ is the external electric field. Replacing this and (2.46) in the energy (2.44) (written in the new gauge) we obtain

$$H_P + H_{Ie} = \sum_{j=1}^{N}(\vec{p}'_j)^2 + V_{coul} - \sum_{j=1}^{N} q_j \vec{E}_e(\vec{0}, t) \cdot \vec{r}_j,$$

and using the definition of the dipole moment (2.45), we have

$$H_P + H_{Ie} = \sum_{j=1}^{N}(\vec{p}'_j)^2 + V_{coul} - \vec{d} \cdot \vec{E}_e(\vec{0}, t). \tag{2.47}$$

We denote by $\hat{d} := \{\hat{d}_x, \hat{d}_y, \hat{d}_z\}$ the quantum mechanical operator corresponding to $\vec{d}$. We have

$$\hat{d} = \sum_{j=1}^{N} q_j \hat{r}_j.$$

Quantizing, the quantum mechanical (dipole) Hamiltonian corresponding to (2.47) is

$$\hat{H}_P + \hat{H}_{Ie} = \sum_{j=1}^{N} -\frac{1}{2m_j}\hbar^2\nabla_j^2 + \hat{V}_{coul} - \hat{d} \cdot \vec{E}_e(\vec{0}, t). \tag{2.48}$$

---

[14]This could have been obtained by a preliminary gauge transformation which leaves $\vec{A}_e$ unchanged.

2.3.2 Controlled Schrödinger wave equation

It is convenient to separate the Hamiltonian of a quantum mechanical system interacting with an external field into a constant part and a time-varying part depending on the external field. We denote the resulting quantum mechanical Hamiltonian (after the appropriate approximations) by $\hat{H}$ and write it as

$$\hat{H} = \hat{H}_0 + \hat{H}_I(t). \tag{2.49}$$

The operator $\hat{H}_0$, which represents sums of kinetic and potential energy in absence of the external transversal field, is called the *internal Hamiltonian*. The operator $\hat{H}_I$ represents the interaction energy between the system and the external field and it is called *interaction Hamiltonian*. For $\hat{H} = \hat{H}_P + \hat{H}_{Ie}$ in (2.48) we have

$$\hat{H}_0 = \hat{H}_P = \sum_{j=1}^{N} -\frac{1}{2m_j}\hbar^2\nabla_j^2 + \hat{V}_{coul},$$

and

$$\hat{H}_I = \hat{H}_{Ie} = -\vec{d}\cdot\vec{E}(\vec{0},t).$$

The *controlled Schrödinger wave equation* reads as

$$i\hbar\frac{\partial\psi(\vec{r},t)}{\partial t} = (\hat{H}_0 + \hat{H}_I)\psi(\vec{r},t). \tag{2.50}$$

In the absence of the external field, the state of the molecule will be described by a wave function $\psi_s(\vec{r},t)$, solution of the Schrödinger wave equation (cf. (2.2))

$$i\hbar\frac{\partial\psi_s(\vec{r},t)}{\partial t} = \hat{H}_0\psi_s(\vec{r},t). \tag{2.51}$$

A wave function $\psi_k(\vec{r},t)$ corresponding to a fixed value of the energy E_k has to satisfy the Schrodinger wave equation

$$\hat{H}_0\psi_k(\vec{r},t) = E_k\psi_k(\vec{r},t). \tag{2.52}$$

There are various possible values for the energy of the system which might form a discrete set or a continuous set of a mixed continuous and discrete set. The state at the lowest value of the energy is called the *ground state* while the other states are called *excited states*.

Placing (2.52) into (2.50) and expressing $\psi_k(\vec{r},t)$ as $\psi_k(\vec{r},t) = \alpha_k(t)\phi_k(\vec{r})$, we can solve the partial differential equation by separation of variables and obtain the solution

$$\psi_k(\vec{r},t) = \phi_k(\vec{r})e^{\frac{-iE_kt}{\hbar}},$$

where $\phi_k(\vec{r})$ is solution of the *time-independent Schrödinger wave equation* (cf. (2.52))

$$\hat{H}_0\phi_k(\vec{r}) = E_k\phi_k(\vec{r}).$$

The (approximate) solution of this equation in molecular dynamics is obtained by methods of quantum chemistry (see, e.g., [266]). This way, one obtains the initial equilibrium state $\phi_k(\vec{r})$. Let us assume that there is a discrete set of energy values E_k along with the eigenstates $\phi_k(\vec{r})$ associated with them. It is also customary to assume that we work in a finite range of energies. This is referred to as **multilevel approximation** or **truncation**.[15] The control problem in molecular dynamics can often be formulated as transferring the state of a system between two eigenstates corresponding to two different values of the energy.

Writing the general form of the wave function solution of (2.2) in terms of the eigenstates ϕ_k of the Hamiltonian operator $\hat{H}_0$, we obtain

$$\psi(\vec{r}, t) = \sum_{k=1}^{n} c_k(t)\phi_k(\vec{r}), \qquad (2.53)$$

with c_k complex coefficients and $\sum_k |c_k|^2 = 1$. $|c_k|^2$ are the probabilities for the system to be in eigenstate $\phi_k(\vec{r})$ and have energy E_k.[16]

2.3.2.1   The control system

By placing (2.53) into (2.50), we find that the complex, norm 1, vector $\vec{c} = \{c_1, c_2, ..., c_d\}$ satisfies the linear ordinary differential equation[17]

$$i\hbar\frac{d\vec{c}}{dt} = (\hat{H}_0 + \hat{H}_I(t))\vec{c}, \qquad (2.54)$$

with $\hat{H}_0$ and $\hat{H}_I(t)$ Hermitian matrices.[18] In particular $\hat{H}_0$ is diagonal and the values on the diagonal are the possible values for the energy. Using the dipole approximation, the matrix $\hat{H}_I$ is the matrix representation of the operator (cf. (2.48)) $-\hat{d} \cdot \vec{\epsilon}E(\vec{0}, t)$, where $\vec{\epsilon}$ is the direction of the electric field.[19] The control problem of transferring the state between two different energy eigenstates can now be naturally formulated for system (2.54). This system has state varying

---

[15]See more on the multilevel approximation in Chapter 8, section 8.1.

[16]This is easily seen by recalling that the probability for a state $|\psi\rangle$ to be found with a value of an observable corresponding to an eigenvector $|\phi_n\rangle$ is $|\langle\phi_n|\psi\rangle|^2$. In our case

$$|\psi\rangle = \int \psi(\vec{r}, t)|\vec{r}\rangle d\vec{r} = \sum_{k}^{n} c_k(t)\int \phi_k(\vec{r})|\vec{r}\rangle d\vec{r} := \sum_{k}^{n} c_k(t)|\phi_k\rangle,$$

where $|\phi_k\rangle := \int \phi_k(\vec{r})|\vec{r}\rangle d\vec{r}$ and taking the product of both sides with $\langle\phi_n|$ we obtain $c_n = \langle\phi_n|\psi\rangle$ (since $\hat{H}_0$ is Hermitian, $|\phi_n\rangle$ are orthonormal).

[17]Also, cf. Exercise 2.1.

[18]Continuing with the notation introduced in Chapter 1, we denote with the same symbol an operator and its matrix representation (in a natural basis).

[19]For simplicity we assume here this direction fixed although what we say can be easily adapted for a direction $\vec{E}$ variable with time.

on a complex sphere of radius 1 and if we treat the electric field as the control u it has the form of a *bilinear system*

$$\frac{d\vec{c}}{dt} = (A + Bu)\vec{c},$$

with A and B Hermitian matrices.

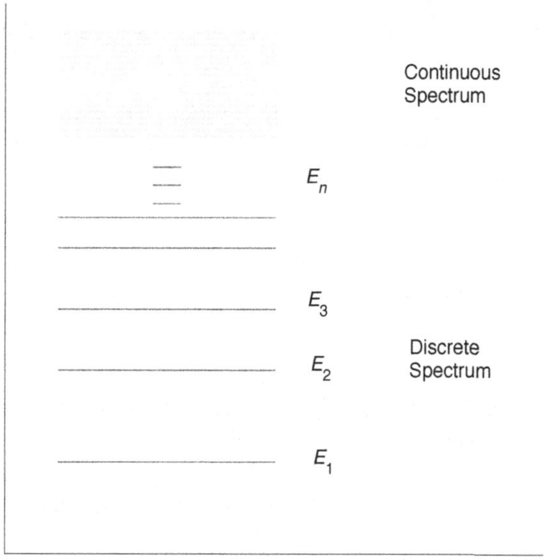

FIGURE 2.5: A typical situation in the dissociation of molecule is when the energy spectrum has the structure outlined in this figure. This is a mixed structure, discrete and continuous. The ground state E_1 and the excited states $E_2, E_3, \ldots, E_n, \ldots$ correspond to a discrete spectrum of energy levels. There then exists a continuum of energy levels. Dissociation of the molecule happens in the passage from a state corresponding to a value of energy in the discrete spectrum to one corresponding to a value in the continuous spectrum.

2.3.2.2 Summary of the modeling procedure

We summarize the procedure for modeling a quantum control system outlined in this chapter so far. The procedure follows the following points:

1. Obtain the classical Hamiltonian (the energy) of the system under

consideration identifying the time-varying parameters corresponding to an external action.

2. Introduce appropriate approximations neglecting small terms (e.g., electrodynamic interactions) in the expression of the Hamiltonian, in order to obtain a simplified form.

3. Express the Hamiltonian in terms of canonical variables and canonical momenta only.

4. Perform canonical quantization to obtain the quantum mechanical Hamiltonian and the relevant Hilbert space for the quantum mechanical system.

5. Identify the internal, $\hat{H}_0$, and interaction, $\hat{H}_I$, Hamiltonian in (2.49) and calculate the energy eigenvalues and eigenstates of $\hat{H}_0$ from the time-independent Schrödinger equation (2.52).

6. Identify the relevant energy levels (multilevel approximation) and expand the wave function in terms of the corresponding eigenstates as in (2.53) (assumed finite).

7. Obtain a finite-dimensional control system for the coefficient c_k in (2.53).

Notice that the modeling procedure remains valid up to the Schrodinger wave equation (2.50) even though we do not assume a *finite* number of levels. In this case, the problem is a control problem for an infinite-dimensional system. Control problems of this type arise for example in molecular dissociation as discussed in Figure 2.5. Control problems for infinite-dimensional quantum systems have attracted much attention in the last decades. The mathematical tools used depart significantly from the ones for finite-dimensional systems which are the focus of this book. One paper that can be used as an entry point to a vast literature in the mathematical theory is [30].

2.4   Spin Dynamics and Control

Several particles such as electrons, some types of nuclei[20] and molecules display an *intrinsic* (as opposed to *orbital*) angular momentum referred to as **spin angular momentum**. We denote the observable spin angular momentum in the x, y, z-direction by $S_{x,y,z}$, respectively. If $S_{x,y,z}$ is measured, there are $2j + 1$ possible results, $-\hbar j, -\hbar(j - 1), \cdots \hbar(j - 1), \hbar j$, where j is either a

---

[20]In particular all isotopes which contain an odd number of nucleons and some of the ones with an even number of nucleons have nuclear spin.

positive integer or a positive half integer whose value depends on the system under consideration (cf. Figure 2.6). It is called the **spin** of the system and it is one of its intrinsic properties.

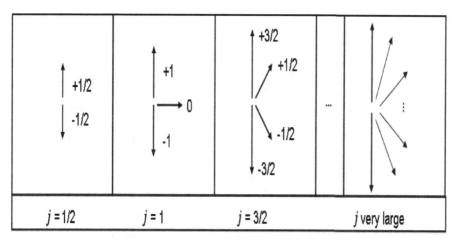

FIGURE 2.6: Pictorial representation of particles with spin for several values of the spin j. For $j = \frac{1}{2}$ only two values of the spin are possible when measured, as in a Stern-Gerlach experiment. For spin $j = 1$ there are three possible values of the spin. As the intrinsic spin of the particle increases, the behavior resembles more the one of classical orbital angular momentum, which may give a continuum of possible values, when measured.

Particles with spin (and other degrees of freedom neglected) are of great theoretical and practical interest in quantum physics. They are the prototypical example of finite-dimensional quantum systems, being similar to classical systems for large values of j (many possible values of the angular momentum) and displaying distinctive quantum mechanical behavior for small values of the spin j. Several experimental scenarios for the manipulation of particles with spin include *nuclear magnetic resonance* (see, e.g., [105]) and *electron paramagnetic resonance* (see, e.g., [119]), with a large variety of applications. Chapter 10, in section 10.1, presents a discussion of control problems in nuclear magnetic resonance. It should be mentioned that techniques for the control of spins using electromagnetic fields have been in place in these areas for several decades. Often these techniques have been suggested by physical intuition and the introduction of control theory in these areas is more recent.

2.4.1 Introduction of the spin degree of freedom in the dynamics of matter and fields

The existence of the spin degree of freedom, which was first demonstrated in the Stern-Gerlach experiment (see Chapter 1), can be derived theoretically using relativistic quantum electrodynamics (see, e.g., [67]). In this way, one can also derive the form of the Hamiltonian modeling the interaction between electromagnetic field and particles with spin. In a heuristic way, this Hamiltonian can be obtained using the correspondence between orbital angular momentum and spin (intrinsic) angular momentum. After having rewritten the Hamiltonian of classical electrodynamics in terms of orbital angular momentum, one 'augments' the orbital angular momentum with an additional spin term (which in some cases is the dominant term). We illustrate this for the term which models the interaction of the particles with the external field, that is for H_{Ie} in (2.43).

The definition of the **orbital angular momentum** $\vec{L}_j$ for the particle j is given in terms of the canonical momentum $\vec{p}_j$, as

$$\vec{L}_j := \vec{p}_j \times \vec{r}_j. \tag{2.55}$$

Let us make the assumption of *low intensity of the field*. Under this condition, in the expression of H_{Ie} in (2.43), the magnitude of the first term on the right is much smaller than the second term, and so we shall neglect it in the following discussion. Let us also make a long-wavelength approximation by assuming that the wavelength of the field is much larger than the size of the system of particles which is assumed localized in a small region of space, just as it was done in 2.3.1.2. This allows us to neglect the dependence of the magnetic field on space. Let $\vec{0}$ be the position of the center of mass of the system of particles. The expression (2.12) for the magnetic field in terms of the vector potential $\vec{A}$ suggests to write $\vec{A}_e$ as

$$\vec{A}_e(\vec{r}, t) = -\frac{1}{2}\vec{r} \times \vec{B}_e(\vec{0}, t),$$

where $\vec{B}_e$ represents the external magnetic field. Notice that this leaves $\vec{A}_e$ in the Coulomb gauge since $\nabla \cdot \vec{A}_e = 0$. Using this formula for the vector potential $\vec{A}_e$ in (2.43), H_{Ie} (with low field intensity approximation) takes the form

$$H_{Ie} = \frac{1}{2}\sum_{j=1}^{N}\frac{q_j}{m_j}\vec{p}_j \cdot (\vec{r} \times \vec{B}_e(\vec{0}, t)) + \sum_{j=1}^{N}q_j\phi_e(\vec{r}_j, t).$$

Using the vector identity

$$\vec{P} \cdot (\vec{R} \times \vec{B}) = \vec{B} \cdot (\vec{P} \times \vec{R}),$$

and the definition of the orbital angular momentum (2.55), we write H_{Ie} as

$$H_{Ie} = \sum_{j=1}^{N}\frac{q_j}{2m_j}\vec{B}_e(\vec{0}, t) \cdot \vec{L}_j + \sum_{j=1}^{N}q_j\phi_e(\vec{r}_j, t). \tag{2.56}$$

The quantum mechanical operator

$$\hat{L}_j := -i\hbar \vec{r}_j \times \nabla_j \tag{2.57}$$

is the *orbital angular momentum* operator for the j-th particle. With this definition we rewrite the operator corresponding to the first term of H_{Ie} in (2.56) as

$$\hat{H}_{IeM} := \sum_{j=1}^{N} \frac{q_j}{2m_j} (\hat{L}_j) \cdot \vec{B}_e(0,t) + \sum_{j=1}^{N} q_j \phi_e(\vec{r}_j, t), \tag{2.58}$$

and this is another way of writing the part of the Hamiltonian which models the interaction between particles and fields. The spin angular momentum is an additional, intrinsic angular momentum of the particle. If the particle has spin, the Hamiltonian in (2.58) is therefore heuristically modified by adding to $\hat{L}_j$ a term $g_j \hat{S}_j$, where $\hat{S}_j$ is the spin angular momentum operator and g_j is an extra factor called the *Lande factor*. Therefore, we replace the term $\sum_{j=1}^{N} \frac{q_j}{2m_j} (\hat{L}_j) \cdot \vec{B}_e(\vec{0}, t)$ in (2.58), with $\sum_{j=1}^{N} \frac{q_j}{2m_j} (\hat{L}_j + g_j \hat{S}_j) \cdot \vec{B}_e(0, t)$. Pictorial representation of the orbital angular momentum and the spin angular momentum is given in Figure 2.7.

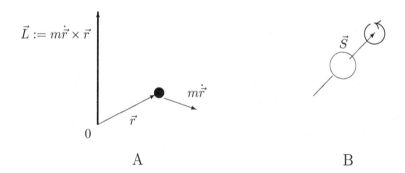

$$\vec{L} := m\dot{\vec{r}} \times \vec{r}$$

$m\dot{\vec{r}}$

$\vec{r}$

0

$\vec{S}$

A   B

FIGURE 2.7: If the conjugate momentum $\vec{p}$ is equal to the kinematical momentum $m\dot{\vec{r}}$ the *orbital angular momentum* $\vec{L}$ is given by the cross product $m\dot{\vec{r}} \times \vec{r}$ (Part A). In many cases, a quantum mechanical treatment has to consider an intrinsic angular momentum, the spin $\vec{S}$ depicted in Part B.

The state of a general system of moving particles that have spin will be described by a vector in a Hilbert space $\mathcal{H}_P \otimes \mathcal{H}_S$, where $\mathcal{H}_P$ refers to the position degrees of freedom and is infinite-dimensional, while $\mathcal{H}_S$ refers to the spin degrees of freedom and is finite-dimensional. $\mathcal{H}_S$, in turn, is the tensor product of Hilbert spaces $\mathcal{H}_{Sk}$, where Sk refers to the spin of the k-th particle. The state $|\psi\rangle$ is then written as

$$|\psi(t)\rangle = \sum_k \int \psi_k(\vec{r}, t)|\vec{r}\rangle \otimes |k\rangle d\vec{r},$$

where the states $|k\rangle$ form an orthogonal basis in $\mathcal{H}_S$. The vector operator $\hat{S}_k$ only acts on the $\mathcal{H}_{Sk}$ part of the Hilbert space.

2.4.2   Spin networks as control systems

2.4.2.1   Spin in an electromagnetic field

The simplest example of a control system involving particles with spin is a *single* particle, with all other degrees of freedom neglected, interacting with a classical magnetic field $\vec{B}_e$. The field $\vec{B}_e$, which plays the role of the control, is assumed constant in space (long-wavelength approximation) and varying with time. In the associated Hamiltonian, the only relevant term is the one which models the interaction of the spin angular momentum with the external field. As we have seen in the previous subsection, under long-wavelength approximation, this term has the form of a dot product of the spin operator with the external field. Given $\vec{B}_e = \vec{B}_e(t) := \{B_x, B_y, B_z\}$ and $\hat{S} := \{\hat{S}_x, \hat{S}_y, \hat{S}_z\}$, the Hamiltonian can therefore be written as

$$\hat{H}(t) := \gamma \hat{S} \cdot \vec{B}_e(t) = \gamma(\hat{S}_x B_x(t) + \hat{S}_y B_y(t) + \hat{S}_z B_z(t)), \qquad (2.59)$$

for some proportionality factor γ called the *gyromagnetic ratio*.[21]

In nuclear magnetic resonance (NMR) [105] and electron paramagnetic resonance (EPR)[119] experiments, a sinusoidal magnetic field at different frequencies is applied to a particle with Hamiltonian $\hat{H}(t)$ as in (2.59). Only sinusoids at a given frequency induce a state transition. The frequency contains the information on the nature of the particle under consideration (as it depends on the parameter γ in (2.59)) and this gives the crucial information on the chemical nature of the atoms and molecules under study.[22] This method has enjoyed for several decades important applications. More recently particles with spin have been proposed as elementary components of quantum computers. In fact, particles with spin $\frac{1}{2}$ are attractive devices to be used as quantum bits in quantum information theory.

---

[21]The value of the factor γ depends on the Lande factor g, but often its value has to be modified to take into account the effect of neighboring particles. This modification is called *chemical shift* (cf. Chapter 10, section 10.1).

[22]cf. section 10.1 in Chapter 10 for more discussion on methods of nuclear magnetic resonance.

Let us consider the Hamiltonian (2.59) and, for sake of simplicity, the case of a particle with spin $j = \frac{1}{2}$. We consider a basis of eigenstates of S_z. The corresponding eigenvalues (possible values of the spin when measured, setting $\hbar = 1$) are $\pm\frac{1}{2}$ and the eigenvectors are denoted by $|\pm\frac{1}{2}\rangle$. Let us express the state $|\psi(t)\rangle$ as $|\psi(t)\rangle = c_1(t)|+\frac{1}{2}\rangle + c_2(t)|-\frac{1}{2}\rangle$. From the Schrödinger equation and the Hamiltonian (2.59), we derive the differential equation for the coefficients $\vec{c} = (c_1, c_2)^T$

$$i\frac{d}{dt}\vec{c} = \gamma\frac{1}{2}(\sigma_x B_x(t) + \sigma_y B_y(t) + \sigma_z B_z(t))\vec{c}. \tag{2.60}$$

The matrices $\frac{1}{2}\sigma_{x,y,z}$ are the matrix representations of the operators $S_{x,y,z}$, where $\sigma_{x,y,z}$ are the Pauli matrices already introduced in (1.20).

$$\sigma_x := \begin{pmatrix} 0 & 1 \\ 1 & 0 \end{pmatrix}, \quad \sigma_y := \begin{pmatrix} 0 & -i \\ i & 0 \end{pmatrix}, \quad \sigma_z := \begin{pmatrix} 1 & 0 \\ 0 & -1 \end{pmatrix}. \tag{2.61}$$

In the case of an ensemble with state described by a density matrix ρ, ρ satisfies Liouville's equation

$$i\frac{d}{dt}\rho = \gamma\frac{1}{2}[\sigma_x B_x(t) + \sigma_y B_y(t) + \sigma_z B_z(t), \rho] := \gamma\frac{1}{2}[\vec{\sigma}\cdot\vec{B}, \rho]. \tag{2.62}$$

It is conceivable to measure the spin in one direction in space, i.e., one of $S_{x,y,z}$ with a selective measurement or to measure its mean value $\mathrm{Tr}(\rho S_{x,y,z})$ with a nonselective measurement.

2.4.2.2 Control systems of several spins

A more general situation is the one of a *network* of particles with spin (with the position degrees of freedom neglected) interacting with each other and with an external magnetic field. This can be realized for example in a molecule when two or more nuclei (and/or electrons) interact. An example is given in Figure 2.8.

In this case, the Hamiltonian operator is composed of a part modeling the interaction among the particles and a part modeling the interaction with the external field. This, under the appropriate approximations, as seen above, has the form

$$\hat{H}_I(t) = \sum_{j=1}^{N}\gamma_j\hat{S}_j\cdot\vec{B}(t),$$

where $\hat{S}_j$ (γ_j) is the spin angular momentum (gyromagnetic ratio) of the j-th particle and N is the number of particles. The total Hilbert space for this system is the tensor product of the Hilbert spaces for the single subsystems, each one having possibly different dimensions according to the values of the spins.[23] The x, y, z-component of the vector operator $\hat{S}_j := \{\hat{S}_{xj}, \hat{S}_{yj}, \hat{S}_{zj}\}$

---

[23]In particular, for a spin j the associated Hilbert space has dimension $2j + 1$.

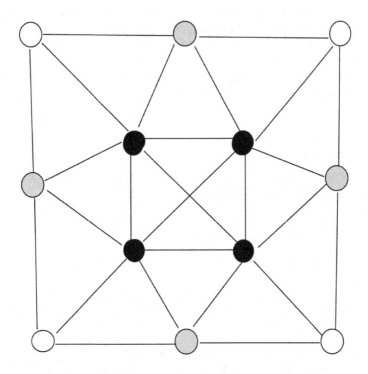

FIGURE 2.8: The magnetic molecule Mn_{12} can be seen as a spin network. Three different types of Manganese ions, which are represented by the white, grey and black circles, all have spin and interact with each other. The lines connecting circles represent interactions among them (cf. [44] and references therein).

acting on the tensor product Hilbert space, $j = 1, ..., N$, is

$$\hat{S}_{xj,yj,zj} = \mathbf{1} \otimes \mathbf{1} \otimes \cdots \otimes \mathbf{1} \otimes \hat{S}_{x,y,z} \otimes \mathbf{1} \otimes \cdots \otimes \mathbf{1}, \qquad (2.63)$$

where $\hat{S}_{x,y,z}$ appears in the j-th position.

The part of the Hamiltonian which models the interaction among the particles may take different forms according to different situations. It is in principle obtained with a procedure similar to the one described in the previous subsection for the interaction with the magnetic field. Starting from the electrodynamic interaction (2.42) one expresses the interaction in terms of angular momentum interactions and then introduces the spin. Various approximations need to be introduced and the subject of modeling spin-spin interaction has a vast literature. The book [2] contains several examples of Hamiltonians modeling the interaction among particles with spin. A common form is the

Heisenberg interaction given by (cf. (2.63))

$$\hat{H}_0 = \sum_{k<j} J_{kj} \hat{S}_k \cdot \hat{S}_j, \tag{2.64}$$

where k and j vary over all the spin particles (modulo the constraint $k < j$) and J_{kj} are the *coupling constants*. Another form is the *Ising interaction* which has the form (choosing the z direction)

$$\hat{H}_0 = \sum_{k<j} J_{kj} \hat{S}_{zk} \hat{S}_{zj}, \tag{2.65}$$

where $\hat{S}_{z(k,j)}$ is the z component of the spin angular momentum for particle k, j. The dynamics is described by the Schrödinger equation with the appropriate Hamiltonian.

To write the Schrödinger equation as a differential equation in matrix form, one considers the Hilbert space of the system of n spins which is the tensor product of the Hilbert spaces of the single spins. With the matrix form for the Hamiltonian, one can write the differential equation for the components of the state, as well as for the density matrix. It is easily seen that in such a differential equation, the control $\vec{B}_e(t)$ multiplies the state.

As the output of a putative experiment, one may consider for example the result of a selective measurement of the spin angular momentum in the z direction of the j-th spin. This is a selective measurement of the observable

$$\hat{M}_{jz} := \mathbf{1} \otimes \cdots \otimes \mathbf{1} \otimes \hat{S}_z \otimes \mathbf{1} \otimes \cdots \otimes \mathbf{1},$$

where $\hat{S}_z$ appears in the j-th position. Another possible output in a nonselective measurement experiment is the expectation value of such an observable, $\mathrm{Tr}(\hat{M}_{jz}\rho)$. Another possible output in a nonselective measurement is the (expectation) value of the total spin in the z direction, which is proportional to $\mathrm{Tr}(\sum_j \hat{M}_{jz}\rho)$, where the sum is taken over all the spins.

More discussion on NMR experiments and spin manipulation is given in Chapter 10.

2.5 Mathematical Structure of Quantum Control Systems

A finite-dimensional **control system** is a system of ordinary differential equations which displays one or more time-varying functions of time, $u := u(t)$, the *controls*, to be chosen in an appropriate set of functions. Therefore, a general control system has the form

$$\dot{x} = f(t, x, u). \tag{2.66}$$

Here x represents the *state* of the system, and $f(t, \cdot, u)$ is a family of vector fields, parametrized possibly explicitly by the time t and by the control u. In some cases, the state of the system varies on a manifold and the system of differential equations (2.66), which is defined in open set of $\mathbf{R}^n$, describes the evolution locally, in appropriate coordinates. The state x completely describes the system under consideration. In most cases, the whole state x is not accessible to an outside observer. Only a (vector) function of x, $y := y(x)$, is accessible and this is called the *output*. It models the interface between the system and the external world.

For **quantum control systems**, the relevant equation describing the dynamics is the Schrödinger equation, in its various forms we have described so far. Let $\vec{\psi}$ denote the vector column corresponding to the ket $|\psi\rangle$ in a given basis (cf. subsection 1.1.5). In this book, we shall focus on finite-dimensional quantum control systems, after appropriate approximations, so that $\vec{\psi}$ is a vector in a finite-dimensional complex vector space. For quantum control systems, the energy of the system depends on some functions of time which play the role of control. In the examples discussed in the previous sections, these functions model time-dependent electromagnetic fields that can be assumed *constant in space*. As a consequence, the Hamiltonian depends on time through the controls and the Schrödinger equation in the column vector form can be written (setting $\hbar = 1$) as

$$\frac{d}{dt}\vec{\psi} = -iH(u(t))\vec{\psi}. \tag{2.67}$$

Here $H(u)$ is a matrix function of u which is Hermitian for every value of u. Equation (2.67) should be compared with (2.66) with $\vec{\psi}$ playing the role of the state. In this case, because of the condition $|||\psi\rangle|| = 1$, the state $\vec{\psi}$ varies on a complex sphere with radius equal to one in the space $\mathbb{C}^n$, which is invariant for (2.67) due to the fact that $H(u)$ is Hermitian for every u. The classes of examples treated in the previous sections show that $H(u)$ often has the form

$$H(u) = H_0 + \sum_k H_k u_k. \tag{2.68}$$

Therefore, quantum control systems can be modeled as **bilinear systems** (i.e., the right-hand side, f, of (2.66) is a linear function of the state and an affine function of the control) with state varying on a complex sphere with radius equal to one. Moreover, since the matrices H_0, H_k are Hermitian, the matrices $-iH_0, -iH_k$ are skew-Hermitian. In some cases the control allows to *switch* among Hamiltonians in a finite set $\{H_1, ..., H_p\}$. This can be obtained for example by adjusting the experimental set-up so as to induce different evolutions.[24] We can write the Hamiltonian as in (2.68) with $H_0 \equiv 0$ and the

---

[24]An example is given by the quantum information processor based on ion traps discussed in section 10.3 of Chapter 10.

u_k's attaining values such that only one of them is equal to 1 and the others are equal to zero. In these cases, the controls do not necessarily have a direct physical meaning (of an electromagnetic field). They are only a mathematical tool to model the switching between different configurations.

2.5.1 Control of ensembles

If the system is an ensemble, the relevant equation describing the dynamics is Liouville's equation with the Hamiltonian depending on the control,

$$\dot{\rho} = [-iH(u(t)), \rho]. \tag{2.69}$$

In this case, the state ρ varies in the set of Hermitian, semidefinite positive matrices with trace equal to one, i.e., in the set of density matrices. This set is not a manifold but rather a manifold with boundary. However, we can assume that the state ρ is just varying in the manifold of Hermitian matrices with trace equal to one as equation (2.69) preserves the property of the matrix ρ being positive semidefinite. In fact, the solution of (2.69) can be written as

$$\rho(t) = X(t)\rho(0)X^\dagger(t), \tag{2.70}$$

where $X(t)$ is the unitary matrix representing the propagator, which is a solution at time t of the Schrödinger operator equation (cf. (1.41)). This is given by

$$\dot{X} = -iH(u)X, \qquad X(0) = \mathbf{1}_{n \times n}, \tag{2.71}$$

where $\mathbf{1}_{n \times n}$ is the $n \times n$ identity matrix.

2.5.2 Control of the evolution operator

System (2.71) may be taken as the model control system and the propagator X as the object of control. The state X is now varying in the Lie group (see Chapter 3) of unitary matrices of dimension n, with n being the dimension of the system. Considering X as the object of control can be more natural and convenient in a number of situations, for the following reasons:

1. Systems of the form (2.71) have been among the most studied in *geometric control theory* [153], [263] and therefore many tools developed there can be directly applied.

2. As we have seen in subsection 1.4.1, in applications to quantum computing, the state $\vec{\psi}$ represents the *information* and the matrix X describes the *operation* to be carried out according to

$$\vec{\psi}(t) = X(t)\vec{\psi}(0).$$

Therefore it is a natural formulation, in this context, to pose the control problem in terms of the matrix X, when the desired evolution has to perform a given (logic) operation. This problem is of interest when we want to design a quantum gate transforming the state ψ in a desired way. The problem is to choose the control u in (2.71) so that the solution X of (2.71) attains a desired value at a given time.

3. If we are able to solve a control problem for X, we are also able to control pure states ψ and ensembles ρ for the given system.

2.5.3   Output of a quantum control system

So far, quantum control systems appear to be just a special case of classical control systems with a specific structure for their dynamics. However, in describing the interaction of a quantum control system with an observer, new elements have to be taken into account which are a consequence of the measurement postulate discussed in subsection 1.2.2.

The output of a quantum control system consists of one or more observables that can be *measured simultaneously* no matter what the current state of the system is, i.e., the output consists of the results of compatible measurements (cf. subsection 1.2.2). The measurement may be selective or nonselective (cf. subsection 1.2.3) and can be modeled differently according to the specific experimental apparatus. In general, for selective measurements, the probability of each result will depend on the state. Therefore, even though the dynamics is described by a deterministic model, one should consider probability theory in the modeling of the output. This is not the case in a nonselective measurement if one considers as the output y the expectation value of one or more compatible observables. The output y can be written, in the case of a Von Neumann-Lüders measurement, as $y = \text{Tr}(S\rho)$, where S is the corresponding observable and ρ the density matrix.

Measurements are (theoretically) instantaneous. After the measurement, the state is modified according to the measurement postulate, and therefore the dynamics has to be considered starting from a different initial condition. For these schemes, the modification of the state due to measurement prevents, in principle, the implementation of feedback control, and this is the main difference between classical and quantum schemes. In fact, all the control schemes to be applied to the systems described in the previous sections must be **open-loop control schemes** and all the methods described in this book will be open-loop schemes. This means that the control law is completely determined prior to the control experiment. It is *not modified* during the experiment.

2.6 Notes and References

2.6.1 An example of canonical quantization: The quantum harmonic oscillator

The quantum harmonic oscillator is one of the most important and most studied quantum systems as it is the model for several phenomena in molecular dynamics, solid-state physics, quantum optics, etc. It is obtained from a procedure of canonical quantization starting from a classical harmonic oscillator. For simplicity, we consider only the one-dimensional harmonic oscillator whose only degree of freedom is the x coordinate. Extensions to higher-dimensional case are natural. For this system the conjugate canonical momentum p, calculated according to Lagrangian mechanics as in Appendix B (B.10), is found to be coinciding with the kinematical momentum, i.e.,[25]

$$p = m\dot{x}. \tag{2.72}$$

The total energy is given by the Hamiltonian

$$H_{ho} = \frac{p^2}{2m} + \frac{1}{2}m\omega^2 x^2, \tag{2.73}$$

which is the sum of the kinetic energy (first term) and the potential energy of the system (second term).

To obtain the quantum mechanical description of the harmonic oscillator, one defines the position, $\hat{x}$, and momentum, $\hat{p}$, operators, associated with x and p, respectively, according to the rules of canonical quantization. In particular[26] $\hat{x}$ corresponds to multiplication of the wave function $\psi(x)$ by x while $\hat{p}$ transforms $\psi(x)$ according to $\psi(x) \rightarrow -i\hbar\frac{d}{dx}\psi(x)$. They satisfy the commutation relation

$$[\hat{x}, \hat{p}] = i\hbar,$$

---

[25]The associated Lagrangian is given by

$$L_{ho} = \frac{1}{2}m\dot{x}^2 - \frac{1}{2}m\omega^2 x^2,$$

where (cf. (B.4)) $T := \frac{1}{2}m\dot{x}^2$ is the kinetic energy while $V := \frac{1}{2}m\omega^2 x^2$ represents the potential energy of the system and ω is a constant having units of a frequency. Using this Lagrangian, the Euler-Lagrange equations (B.2) give the equation of motion

$$\ddot{x} - \omega^2 x = 0.$$

The Hamiltonian (2.73) is found from the definition (B.12) using (2.72).
[26]Recall we are working in the position representation, i.e., we write the state as $|\psi\rangle = \int_{-\infty}^{\infty} \psi(x)|x\rangle dx$.

which is a special case of (2.39). From (2.73), the Hamiltonian operator is given by

$$\hat{H}_{ho} = \frac{\hat{p}^2}{2m} + \frac{1}{2}m\omega^2\hat{x}^2.$$

We can express the state in the *energy representation*, i.e., with an expansion in terms of the energy eigenstates.[27] The eigenvalues of the energy operators can be obtained by solving the time-independent Schrödinger wave equation (cf. (1.49)) which in this case is

$$\hat{H}\psi(x) = -\frac{\hbar^2}{2m}\frac{d^2\psi(x)}{dx^2}\psi(x) + \frac{1}{2}m\omega^2 x^2\psi(x) = E\psi(x).$$

This equation can be solved explicitly and gives the values of the energy eigenvalues and the corresponding wave functions. In particular the requirement that $\psi(x)$ be a probability density leads to only a discrete set of possible values for the energy which are equally spaced by a quantity $\hbar\omega$. We shall illustrate this result using Dirac's method which works in terms of state kets rather than wave functions.

Define the **destruction operator**, $\hat{a}$, as

$$\hat{a} := (2m\hbar\omega)^{-\frac{1}{2}}(m\omega\hat{x} + i\hat{p}), \tag{2.74}$$

and the **creation operator**, $\hat{a}^\dagger$, as

$$\hat{a}^\dagger := (2m\hbar\omega)^{-\frac{1}{2}}(m\omega\hat{x} - i\hat{p}). \tag{2.75}$$

They satisfy the commutation relation

$$[\hat{a}, \hat{a}^\dagger] = \mathbf{1}, \tag{2.76}$$

where $\mathbf{1}$ is the identity operator (cf. Exercise 2.2).

The Hamiltonian $\hat{H}_{ho}$ can be rewritten in terms of these operators as

$$\hat{H}_{ho} = \frac{1}{2}\hbar\omega\left(\hat{a}\hat{a}^\dagger + \hat{a}^\dagger\hat{a}\right) = \hbar\omega\left(\hat{a}^\dagger\hat{a} + \frac{1}{2}\right). \tag{2.77}$$

Consider now the eigenstates of $\hat{H}_{ho}$ and denote by $|n\rangle$ a generic eigenstate and E_n the corresponding value of the energy. This notation is chosen because we shall show below that we can put energy values and eigenvectors in one-to-one correspondence with nonnegative integer values. The energy eigenvalue equation (which corresponds to the Schrödinger time-independent equation) reads

$$\hbar\omega\left(\hat{a}^\dagger\hat{a} + \frac{1}{2}\right)|n\rangle = E_n|n\rangle.$$

---

[27]For the relation between the two representations as well as for a more extensive study of the quantum harmonic oscillator see, e.g., [245] section 2.3.

Applying $\hat{a}^\dagger$ to both terms and replacing $\hat{a}^\dagger\hat{a}$ with $\hat{a}\hat{a}^\dagger - \mathbf{1}$ according to (2.76), we obtain

$$\hbar\omega\left(\hat{a}^\dagger\hat{a}\hat{a}^\dagger - \hat{a}^\dagger + \frac{1}{2}\hat{a}^\dagger\right)|n\rangle = E_n|n\rangle.$$

Taking the term $\hbar\omega\hat{a}^\dagger|n\rangle$ to the right-hand side and recalling the expression of the Hamiltonian $\hat{H}_{ho}$, we obtain

$$\hat{H}_{ho}\hat{a}^\dagger|n\rangle = (E_n + \hbar\omega)\hat{a}^\dagger|n\rangle.$$

This shows that $|n+1\rangle := \hat{a}^\dagger|n\rangle$ is also an eigenvector of $\hat{H}_{ho}$ with eigenvalue $E_n + \hbar\omega$. Analogously one can see that $|n-1\rangle := \hat{a}|n\rangle$ is an eigenvector of $\hat{H}_{ho}$ with eigenvalue $E_n - \hbar\omega$. The elementary energy quantity $\hbar\omega$, which represents the gap between two successive energy levels, is called a *phonon* in the case where the harmonic oscillator represents a mechanical system.[28]

Let us now study more in depth the nature of the set of eigenvalues and eigenvectors. Let us assume we start from a certain eigenvalue E and eigenvector $|e\rangle$. Applying $\hat{a}$ n times we obtain the eigenvector $(\hat{a})^n|e\rangle$, with eigenvalue $\bar{E} - n\hbar\omega$. This cannot be negative and denote by $\bar{n}$ the largest value n so that $\bar{E} - n\hbar\omega$ is nonnegative. Call $|0\rangle := (\hat{a})^{\bar{n}}|e\rangle$. Since $\hat{H}_{ho}\hat{a}|0\rangle = (E - (\bar{n}+1)\hbar\omega)\hat{a}|0\rangle$ and $\bar{E} - (\bar{n}+1)\hbar\omega < 0$, $\hat{a}|0\rangle$ cannot be an eigenvector of $\hat{H}_{ho}$, therefore we must have

$$\hat{a}|0\rangle = 0,$$

where 0 represents the zero vector of the associated Hilbert space. Call E_0 the eigenvalue corresponding to $|0\rangle$. Using the expression of the energy operator $\hat{H}_{ho}$ (2.77) and $\hat{a}|0\rangle = 0$ we obtain

$$\hbar\omega(\hat{a}a + \frac{1}{2})|0\rangle = \frac{\hbar\omega}{2}|0\rangle = E_0|0\rangle,$$

which shows that $E_0 = \frac{\hbar\omega}{2}$. We have then shown that $E_n := E_0 + n\hbar\omega$, $n = 0, 1, 2, ...$ are eigenvalues of $\hat{H}_{ho}$ and called $|n\rangle$ the corresponding eigenvectors. We want to show now that these are the *only* possible eigenvalues. If there was any value different from these there would be associated a sequence of energy values. Call $|0'\rangle$ the one corresponding to the lowest possible value of the energy E_0'. As above we have $\hat{a}|0'\rangle = 0$ and therefore $E_0' = \frac{1}{2}\hbar\omega = E_0$. Therefore, the sequence of eigenvalues coincides with $E_n := E_0 + n\hbar\omega$, $n = 0, 1, 2, ...$ and the minimum value is E_0. It is an important result that the minimum value of the energy is not zero. The state $|0\rangle$ is called the *vacuum state* and it corresponds to no phonon. The creation (destruction) operator[29] transforms an eigenstate of the Hamiltonian into an eigenstate corresponding to energy higher (lower) by a quantity $\hbar\omega$. It 'creates' ('destructs') a *quantum of energy* $\hbar\omega$, that is, a phonon.

---

[28]Harmonic oscillators also model an electromagnetic field as a quantum system (see, e.g., [67], [129], [185]). In that context $\hbar\omega$ is called a *photon*.

[29]These operators are also called the *raising* and *lowering* operators, respectively.

2.6.2   On the models introduced for quantum control systems

The models of quantum control systems described in this chapter are simplified, not only because of the various approximations we have introduced (semiclassical approximation, nonrelativistic treatment, etc.) but also because we have neglected the presence of the environment and assumed the *system* to be isolated. In fact, the design of controls which would render the evolution robust to the presence of an external environment is one of the main theoretical and practical problems in quantum control. An excellent treatment of dynamical models of open quantum systems is given in [47]. More generally, the set of models and potential applications presented in this chapter are only a subset of the models of quantum control systems of interest in applications. However, it is a large subset which encompasses several important applications, as we have pointed out. It is amenable to applications of techniques from geometric control theory, which will be one of the main subjects of the following chapters.

The book [53] contains several examples of quantum control systems, while the two recent books [233], [254] focus on the control of molecular dynamics. The papers [22], [186], [116] present further motivations and applications to study quantum control systems and contains an extensive literature. In recent years, several monographs have been written on the neighboring field of quantum information and computation. In this area, we refer to [207], [167], [49], [31].

As we have already mentioned, the measurement postulate prevents the direct implementation of feedback for the models described here. One can circumvent this limitation in several ways. For example, one could imagine having a very large number of identical quantum systems evolving (simultaneously) in the same conditions starting from the same initial value. Measuring each system at slightly spaced instants of time simulates a continuous measurement. At each time the systems on which the measurement has already been performed are not considered anymore and the control is modified to influence the dynamics of the remaining systems, so as to simulate feedback control. More genuine quantum feedback schemes on a single system can be implemented in the case of continuous weak measurement, where the modification of the state (called '*quantum state reduction*') is slow enough so that it can be compensated by appropriate control. In these cases, the model of the dynamics has to be modified to include the continuous back-action of the measurement. The resulting model is a stochastic differential equation and the tools for control methods are based on the methods of stochastic control theory. This approach has received great interest in recent years starting from the seminal work of Belavkin until the more recent investigations and experimental realizations by Wiseman, Mabuchi et al. (see, e.g., [282], [295] and references therein). Quantum feedback is not the main topic of this book, which focuses on deterministic quantum models and open-loop control design.

A very informative review paper is [308] and a more extensive introductory monograph is [296].

2.7 Exercises

Exercise 2.1 Use the definition of the position and canonical momentum operator to prove the commutation relation (2.39). Which ones are the compatible observables and which ones are incompatible? Observe that from this formula and the discussion in 1.2.2.3 one obtains Heisenberg uncertainty relation for the position $\hat{x}$ and momentum $\hat{p}_x$ (consider a single particle moving in the x direction, vector potential equal to zero), that is

$$\langle \Delta \hat{x}^2 \rangle_\psi \langle \Delta \hat{p}_x^2 \rangle_\psi \geq \frac{1}{4} \hbar^2.$$

Exercise 2.2 Using the definitions of the destruction and creation operator $\hat{a}$ and $\hat{a}^\dagger$, (2.74), (2.75), and the commutation relation between $\hat{x}$ and $\hat{p}$ prove (2.76) and (2.77).

Exercise 2.3 A sufficient condition for the validity of *Helmholtz theorem* (Theorem 2.25) is the existence of the *spatial Fourier transform* of the field $\vec{V} = \vec{V}(\vec{r})$, defined as

$$\vec{\mathcal{V}}(\vec{k}) = \frac{1}{(2\pi)^{\frac{3}{2}}} \int_{\mathbf{R}^3} \vec{V}(\vec{r}) e^{-i\vec{k}\cdot\vec{r}} d\vec{r},$$

with inversion formula

$$\vec{V}(\vec{r}) = \frac{1}{(2\pi)^{\frac{3}{2}}} \int_{\mathbf{R}^3} \vec{\mathcal{V}}(\vec{k}) e^{i\vec{k}\cdot\vec{r}} d\vec{k}.$$

To prove Helmoltz decomposition first prove the vector identity

$$\vec{\mathcal{V}}(\vec{k}) = \frac{1}{\|\vec{k}\|^2} \left((\vec{k} \cdot \vec{\mathcal{V}}) \vec{k} + (\vec{k} \times \vec{\mathcal{V}}) \times \vec{k} \right).$$

Then take the anti-transform of both sides and show that the anti-transform of the first term on the right-hand side is irrotational (longitudinal) and the anti-transform of the second term is divergence free (transversal).

Chapter 3

Controllability

Given a quantum control system describing an experimental set-up, a natural question concerns the type of feasible experiments that can be performed. This question can be formulated in terms of the kind of state transfers admissible with the given system, namely, it is a question of *controllability*.

In this chapter, we shall first consider this question for the Schrödinger operator equation in the general form[1]

$$\dot{X} = -iH(u)X, \qquad X(0) = \mathbf{1}_{n \times n}. \tag{3.1}$$

We shall study the set of unitary matrices that can be obtained by changing the control for this system, i.e., the *reachable set* of states for this system. This set is always a subset of the group of unitary matrices of dimension n, $U(n)$. Since $-iH(u)$ is skew Hermitian for every u, X must be unitary. If the set of possible matrices that can be obtained for system (3.1) is the set of all the unitary matrices, the system is said to be *controllable*.

The study of controllability for the Schrödinger operator equation gives the crucial information on the controllability of the state vector for Schrödinger equation and on the controllability of the density matrix for Liouville's equation. We shall define notions of controllability for these equations and give criteria for controllability.

We first give the definition of reachable sets. Consider system (3.1), where the control u is assumed to belong to a space of functions $\bar{\mathcal{U}}$. Denote by $X(t, u)$ the solution of (3.1), with control u, at time t. The **reachable set at time** $T > 0$ for system (3.1), $\mathcal{R}(T)$, is the set of all the unitary matrices $\bar{X}$ such that there exists a control $u \in \bar{\mathcal{U}}$ with $X(T, u) = \bar{X}$. The reachable set $\mathcal{R}(\leq T)$ is defined as $\mathcal{R}(\leq T) := \cup_{0 \leq t \leq T} \mathcal{R}(t)$. The **reachable set** $\mathcal{R}$ is defined as[2]

$$\mathcal{R} = \cup_{T \geq 0} \mathcal{R}(T). \tag{3.2}$$

In this chapter, we shall describe the reachable set for system (3.1) in the case where $\bar{\mathcal{U}}$ is the set of piecewise constant functions with values in a

---

[1]We set $\hbar = 1$ (This is done without loss of generality by changing units, or incorporating $\hbar$ into the Hamiltonian H) and do not assume, in general, any special dependence of the Hamiltonian H on the control u.

[2]These definitions extend naturally to other types of systems although one typically should specify the initial state, which here is naturally taken to be the identity.

DOI: 10.1201/9781003051268-3

set $\mathcal{U}$. Notice that if $X(t, u)$ is the solution of (3.1) with initial condition equal to the identity, then the solution with initial condition equal to S is given by $X(t, u)S$. Therefore, if we can control the system to a state X_1 with control u_1 and to a state X_2 with control u_2, we can also control it to a state $X_2 X_1$, by using a control which is the concatenation of u_1 and u_2. In other words, $\mathcal{R}$ is a *semigroup*. It turns out that $\mathcal{R}$ is often a *Lie group*, namely a group with an additional differentiable structure. We will be able to characterize this Lie group exactly. Before we do that, we need to recall the basic definitions and facts about Lie algebras and Lie groups. The theory of Lie algebras and Lie groups gives the mathematical basis to much of the theory of quantum control as we shall see in the following chapters.

3.1 Lie Algebras and Lie Groups

We summarize in this section the basic definitions concerning Lie groups and Lie algebras. In the process, we shall recall some elementary concepts of linear algebra and differential geometry. There are many introductory texts on the subject. Some of the most popular are [104], [130], [168], [244].

3.1.1 Basic definitions for Lie algebras

3.1.1.1 Lie algebras

Definition 3.1.1 A **Lie algebra** $\mathcal{L}$ over a field $\mathcal{F}$ is a vector space over $\mathcal{F}$ with an additional binary operation $\mathcal{L} \times \mathcal{L} \to \mathcal{L}$. This operation associates with an ordered pair of elements $\{x, y\}$ in $\mathcal{L}$ an element $[x, y]$. It is called the *Lie bracket* or *commutator* and it is required to satisfy the following axioms:

1. Bilinearity:

$$[x + y, z] = [x, z] + [y, z], \qquad [x, y + z] = [x, y] + [x, z], \qquad (3.3)$$

$$[\alpha x, y] = \alpha[x, y], \quad \forall \alpha \in \mathcal{F}.$$

2.

$$[x, x] = 0, \quad \forall x \in \mathcal{L}. \qquad (3.4)$$

3. Jacobi identity:

$$[x, [y, z]] + [y, [z, x]] + [z, [x, y]] = 0.$$

In the cases where the field $\mathcal{F}$ is the field of real numbers $\mathbf{R}$ or the field of complex numbers $\mathbb{C}$, condition (3.4) can be replaced (cf. Exercise 3.5) by the *skew-symmetry condition*, that is:

$$[x, y] = -[y, x].$$

In our treatment we will mostly deal with *real* Lie algebras, that is, the field $\mathcal{F}$ above is the field of real numbers.

The structure of a Lie algebra is unambiguously determined by the commutation relations on a basis, since these determine the value of the commutator of any pair of elements in the Lie algebra.

Perhaps the simplest nontrivial example of a Lie algebra is the set of vectors in three-dimensional space, $\mathbf{R}^3$, with the cross product playing the role of the commutator. If we choose a basis $\{\vec{i}, \vec{j}, \vec{k}\}$, then the commutation relations are given by

$$\vec{i} \times \vec{j} = \vec{k}, \qquad \vec{j} \times \vec{k} = \vec{i}, \qquad \vec{k} \times \vec{i} = \vec{j}. \tag{3.5}$$

We shall be concerned with Lie algebras of matrices which are called *linear Lie algebras*, where the commutator $[A, B]$ is the standard matrix commutator

$$[A, B] := AB - BA.$$

The Lie algebra of all the $n \times n$ matrices with real (complex) entries is denoted by $gl(n, \mathbf{R})$ ($gl(n, \mathbb{C})$) and it is called the *general linear* algebra over the real (complex) numbers.

3.1.1.2 Subalgebras

Definition 3.1.2 Given a Lie algebra $\mathcal{L}$, consider a subspace $\mathcal{A} \subseteq \mathcal{L}$. If $\mathcal{A}$ with the commutator defined on $\mathcal{L}$ is a Lie algebra, then $\mathcal{A}$ is called a **subalgebra** of $\mathcal{L}$.

A subspace $\mathcal{A} \subseteq \mathcal{L}$ is therefore a Lie subalgebra if and only if it is closed under commutation $[\mathcal{A}, \mathcal{A}] \subseteq \mathcal{A}$.

Subalgebras of $gl(n, \mathbf{R})$ or $gl(n, \mathbb{C})$ are called *linear Lie algebras*. An example of a subalgebra of $gl(n, \mathbf{R})$ is the *special linear algebra* $sl(n, \mathbf{R})$ (also denoted by $sl(n)$) which is the Lie algebra of $n \times n$ real matrices with trace equal to zero. Another example is the *orthogonal Lie algebra* $so(n, \mathbf{R})$ (also denoted by $so(n)$) which is the Lie algebra of $n \times n$ skew-symmetric matrices. In the following, we shall be interested in the **Lie algebra** $u(n)$ **of skew-Hermitian** $n \times n$ **matrices** considered as a Lie algebra over the real field. This is because, all matrices $-iH(u)$ in (3.1) are in $u(n)$. The subalgebra $su(n)$ of $u(n)$ will play an important role. It consists of the matrices in $u(n)$ with zero trace. The nontrivial simplest but very important example is $su(2)$ which is spanned by the multiples of the Pauli matrices (1.20)

$$\bar{\sigma}_x := \frac{1}{2}\begin{pmatrix} 0 & i \\ i & 0 \end{pmatrix}, \qquad \bar{\sigma}_y := \frac{1}{2}\begin{pmatrix} 0 & -1 \\ 1 & 0 \end{pmatrix}, \qquad \bar{\sigma}_z := \frac{1}{2}\begin{pmatrix} i & 0 \\ 0 & -i \end{pmatrix}. \tag{3.6}$$

The matrices $\bar{\sigma}_{x,y,z}$ satisfy the commutation relations (cf. Figure 3.1)

$$[\bar{\sigma}_x, \bar{\sigma}_y] = \bar{\sigma}_z, \qquad [\bar{\sigma}_y, \bar{\sigma}_z] = \bar{\sigma}_x, \qquad [\bar{\sigma}_z, \bar{\sigma}_x] = \bar{\sigma}_y, \tag{3.7}$$

which completely describe the Lie algebra $su(2)$.

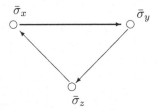

FIGURE 3.1: This graph is useful to remember the commutation relations among $\{\bar\sigma_x, \bar\sigma_y, \bar\sigma_z\}$. If we want to find the commutator $[\bar\sigma_A, \bar\sigma_B]$, we follow a branch from $\bar\sigma_A$ to $\bar\sigma_B$ on the graph. The following node of the graph is the result except possibly for the sign. If our path follows the same direction as the arrows then the sign is positive otherwise it is negative.

3.1.1.3 Homomorphisms and isomorphisms

Definition 3.1.3 Given two Lie algebras $\mathcal{L}_1$, $\mathcal{L}_2$, a **homomorphism** ϕ, $\phi : \mathcal{L}_1 \to \mathcal{L}_2$, is a linear map which preserves the Lie bracket of $\mathcal{L}_1$, namely

$$\phi([x,y]_1) = [\phi(x), \phi(y)]_2, \qquad (3.8)$$

where $[\cdot, \cdot]_{1,2}$ denote the Lie bracket on $\mathcal{L}_{1,2}$. A bijective homomorphism is called an **isomorphism**, and the two Lie algebras are called isomorphic if such an isomorphism exists. A homomorphism of $\mathcal{L}_1$ onto a subalgebra of $gl(n, \mathbf{R})$ or $gl(n, \mathbb{C})$ is called a *representation* of $\mathcal{L}_1$.

It is easily seen that, for a linear map to be a homomorphism, (3.8) needs to be verified only for the elements of a basis of $\mathcal{L}_1$.

Example 3.1.4 As a simple example of two isomorphic Lie algebras, consider the Lie algebras $\mathcal{L}_1$ and $\mathcal{L}_2$ defined as follows. $\mathcal{L}_1$ is the Lie algebra of the vectors in $\mathbf{R}^3$ with the cross product as the commutator. It is spanned by $\vec{i}, \vec{j}, \vec{k}$. The Lie algebra $\mathcal{L}_2 := su(2)$ is spanned by $\bar\sigma_{x,y,z}$ in (3.6). Because of the commutation relations (3.7), (3.5), the following correspondence gives an isomorphism:

$$\vec{i} \leftrightarrow \bar\sigma_x, \qquad \vec{j} \leftrightarrow \bar\sigma_y, \qquad \vec{k} \leftrightarrow \bar\sigma_z.$$

3.1.1.4 Lie algebra generated by a set of elements

Given a set of vectors of a Lie algebra $\mathcal{L}$, $\{x_1, ..., x_n\}$, the set of all the (possibly repeated) commutators of $\{x_1, ..., x_n\}$ spans a subalgebra of $\mathcal{L}$ which is called **the Lie algebra generated by** $\{x_1, ..., x_n\}$. It will be denoted here by $\{x_1, ..., x_n\}_{\mathcal{L}}$. The Lie algebra $\{x_1, ..., x_n\}_{\mathcal{L}}$ is the smallest subalgebra of $\mathcal{L}$ containing $\{x_1, ..., x_n\}$. In general, in order to prove that $\{x_1, ..., x_n\}_{\mathcal{L}}$ is equal to a given subalgebra $\mathcal{A}$, we only need to produce, as (possibly repeated) commutators of $\{x_1, ..., x_n\}$, a basis of $\mathcal{A}$. For example, consider $\mathcal{L}$ the Lie algebra of complex 2×2 matrices over the real numbers. The Lie algebra generated by $\{\bar{\sigma}_x, \bar{\sigma}_y, \bar{\sigma}_z\}$ in $\mathcal{L}$ is equal to $su(2)$. Because of (3.7), the Lie algebra generated by $\{\bar{\sigma}_x, \bar{\sigma}_y\}$, $\{\bar{\sigma}_x, \bar{\sigma}_y\}_{\mathcal{L}}$, is equal to $su(2)$, as well.

3.1.1.5 Ideals; semisimple, abelian and reductive Lie algebras

A Lie algebra $\mathcal{L}$ is called *Abelian* if the Lie bracket between any two elements in $\mathcal{L}$ is zero. The direct sum of two Lie algebras $\mathcal{L}_1$ and $\mathcal{L}_2$ $\mathcal{L}_1 \bar{\oplus} \mathcal{L}_2$ is the Lie algebra obtained as the direct sum of the vector spaces $\mathcal{L}_1$ and $\mathcal{L}_2$, with, moreover, $[\mathcal{L}_1, \mathcal{L}_2] = 0$.

Definition 3.1.5 An **ideal** $\mathcal{I}$ of a Lie algebra $\mathcal{L}$ is a subspace of $\mathcal{L}$ such that

$$[\mathcal{I}, \mathcal{L}] \subseteq \mathcal{I}.$$

It is an immediate consequence of the definition that an ideal of a Lie algebra $\mathcal{L}$ is a Lie subalgebra of $\mathcal{L}$. Strictly related to the concept of ideal is the concept of *normal subgroup*. This is discussed in Appendix C.

Definition 3.1.6 A Lie algebra $\mathcal{L}$ is **simple** if it has dimension > 1 and it has no ideals other than the trivial ones: $\{0\}$ and $\mathcal{L}$ itself. A **semisimple** Lie algebra $\mathcal{L}$ is the direct Lie algebra sum $\bar{\oplus}$ of simple Lie algebras $\mathcal{L}_j$, $j = 1, ..., r$, i.e.,

$$\mathcal{L} = \mathcal{L}_1 \bar{\oplus} \mathcal{L}_2 \bar{\oplus} ... \bar{\oplus} \mathcal{L}_r.$$

Whether or not a Lie algebra is semisimple can be checked using *Cartan semisimplicity criterion* (cf. Appendix C).

Definition 3.1.7 A Lie algebra $\mathcal{L}$ is called **reductive** if it is the direct sum of a semisimple and an Abelian Lie algebra.

Since we will be interested in Lie subalgebras of $u(n)$, the following result proved for instance in Theorem 1 of [80] will be important. It is a special case of the general *Levi decomposition* discussed in Appendix C.

Theorem 3.1.8 *Every Lie subalgebra of $u(n)$ is reductive.*

3.1.2 Lie groups

3.1.2.1 General definition of a Lie group

Definition 3.1.9 A **Lie group** is a group which is also an analytic differentiable manifold and such that the group operations, $\{x, y\} \to xy$, and $x \to x^{-1}$, are analytic.

An important example of a Lie group is the *general linear group*, $Gl(n, \mathbb{C})$, defined as the group of nonsingular matrices of dimension n with complex entries. This is indeed a group under matrix multiplication since $\det(AB) = \det(A)\det(B)$ and we can give it the structure of a differentiable manifold by mapping the real and imaginary parts of each entry of an element to open sets in $\mathbf{R}^{2n^2}$. Because the matrix multiplication and inversion only require analytic operations, this is a Lie group.

Definition 3.1.10 Given two Lie groups G_1 and G_2, an analytic map

$$\phi : G_1 \to G_2$$

is a **homomorphism of Lie groups** if it preserves the group operations of G_1, i.e., for every $g_1, g_2 \in G_1$,

$$\phi(g_1 g_2) = \phi(g_1)\phi(g_2), \qquad \phi(g_1^{-1}) = [\phi(g_1)]^{-1}.$$

The operations on the left- and right-hand sides of these conditions refer to the operations in G_1 and G_2, respectively.

A homomorphism of Lie groups which is also bijective is called an **isomorphism of Lie groups**.

3.1.2.2   Lie subgroups

The Lie groups we shall be most interested in are the Lie group of unitary matrices of dimension n and the Lie group of unitary matrices of dimension n with determinant equal to 1. The first is called the **unitary group** and it is denoted by $U(n)$; the second is called the **special unitary group** and it is denoted by $SU(n)$. These are Lie subgroups of $Gl(n, \mathbb{C})$ in the following sense:

Definition 3.1.11 A subgroup H of a Lie group G which is also an analytic submanifold of G and a Lie group with the structure inherited from G is called a **Lie subgroup** of G.

There are various definitions of submanifold in differential geometry. We adopt the definition of [39] pg. 75 ff. (see also Theorem 6.6 pg. 83, therein). For us a submanifold of M is a subset N of M with the relative topology and such that, for every point $p \in N$, there is a chart (U, ϕ) of M such that i) $\phi(p) = 0$, ii) ϕ maps U onto an open ball of radius ϵ with center at the origin in $\mathbf{R}^m$, B_ϵ^m, m being the dimension of M, iii) $\phi(U \cap N)$ is the set of points in B_ϵ^m with the last $m - n$ coordinates equal to zero, n being the dimension of N.

3.1.2.3   Lie algebra of a Lie group

Consider a Lie group G and a neighborhood N of the identity. A *tangent vector at the identity I* is defined as a class of equivalent curves $\gamma : (-\epsilon, \epsilon) \to$

N, $\epsilon > 0$. Two curves are equivalent if

$$\gamma_1(0) = \gamma_2(0) = I$$

and, for every function $f : N \to \mathbf{R}$,

$$\frac{d}{dt}\bigg|_{t=0}(f \circ \gamma_1(t)) = \frac{d}{dt}\bigg|_{t=0}(f \circ \gamma_2(t)). \tag{3.9}$$

There is a one-to-one correspondence between classes of equivalent curves and *derivations* on the space of C^∞ functions $N \to \mathbf{R}$. The derivation corresponding to the curve γ maps f to $\frac{d}{dt}|_{t=0}(f \circ \gamma(t))$. Using the vector space structure on the space of derivations we can give the space of tangent vectors the structure of a vector space, the *tangent space*. This has the same dimension as a vector space as the dimension of the Lie group G as a manifold.

The tangent space at the identity of a *Lie group of matrices* can be given the structure of a Lie algebra over the reals as follows. Consider a parametrization of the group G as a subgroup of $Gl(n, \mathbf{C})$ or $Gl(n, \mathbf{R})$, namely with $2n^2$ or n^2 (possibly redundant) parameters. Then a curve in G can be written, in these coordinates, as a map $(-\epsilon, \epsilon) \to Gl(n, \mathbf{C})$ or $(-\epsilon, \epsilon) \to Gl(n, \mathbf{R})$, $t \to \tilde{X}(t)$. The derivative $\frac{d}{dt}$ of $\tilde{X}(t)$ at zero is a matrix, $\tilde{A}$. It uniquely determines the derivatives in (3.9). The curve $\{e^{\tilde{A}t}|t \in \mathbf{R}\}$ is in the same equivalence class as $\{\tilde{X}(t)|t \in (-\epsilon, \epsilon)\}$. It is called a *one parameter subgroup* of G and it can be proved to belong to G for every $t \in \mathbf{R}$. In fact, this defines a one-to-one correspondence between tangent vectors and one parameter subgroups. Consider now the Lie algebra generated by the matrices $\tilde{A}$ obtained as above. To each matrix $\tilde{A}$ there corresponds a one parameter subgroup and therefore a tangent vector. In fact one can prove that if e^{At} and e^{Bt} are in G so are $e^{(A+B)t}$ and $e^{[A,B]t}$. Proofs of the above statements can be found in any of the standard introductions to Lie groups and Lie algebras such as [130], [244], [290]. The important facts for our purposes are summarized in the following.

FACT. (**Correspondence between Lie groups and Lie algebras**) Given a Lie group of matrices, the matrices obtained by differentiating curves at the identity at time $t = 0$ generate the corresponding Lie algebra. This Lie algebra is isomorphic (in the sense of vector spaces) to the tangent space of the Lie group at the identity. In particular, it has the same dimension as the tangent space which is the dimension of the Lie group seen as an analytic manifold. Conversely, given a Lie algebra of matrices $\mathcal{L}$, the associated one-dimensional subgroups generate a Lie group

$$e^{\mathcal{L}} := \{e^{A_1}e^{A_2} \cdots e^{A_m}, \quad A_1, A_2, ..., A_m \in \mathcal{L}\} \tag{3.10}$$

which is the unique connected Lie subgroup of $Gl(n, \mathbf{R})$ $(Gl(n, \mathbf{C}))$ with Lie algebra $\mathcal{L}$.

The correspondence in the FACT is often called *Lie correspondence* (e.g., section 2.5, [238]). More notions of Lie algebras and Lie group theory will be presented in this book as they are needed. In particular see Chapter 4, Chapter 6 and Chapter 9. Throughout the book, we shall denote by $e^{\mathcal{L}}$ the Lie group associated with Lie algebra $\mathcal{L}$. The map from the Lie algebra $\mathcal{L}$ to the corresponding Lie group $e^{\mathcal{L}}$ which associate to $A \in \mathcal{L}$, $e^A \in e^{\mathcal{L}}$ is called the *exponential map*. It plays a prominent role in all the theory of Lie groups and Lie algebras and in what follows. In general $e^A e^B \neq e^{A+B}$ unless A and B commute. There exist, however, several known formulas relating matrix exponentials known as *exponential formulas*. Some of the most useful are collected in Appendix E (some more formulas can be found for example in Section 3, of Chapter 5 of [244]).

Remark 3.1.12 Another way to describe elements of a Lie group $e^{\mathcal{L}}$ is as *finite products* (cf. (3.10))

$$e^{B_1 t_1} e^{B_2 t_2} \cdots e^{B_l t_l}, \quad t_1, t_2, ..., t_l \in \mathbf{R},$$

where $B_1, ..., B_l$ are taken from a *basis* of the Lie algebra $\mathcal{L}$. To see this, consider the expression of a given $X \in e^{\mathcal{L}}$ as $X = e^{A_1} e^{A_2} \cdots e^{A_m}$ with $A_1, ..., A_m \in \mathcal{L}$. Take one of the exponentials e^{A_j}. By writing $e^{A_j} = (e^{\frac{A_j}{M}})^M$ for M integer and large, we can assume that $e^{\frac{A_j}{M}}$ is in an arbitrarily small neighborhood of the identity in $e^{\mathcal{L}}$. From the inverse function theorem it follows that the set $\{e^{B_1 t_1} e^{B_2 t_2} \cdots e^{B_l t_l} \mid t_1, ... t_l \in \mathbf{R}\}$ includes all elements in a neighborhood of the identity in $e^{\mathcal{L}}$. In fact, we shall see below (cf. subsection 3.2.3 and Appendix D) that it is enough for the elements $\{B_1, ..., B_l\}$ to *generate* the Lie algebra $\mathcal{L}$.

We conclude this section by discussing the correspondence between Lie subalgebras $\mathcal{A}$ of $\mathcal{L}$ and Lie subgroups $e^{\mathcal{A}}$ of $e^{\mathcal{L}}$. In order for the Lie group corresponding to $\mathcal{A} \subseteq \mathcal{L}$ to be a Lie subgroup of $e^{\mathcal{L}}$, we need that the topology and differentiable structure of $e^{\mathcal{A}}$ to be the one induced as a subset of $e^{\mathcal{L}}$. In other terms, the open sets in $e^{\mathcal{A}}$ must be of the form $e^{\mathcal{A}} \cap N$ where N is an open set in $e^{\mathcal{L}}$. For example, $U(n)$ is a Lie subgroup of $Gl(n, \mathbb{C})$ because on $U(n)$ we take the topology generated by open balls $B_\epsilon(U_0) := \{U \in U(n) \mid \|U - U_0\| < \epsilon\}$, where the $\|A\|$ is defined by $\|A\| := \text{Tr}(AA^\dagger)$ and such a definition is the same as the one for the topology on $Gl(n, \mathbb{C})$. With this topology the operations of group multiplications and inversion are continuous, in fact, analytic. On the other hand, consider the one-dimensional (Abelian) Lie algebra $\mathcal{A}$ spanned by

$$A := \begin{pmatrix} 0 & 1 & 0 & 0 \\ -1 & 0 & 0 & 0 \\ 0 & 0 & 0 & x \\ 0 & 0 & -x & 0 \end{pmatrix}, \tag{3.11}$$

with x an irrational number. Then $e^{\mathcal{A}} := \{e^{At} \,|\, t \in \mathbf{R}\}$. The topology which makes this a Lie group is the one inherited by $\mathbf{R}$. That is, the open sets in $e^{\mathcal{A}}$ are the sets $\{e^{At} \,|\, t \in N\}$, where N is an open set in $\mathbf{R}$. Consider the set $N_a := \{e^{At} \,|\, t \in (-a, a)\}$, for $a > 0$, which is an open set and, in fact, a coordinate neighborhood, in $e^{\mathcal{A}}$. If we try to realize N_a as $e^{\mathcal{A}} \cap N$, for some neighborhood N of the identity in $U(n)$ we have the problem that e^{At} goes arbitrarily close to the identity an infinite number of times as $t \to \infty$. Therefore, since a is finite, N_a cannot be written as $e^{\mathcal{A}} \cap N$, for N a neighborhood of the identity in $U(n)$.

3.2 Controllability Test: The Dynamical Lie Algebra

The main test for controllability of quantum systems (3.1) is the following:

Theorem 3.2.1 *The set of reachable states for system (3.1) is dense (in the subset topology induced by $U(n)$) in the connected Lie group $e^{\mathcal{L}}$ associated with the Lie algebra $\mathcal{L}$ generated by $\{-iH(u) \,|\, u \in \mathcal{U}\}$. Furthermore, if $e^{\mathcal{L}}$ is a Lie subgroup of $U(n)$ then*

$$\mathcal{R} = e^{\mathcal{L}}. \tag{3.12}$$

The Lie algebra $\mathcal{L}$ is called the **dynamical Lie algebra** associated with the system. This is always a subalgebra of $u(n)$. In the case $\dim(\mathcal{L}) = n^2 = \dim(u(n))$, which is equivalent to $\mathcal{L} = u(n)$ and $e^{\mathcal{L}} = U(n)$, the system is said to be **controllable**. In this case $\mathcal{R} = U(n)$, which means that every unitary matrix can be obtained by choosing an appropriate control in (3.1). We shall say that the system is controllable even in the case where $\dim(\mathcal{L}) = n^2 - 1 = \dim(su(n))$ which is equivalent to $\mathcal{L} = su(n)$ and $e^{\mathcal{L}} = SU(n)$. Sometimes we shall use the terminology **operator controllability** or **complete controllability** to distinguish this case (controllability of system (3.1)) from the case where we have controllability of the state $|\psi\rangle$ which will be treated later in this chapter. The condition of Theorem 3.2.1 is often referred to as the **Lie Algebra Rank Condition**.

The theorem states that the dynamical Lie algebra $\mathcal{L}$ essentially gives all the information on the set of reachable states $\mathcal{R}$ for the system given by the Schrödinger operator equation (3.1). It is important to discuss in which case the equality $\mathcal{R} = e^{\mathcal{L}}$ holds. First of all, we notice that the assumption that $e^{\mathcal{L}}$ is a Lie subgroup of $U(n)$ cannot, in general, be removed. For example, consider the 'uncontrolled' system $\dot{X} = -iH_0 X$ where H_0 is given by (cf. equation (3.11))

$$H_0 = \begin{pmatrix} 0 & i & 0 & 0 \\ -i & 0 & 0 & 0 \\ 0 & 0 & 0 & ix \\ 0 & 0 & -ix & 0 \end{pmatrix}, \tag{3.13}$$

with x an irrational number. In this case, $\mathcal{R} = \{e^{-iH_0t}|t \geq 0\}$ while $e^{\mathcal{L}} = \{e^{-iH_0t}|t \in \mathbf{R}\}$ and the two sets do not coincide. As discussed for the example of (3.11) the Lie group $e^{\mathcal{L}}$ is not a Lie subgroup of $U(n)$ ($U(4)$ in this case).

The problem of whether a Lie group is a Lie subgroup of another Lie group is a classical question in Lie theory which is often referred to as the *correspondence between Lie subgroups and Lie subalgebras*. A classical result which is known as the **closed subgroup theorem** (e.g., [238] Corollary 3 of §2.7 and Exercise 2, therein) tells us that $e^{\mathcal{L}}$ is a Lie subgroup if and only if it is closed in the topology of $U(n)$. From a practical point of view, it is of interest to have a way to check this property on the Lie algebra $\mathcal{L}$. The *Killing form*[3] gives a way to do this. It is defined as follows.

Definition 3.2.2 Consider a Lie algebra $\mathcal{L}$. If ad_x is the matrix representation of the linear operator on $\mathcal{L}$, $L \to [x, L]$ for a given $x \in \mathcal{L}$, the Killing form $\langle x, y \rangle_K$ is defined as $\langle x, y \rangle_K := \text{Tr}(ad_x ad_y)$. In the case of $su(n)$, the Killing form is proportional to $\text{Tr}(xy^\dagger)$. See, e.g., [130] for further details. See also Appendix C.

We have (see, e.g., Proposition 4.27 in [168]).

Proposition 3.2.3 If the Killing form of $\mathcal{L}$ is negative definite, then $e^{\mathcal{L}}$ is a compact Lie group (and therefore is closed).

This is an important result which allows us to check a topological property using an algebraic test. However, it only gives a sufficient condition and does not use any possible further structure of the Lie algebra $\mathcal{L}$. In alternative we can use more specific properties of the Lie group $U(n)$ and the Lie algebra $u(n)$. Recall from Theorem 3.1.8 that $\mathcal{L}$ being a Lie subalgebra of $u(n)$ is reductive and therefore it can be written as the direct sum

$$\mathcal{L} = \mathcal{S} \bar{\oplus} \mathcal{Z},$$

where $\mathcal{S}$ is semisimple and $\mathcal{Z}$ is Abelian. The subalgebra $\mathcal{Z}$ is the *center*, i.e., the subspace of all elements that commute with all of $\mathcal{L}$. Therefore, every element in $e^{\mathcal{L}}$ is the commuting product of an element in the Lie group $e^{\mathcal{S}}$ associated with $\mathcal{S}$ and an element in the Abelian Lie group $e^{\mathcal{Z}}$ associated with $\mathcal{Z}$. Since $e^{\mathcal{S}}$ is a semisimple Lie group and a subgroup of the compact Lie group $U(n)$ we can apply Theorem 6.3.13 of [220] which says:[4]

Theorem 3.2.4 *If $\mathcal{S}$ is a semisimple Lie subalgebra of a Lie algebra $\mathcal{G}$ and $e^{\mathcal{G}}$ is a compact Lie group (such as $U(n)$) then $e^{\mathcal{S}}$ is a Lie subgroup of $e^{\mathcal{G}}$. In particular it is compact.*

---

[3]This will also be used in other parts of this book and it is one of the fundamental tools in Lie algebra theory.

[4]Notice that the definition of Lie subgroup in [220] is different from the one adopted in this book. However, we state the theorem in a form coherent with our notations and definitions.

Therefore, problems may only come from the Abelian part $\mathcal{Z}$ which has usually small dimension and it is easy to analyze as it is commuting with the whole group. In this sense, the example of the Abelian (one-dimensional) Lie group associated with H_0 described in (3.13) is somehow prototypical. Being Abelian, $e^{\mathcal{Z}}$ is the same as $T^k \times \mathbf{R}^j$ where T^k is a k-torus (which is compact) (see, e.g., Theorem 2.6.2 in [194]). The Lie group $e^{\mathcal{L}}$ is compact if $j = 0$.

Another linear algebra type of test is suggested in Problem 4 of §2.7 of [238]. If for $X \in u(n)$, $[X, \mathcal{L}] \subseteq \mathcal{L}$ implies $X \in \mathcal{L}$, then $e^{\mathcal{L}}$ is a closed subgroup and one can apply the *closed subgroup theorem* above.

3.2.1 On the proof of the controllability test

The proof of Theorem 3.2.1 is presented in Appendix D. To a certain extent such a proof is constructive, in the sense that it shows how to find a sequence of piecewise constant controls, that is, a sequence of Hamiltonians, which drives the state X of (3.1) to any desired final value. In particular, assume we wish to drive the evolution operator X from the identity to $X_f \in e^{\mathcal{L}}$, and notice that the exponential map $\mathcal{L} \to e^{\mathcal{L}}$ is onto in our cases[5], that is, there exists a matrix $A \in \mathcal{L}$ such that $e^A = X_f$, for every X_f. This also means that given any neighborhood N of the identity in $e^{\mathcal{L}}$, we can choose K sufficiently large such that $e^{\frac{A}{K}} = X_f^{\frac{1}{K}} \in N$. Now, assume first that we have available in the set $\{-iH(u) \,|\, u \in \mathcal{U}\}$ a basis of elements (Hamiltonians) for $\mathcal{L}$. That is, no Lie bracket is necessary to obtain a basis of $\mathcal{L}$. Let $m := \dim \mathcal{L}$. There exist m values of the control u_j, and corresponding m matrices $A_j := -iH(u_j)$, $j = 1, \ldots, m$, (cf. Lemma 1 in Appendix D) which form a basis for $\mathcal{L}$. By varying $t_1, \ldots, t_m$ in a neighborhood of the origin in $\mathbf{R}^m$, $e^{A_m t_m} e^{A_{m-1} t_{m-1}} \cdots e^{A_1 t_1}$ gives a neighborhood of the identity in $e^{\mathcal{L}}$ and, in particular, it contains $e^{\frac{A}{K}}$. Therefore, there exist values $\bar{t}_1, \ldots, \bar{t}_m$ such that setting the control equal to u_1 for time $\bar{t}_1$ and then equal to u_2 for time $\bar{t}_2$, and so on, up to, equal to u_m for time $\bar{t}_m$, we drive the value of the evolution operator to $e^{\frac{A}{K}}$. Obviously this argument ignores the fact that some of the values $\bar{t}_j$, $j = 1, \ldots, m$, may be negative. However, this problem can be overcome. In fact, in many quantum mechanical models, it is possible to choose the controls u_j and therefore the matrices A_j so that the orbits $\{e^{A_j t} | t \in \mathbf{R}\}$ are periodic, which allows us to assume all the $\bar{t}_j$'s positive, without loss of generality. If this is not the case one can show that it is possible to reach a state arbitrarily close to the desired one and reach it exactly if $e^{\mathcal{L}}$ is a Lie subgroup of $U(n)$.

In the more common case where it is not possible to choose controls to obtain a basis of $\mathcal{L}$, we can proceed as in the proof of Lemma 1, Appendix D.

---

[5]This is in particular true if $e^{\mathcal{L}}$ is compact cf. Corollary 4.48 in [168], and true for Abelian Lie subgroups of $U(n)$ and therefore true in the reductive case which is the one of interest here since $e^{\mathcal{S}}$ is compact for $\mathcal{S}$ semisimple.

Let $\{A_1, \ldots, A_s\}$ be a set of linearly independent available elements (Hamiltonians) which generate $\mathcal{L}$. There exist two values $1 \leq k, l \leq s$ such that the commutator $[A_l, A_k]$ is linearly independent of $\{A_1, \ldots, A_s\}$. This implies that there exists a value $t \in \mathbf{R}$ such that $F := e^{A_l t} A_k e^{-A_l t}$ is also linearly independent, so that we can add F to $\{A_1, \ldots, A_s\}$, to obtain $s + 1$ linearly independent matrices in $\mathcal{L}$. Notice that it is very easy to obtain the exponential of F, since $e^{F\tilde{t}} = e^{A_l t} e^{A_k \tilde{t}} e^{-A_l t}$. If the dimension of $\mathcal{L}$ is $s + 1$ we have now a basis of $\mathcal{L}$ and we can proceed as above. If this is not the case, we can obtain more linearly independent matrices in the same way and eventually obtain a basis of $\mathcal{L}$. From this basis, using only exponentials of the available elements, we obtain all the elements of a neighborhood of the identity in $e^{\mathcal{L}}$ including $e^{\frac{A}{K}}$. Repeating the associated control sequence K times we obtain e^A.

This approach allows us to drive the evolution operator X from the identity (arbitrarily close) to any point in a neighborhood of the identity and therefore to $e^{\frac{A}{K}}$ for K sufficiently large. The main problem with this approach is to find a suitable way to estimate the 'size' of the neighborhood of the identity which is obtained so as to know how to choose K so that $e^{\frac{A}{K}}$ is in it. More discussion on control algorithms based on the proof of the controllability test of Theorem 3.2.1 can be found in [79]. In the following chapters we will give many techniques to construct control algorithms for quantum mechanical systems.

3.2.2   Procedure to generate a basis of the dynamical Lie algebra

To identify the nature of the dynamical Lie algebra $\mathcal{L}$ one must obtain a basis of $\mathcal{L}$. To do that, one first takes a basis in $\{-iH(u) \mid u \in \mathcal{U}\}$, $\{A_1, ..., A_s\}$. A *repeated Lie bracket* is an element of $\mathcal{L}$ of the form $[X_1, [X_2, [\ldots [X_{p-1}, X_p]]]]$, with $X_1, ..., X_p \in \{A_1, ..., A_s\}$. One defines the *depth* of a repeated Lie bracket B obtained with $\{A_1, ..., A_s\}$ as the number of Lie brackets performed to obtain B, i.e., $p - 1$ in the above definition. The recursive procedure is as follows:

Step 0: List the linearly independent vectors of depth 0, $\{A_1, ..., A_s\}$.

Step k:

1. Calculate the Lie brackets of the elements of depth $k - 1$ obtained at step $k - 1$, with the elements of depth 0. This way one obtains elements of depth k.

2. Out of the elements obtained in the previous point, select the ones that form a linearly independent set along with the ones obtained up to step $k - 1$.

3. Stop the procedure if there is no new vector or if the dimension of the linearly independent set is $n^2 - 1$ or n^2.

The set of matrices obtained this way forms a basis of the dynamical Lie algebra $\mathcal{L}$.

A justification of this procedure is the subject of Exercise 3.1. We remark that the procedure above described will always terminate since the dynamical Lie algebra $\mathcal{L}$ is a subalgebra of $u(n)$ and therefore it has at most dimension n^2.

Example 3.2.5 Consider the system of two spin-$\frac{1}{2}$ particles with Ising interaction (2.65), interacting with a magnetic field which has zero component only in the x-direction. The magnetic field plays the role of the control. Assuming two equal gyromagnetic ratios for the two particles, the set of matrices one obtains at Step 0 of the above procedure is given by

$$\{A_1, A_2\} = \{i\sigma_z \otimes \sigma_z, i\sigma_x \otimes 1_{2\times 2} + i1_{2\times 2} \otimes \sigma_x\},$$

where the matrices σ_x, σ_y, σ_z are the Pauli matrices defined in (1.20). At Step 1, using (1.21), we find the matrix $A_3 = [A_1, A_2]$, with

$$A_3 = 2i\sigma_y \otimes \sigma_z + 2i\sigma_z \otimes \sigma_y.$$

At Step 2 we calculate $[A_3, A_1]$ and $[A_3, A_2]$. The commutator $[A_3, A_1]$ gives a matrix proportional to $i\sigma_x \otimes 1 + i1 \otimes \sigma_x$, which therefore is not linearly independent from the ones we already have and we discard. The commutator $[A_3, A_2]$ gives, up to an unimportant proportionality factor, $A_4 := i\sigma_y \otimes \sigma_y - i\sigma_z \otimes \sigma_z$. At Step 3 we have $[A_4, A_1] = 0$ and $[A_4, A_2]$ proportional to A_3. Therefore, no new linearly independent matrix is found and the procedure stops. The dynamical Lie algebra $\mathcal{L}$ is spanned by $\{A_1, A_2, A_3, A_4\}$, with

$$A_1 := i\sigma_z \otimes \sigma_z, \qquad A_2 := i\sigma_x \otimes 1 + i1 \otimes \sigma_x, \qquad A_3 := i\sigma_y \otimes \sigma_z + i\sigma_z \otimes \sigma_y,$$

$$A_4 := i\sigma_y \otimes \sigma_y - i\sigma_z \otimes \sigma_z.$$

The matrix $i\sigma_z \otimes \sigma_z + i\sigma_y \otimes \sigma_y$ commutes with the generators A_1 and A_2 and therefore with the whole dynamical Lie algebra $\mathcal{L}$. It spans the center of such Lie algebra. The set $\{A_2, A_3, A_4\}$ spans a Lie subalgebra which is isomorphic to $su(2)$ and therefore simple (since $su(2)$ is simple from Proposition 3.3.7 below). The corresponding Lie group is compact because of Theorem 3.2.4. Since the matrix $i\sigma_y \otimes \sigma_y + i\sigma_z \otimes \sigma_z$ has eigenvalues $\{0, 0, \pm 2i\}$ the one-dimensional Lie group $\{e^{(i\sigma_y \otimes \sigma_y + i\sigma_z \otimes \sigma_z)t} \,|\, t \in \mathbf{R}\}$ is closed and therefore compact. Therefore, $e^{\mathcal{L}}$ is a compact Lie subgroup of $U(4)$ and the reachable set is exactly $e^{\mathcal{L}}$ in this case, from Theorem 3.2.1.

3.2.3   Uniform finite generation of compact Lie groups

The proof of the controllability test presented in Appendix D shows, among other things, the two facts in the following corollary. The crucial assumption is that $e^{\mathcal{L}}$ is a Lie subgroup of $U(n)$ and $U(n)$ is compact.

Corollary 3.2.6 Consider a connected Lie group $e^{\mathcal{L}}$ corresponding to a Lie algebra $\mathcal{L}$. Then
a) Every element X_f in $e^{\mathcal{L}}$ can be written in the form

$$X_f = e^{At_1} \cdots e^{At_r}, \tag{3.14}$$

with the indeterminates A in the set $\mathcal{S} := \{A_1, ..., A_s\}$ of generators of $\mathcal{L}$ and real $t_1, ..., t_r$. The number r will depend on X_f.
b) Every element X_f in a neighborhood N of the identity in $e^{\mathcal{L}}$ can be expressed as a product of the form (3.14) with uniformly bounded r.

If the Lie group $e^{\mathcal{L}}$ is *compact*, for example $e^{\mathcal{L}} = U(n)$ or $e^{\mathcal{L}} = SU(n)$, then we see that the number of factors r is uniformly bounded as X_f varies in $e^{\mathcal{L}}$. To see this, notice that, for an open neighborhood N of the identity in $e^{\mathcal{L}}$, we can write $e^{\mathcal{L}} = \cup_{K \in e^{\mathcal{L}}} KN$. This is an open cover of $e^{\mathcal{L}}$ and, by compactness of $e^{\mathcal{L}}$, contains a finite subcover. Therefore, we can write $e^{\mathcal{L}} = \cup_{i=1}^{l} K_i N$, for finite l. Since, from part a) of the above corollary, K_i is itself the finite product of elements of the form e^{At}, the result follows. In this situation, the Lie group $e^{\mathcal{L}}$ is said to be *uniformly finitely generated* (see, e.g., [169]) by the set of generators $\mathcal{S}$. Therefore, every compact Lie group is uniformly generated by a set of generators of the corresponding Lie algebra.

A physical interpretation of the above uniform generation result concerns universality of quantum gates. Consider two Hamiltonians corresponding to the skew-Hermitian matrices A_1 and A_2. Each of them represents a possible evolution of the quantum system which modifies the state. We have interpreted each evolution $e^{A_1 t}$ or $e^{A_2 t}$ as a quantum logic gate (see subsection 1.4.1). Assume it is possible to switch between these two evolutions. We can call the set of gates $\{A_1, A_2\}$ *universal* if, by switching between them, it is possible to generate all the (special) unitary evolutions. According to the above results, this is possible if and only if $\{A_1, A_2\}$ generate the whole Lie algebra $u(n)$ or $su(n)$. Moreover, since $U(n)$ and $SU(n)$ are compact Lie groups, the number of gates needed to generate every (special) unitary evolution is uniformly bounded.

3.2.4   Controllability as a generic property

Assume we have two generators of the dynamical Lie algebra $\mathcal{L}$, A_1 and A_2 in $su(n)$. We might ask the question of how likely it is that A_1 and A_2 generate $su(n)$, i.e., $\mathcal{L} = su(n)$. In order to study this question we put a measure on $su(n) \times su(n)$ which is the one inherited by $\mathbf{R}^{n^2-1} \times \mathbf{R}^{n^2-1}$. The (real) dimension of $su(n)$ is $n^2 - 1$. The set of pairs which give a Lie algebra smaller than $su(n)$ is a set of measure zero in $su(n) \times su(n)$. In order to see this take a basis in $su(n)$, $\{E_j \mid j = 1, ..., n^2 - 1\}$ so that any element $A \in su(n)$ can be written as $A = \sum_j a_j E_j$. Analogously we write for $B \in su(n)$, $B = \sum_j b_j E_j$. Because of linearity, the commutators $[E_i, E_j] = \sum_k c_{ij}^k E_k$ determine the whole structure of the Lie algebra $su(n)$. The constants c_{ij}^k describe such a structure and they are called the *structure constants*. They are fixed once the basis $\{E_j\}$ has been chosen. Now take A, B and an arbitrary sequence of commutators of A and B, for a total of $n^2 - 1$ matrices. By a straightforward induction argument (on the depth of the commutator), it follows that each element of such a sequence is of the form $\sum_l w_l E_l$, where the coefficients w_l are *analytic* functions of the coefficients $\{a_j, b_k\} \in \mathbf{R}^{n^2-1} \times \mathbf{R}^{n^2-1}$, depending on the structure constants. The determinant function det giving the rank of the space spanned by such matrices is also an analytic function of $\{a_j, b_k\} \in \mathbf{R}^{n^2-1} \times \mathbf{R}^{n^2-1}$. By a classical result in analysis (see, e.g., [204]) the set of zeros of det has zero measure as det is not identically zero. Therefore, to prove the claim it is enough to show that there always (that is, for every n) exist two matrices A and B which generate $su(n)$. This is also a classical result in Lie algebra theory [146], [172], for any semisimple Lie algebra such as $su(n)$ (cf. Proposition 3.3.7 below). Therefore, we have.

Theorem 3.2.7 *A generic pair of Hamiltonians $A, B \in su(n)$ generates $su(n)$.*

In the quantum information theory literature, the two Hamiltonians A and B are seen as performing quantum logic gates and they are called *universal* if every unitary can be expressed as a sequence of exponentials e^{At} and e^{Bt} for $t \geq 0$. Therefore, the above result (together with the controllability test of Theorem 3.2.1) is often quoted by saying that 'almost every quantum logic gate is universal'. We refer to [184], [291] for more discussion on this.

3.2.5 Reachable set from some time onward

We conclude this section by stating a property of controllable quantum systems which is often useful. Theorem 3.2.1 does not give details about the dependence of the reachable sets ($\mathcal{R}(t)$) on t. It only characterizes the union over all the t's of such sets. A more detailed description is given by the following property.

Theorem 3.2.8 *Assume that the dynamical Lie algebra $\mathcal{L}$ is semisimple (in particular $\mathcal{L} = su(n)$ when the system is controllable). Then there exists a critical time T_c such that for every $T \geq T_c$,*

$$\mathcal{R}(T) = e^{\mathcal{L}}. \tag{3.15}$$

Therefore, under the controllability assumption, if we take the final time T sufficiently large we can find a control steering to any $X_f \in SU(n)$ in exactly time T. Finding estimates of the critical time for various systems of interest is an important topic of research. For a proof of this result we refer to Theorem 7.2 in [154].[6]

3.3 Notions of Controllability for the State

In the previous section, we saw that a calculation of the dynamical Lie algebra gives information on the set of all the possible transformations between states achieved with a physical apparatus. In this section, we investigate in what cases this set of transformations allows any desired transfer between two quantum mechanical states. In particular, we shall investigate the following notions of controllability.

Definition 3.3.1 The Schrödinger equation

$$\frac{d}{dt}\vec{\psi}(t) = -iH(u(t))\vec{\psi}(t) \tag{3.16}$$

is **pure state controllable (PSC)** if for every pair of initial and final states, $\vec{\psi}_0$ and $\vec{\psi}_1$, there exist control functions u and a time $T > 0$ such that the solution of (3.16) at time T, with initial condition $\vec{\psi}_0$, satisfies $\vec{\psi}(T) = \vec{\psi}_1$. Here $\vec{\psi}_0$ and $\vec{\psi}_1$ are two vectors on the complex sphere of radius 1, $S_{\mathbb{C}}^{n-1}$.

Since, for any vector $\vec{\psi}_1$, $\vec{\psi}_1$ and $e^{i\phi}\vec{\psi}_1$ represent the same physical state, for any $\phi \in \mathbb{R}$, from a physics viewpoint, the following property is equivalent to PSC.

Definition 3.3.2 The system (3.16) is **equivalent-state-controllable (ESC)** if, for every pair of initial and final states, $\vec{\psi}_0$ and $\vec{\psi}_1$ in $S_{\mathbb{C}}^{n-1}$, there exist controls u and a phase factor ϕ such that the solution $\vec{\psi}$ of (3.16), with $\vec{\psi}(0) = \vec{\psi}_0$, satisfies $\vec{\psi}(T) = e^{i\phi}\vec{\psi}_1$, at some $T > 0$.

Assume the system is an ensemble and the dynamics is described by Liouville equation for the density matrix ρ (cf. (1.51))

$$\frac{d}{dt}\rho(t) = [-iH(u(t)), \rho(t)]. \tag{3.17}$$

---

[6]The statement of this theorem is a bit different but equivalent in [154]. It is stated in the paper that there is a critical time T_c such that every transfer $X_1 \to X_2$ is possible in exactly time T_c. For times T larger than T_c we can drive from the identity to any arbitrary $\hat{X} \in e^{\mathcal{L}}$ in time $T - T_c$ and then use the theorem to drive from the identity to $X_f \hat{X}^{-1}$ in time T_c so that the complete transfer is from the identity to X_f in time T.

Then if the initial condition is ρ_0, $\rho(t)$ is unitarily equivalent[7] to ρ_0 for every t (cf. (1.50)). Therefore, the following definition is appropriate.

Definition 3.3.3 The system (3.17) is **density matrix controllable (DMC)** if, for each pair of unitarily equivalent density matrices ρ_1 and ρ_2, there exists a control u and a time $T > 0$, such that the solution $\rho(t)$ of (3.17) with initial condition equal to ρ_1 at time T satisfies

$$\rho(T) = \rho_2.$$

In the rest of this section, we will show how to use the dynamical Lie algebra to check whether or not the system under consideration possesses any of the controllability properties above defined.

3.3.1 Pure state controllability

The solution of (3.16) with initial condition $\vec{\psi}_0$ can be written in matrix-vector form as

$$\vec{\psi}(t) = X(t)\vec{\psi}_0,$$

where $X(t)$ is the matrix solution of the Schrödinger operator equation (3.1). The matrix $X(t)$ attains a set of values dense in the Lie group $e^{\mathcal{L}}$, where $\mathcal{L}$ is the dynamical Lie algebra associated to the system. Therefore, the set of states that can be obtained as a solution of (3.16) with initial condition equal to $\vec{\psi}_0$ is in the set $\mathcal{O}_{\psi_0} := \{X\vec{\psi}_0 | X \in e^{\mathcal{L}}\}$. In order for the system to be pure state controllable we need that, for every pair of complex vectors with radius equal to 1, $\{\vec{\psi}_0, \vec{\psi}_1\}$, that is, for every pair of elements of the complex sphere $S_{\mathbb{C}}^{n-1}$ of radius 1 in $\mathbb{C}^n$, there exists a matrix X in $e^{\mathcal{L}}$, such that $\vec{\psi}_1 = X\vec{\psi}_0$. In view of this observation, the question of pure state controllability reduces to an analysis of the properties of $e^{\mathcal{L}}$ acting as a **Lie transformation group** on $S_{\mathbb{C}}^{n-1}$. The Lie group $e^{\mathcal{L}}$ is called **transitive** if the above property holds. In the next two subsections, we give the main definitions concerning Lie transformation groups.

3.3.1.1 Lie transformation groups

Given a Lie group G and a manifold M, a (left) *action* of G on M is defined as an analytic map

$$\theta : G \times M \to M,$$

satisfying the conditions

---

[7]Two Hermitian matrices A, B are said to be unitarily equivalent if there exists a matrix $C \in U(n)$ such that $CAC^\dagger = B$. This is equivalent to the two matrices having the same spectrum.

1.

$$\theta(I, x) = x$$

for every $x \in M$, where I is the identity element of the group G;

2.

$$\theta(g_2, \theta(g_1, x)) = \theta(g_2 g_1, x). \qquad (3.18)$$

It is more convenient to use the notation gx instead of $\theta(g, x)$, so that property (3.18) reads $g_2(g_1 x) = (g_2 g_1)x$. If an action exists, G is called a Lie transformation group on M. It is said to be **transitive** if, for every pair of points in M, x_1, x_2, there exists a g in G such that $x_2 = gx_1$. If G is transitive on M, M is called a **homogeneous space** of G. The terminology derives from the fact that if P is a *property* on M that is invariant for any element g of G, namely a function

$$P : M \to \mathbf{R}, \quad Pg = P, \quad \forall g \in G,$$

then P is constant over M. In fact, if for every two points, x_1 and x_2, we have $x_2 = gx_1$ for some $g \in G$, then $P(x_2) = P(gx_1) = P(x_1)$. We shall see in subsection 3.3.1.3 that the special unitary group $SU(n)$, with action given by the standard matrix-vector multiplication, is transitive on the sphere. The norm of a vector v is a property which is invariant for any element $X \in SU(n)$, since $\|Xv\| = \|v\|$, and it is constant on the sphere.

3.3.1.2   Coset spaces and homogeneous spaces

Consider a Lie group G and a closed Lie subgroup H, obtained as the connected Lie subgroup corresponding to a Lie subalgebra of the Lie algebra associated with G. Given $g \in G$, all the elements of G that can be written as gX_h, with $X_h \in H$, form an equivalence class in G which is called a **left coset** and is denoted by $\{gH\}$. Analogously one defines a **right coset** as the equivalence class of all the elements in G that can be written as $X_h g$ with $X_h \in H$. The right coset containing g is denoted by $\{Hg\}$. In the following definitions we shall refer to left cosets but only natural modifications are needed for right cosets. The **coset space** G/H is the set of left cosets. There is a natural mapping π,

$$\pi : G \to G/H, \qquad (3.19)$$

which maps $g \in G$ to $\{gH\}$ and a topology can be given to G/H from the requirement that π is continuous and open. In fact, G/H can be given the structure of an analytic manifold (cf., e.g., [157] Theorem 3.37 and [290] Theorem 3.58).

There is a natural way of defining an action of G on G/H which associates to $g \in G$ and $\{bH\} \in G/H$ the coset $\{gbH\}$. With this action and the above analytic structure of G/H, G acts as a *Lie transformation group* on G/H. It is clear that G is transitive and therefore G/H is a homogeneous space of G.

Coset spaces are canonical examples of homogeneous spaces in a sense that will now be explained. Consider a connected transitive Lie transformation group G on a connected manifold M which is therefore a homogeneous space of G. Select a point p in M. The group of elements of G that leaves p fixed is called the **isotropy group** or **the stabilizer** of p and is denoted by G_p. It is a closed Lie subgroup of G. The correspondence

$$\{gG_p\} \leftrightarrow gp \qquad (3.20)$$

for $g \in G$ is a diffeomorphism and therefore M and G/G_p are diffeomorphic. In particular, they have the same dimension. Therefore, every homogeneous space is diffeomorphic to a coset space.

In the following we shall be interested in matrix Lie subgroups $e^{\mathcal{L}}$ of the unitary group acting on the complex sphere $S_{\mathbb{C}}^{n-1}$ via the standard matrix vector multiplication. To show that a Lie group $e^{\mathcal{L}}$ is transitive, it is enough to fix a $\vec{\psi}_0 \in S_{\mathbb{C}}^{n-1}$ and show that, for every $\vec{\psi}_1 \in S_{\mathbb{C}}^{n-1}$, there exists an $X \in e^{\mathcal{L}}$ such that $\vec{\psi}_1 = X\vec{\psi}_0$. This implies, in fact, that transfer is possible between any two states.

3.3.1.3 The special unitary group and its action on the unit sphere

The **special unitary group** of dimension n, $SU(n)$, is the Lie group of unitary $n \times n$ matrices with determinant equal to one. It is the connected Lie group associated to the Lie algebra $su(n)$ of skew-Hermitian $n \times n$ matrices, with trace equal to zero. Its dimension as a manifold, which is equal to the dimension of its Lie algebra as a vector space, is $n^2 - 1$.

The Lie group $SU(n)$ is transitive on the sphere $S_{\mathbb{C}}^{n-1}$. In order to see this, consider the point $\vec{\psi}_0 := [1, 0, ..., 0]^T$. Every matrix X in $SU(n)$ whose first column is equal to $\vec{\psi}_f$ is such that

$$\vec{\psi}_f = X\vec{\psi}_0. \qquad (3.21)$$

We can always construct such a matrix by choosing the remaining columns so as to form an orthonormal set with $\vec{\psi}_f$. This gives a unitary matrix whose determinant will in general be equal to $e^{i\phi}$ for some $\phi \in \mathbf{R}$. By multiplying one of the columns 2 through n by $e^{-i\phi}$, we still have a unitary matrix X with the property (3.21) and with determinant equal to one. Of course we always assume $n \geq 2$.

The isotropy group of $\vec{\psi}_0 = [1, 0, ..., 0]^T$ is the Lie group of matrices $\tilde{X} = \begin{pmatrix} 1 & 0 \\ 0 & X \end{pmatrix}$, with $X \in SU(n-1)$, clearly isomorphic to $SU(n-1)$. The diffeomorphism (3.20) between $SU(n)/SU(n-1)$ and the sphere $S_{\mathbb{C}}^{n-1}$ maps a representative g of $\{gSU(n-1)\}$ to $g\vec{\psi}_0$. It is easily seen that this map does not depend on the representative chosen. The inverse map associates to $\vec{\psi}_f$ the equivalence class of elements of $SU(n)$ which map $\vec{\psi}_0$ to $\vec{\psi}_f$. The dimension of $SU(n)/SU(n-1)$ is given by $n^2 - 1 - ((n-1)^2 - 1) = 2n - 1$ which is indeed the (real) dimension of $S_{\mathbb{C}}^{n-1}$.

3.3.1.4   The symplectic group and its action on the unit sphere

Definition 3.3.4 The **symplectic group** $Sp(k)$ is defined as the Lie subgroup of $SU(2k)$ of matrices X satisfying

$$XJX^T = J, \tag{3.22}$$

where the $2k \times 2k$ matrix J is defined as

$$J = \begin{pmatrix} 0 & -\mathbf{1}_{k \times k} \\ \mathbf{1}_{k \times k} & 0 \end{pmatrix}, \tag{3.23}$$

with $\mathbf{1}_{k \times k}$ the $k \times k$ identity matrix.

The set of matrices in $SU(2k)$ satisfying (3.22) is, in fact, a group. In particular if X satisfies (3.22) by multiplying (3.22) on the left by $X^\dagger = X^{-1}$ and on the right by $(X^\dagger)^T$, we obtain that $X^\dagger$ also satisfies (3.22), i.e., $X^\dagger J (X^\dagger)^T = J$. Taking the complex conjugate of this expression, we obtain

$$X^T J X = J, \tag{3.24}$$

which can be used in alternative to (3.22) as the definition.

The Lie group $Sp(k)$ is the group of matrices in $SU(2k)$ leaving invariant the bilinear form

$$(y, z)_J := y^T J z$$

on $\mathbb{C}^{2k}$, since, for every $X \in Sp(k)$, (3.22) implies $(Xy, Xz)_J = (y, z)_J$.

3.3.1.5   Structure of the matrices in $Sp(k)$

From equation (3.22), since the conjugate of X, $\bar{X}$, is the inverse of X^T, we have

$$XJ = J\bar{X}. \tag{3.25}$$

If we write X by partitioning its columns into sub-columns of dimension k,

$$X := \begin{pmatrix} x_1 & x_2 & \dots x_k & x_{k+1} & x_{k+2} & \cdots & x_{2k} \\ y_1 & y_2 & \dots y_k & y_{k+1} & y_{k+2} & \cdots & y_{2k} \end{pmatrix}, \tag{3.26}$$

with $x_l, y_l \in \mathbb{C}^k$, $l = 1, ..., 2k$, then equation (3.25) implies

$$x_{k+j} = -\bar{y}_j, \qquad y_{k+j} = \bar{x}_j, \qquad j = 1, ..., k. \tag{3.27}$$

Therefore, symplectic matrices are special unitary matrices (cf. Exercise 3.6) of the form

$$X = \begin{pmatrix} A & -\bar{B} \\ B & \bar{A} \end{pmatrix}. \tag{3.28}$$

3.3.1.6 The symplectic Lie algebra

The Lie algebra of the symplectic Lie group $Sp(k)$ is denoted by $sp(k)$ and it is called the **symplectic Lie algebra**. Its elements have to satisfy a condition obtained from (3.22). By letting X depend on t, with $X(0) = 1_{n \times n}$, and differentiating (3.22), we obtain

$$\dot{X}(0)J + J\dot{X}^T(0) = 0.$$

Therefore, the Lie algebra of *skew-Hermitian* matrices A satisfying

$$AJ + JA^T = 0 \tag{3.29}$$

is the Lie algebra of $Sp(k)$. This is in fact a Lie algebra since if A and B satisfy (3.29) so does $[A, B]$.

If we partition A in $k \times k$ blocks as $A := \begin{pmatrix} L_1 & L_2 \\ -L_2^\dagger & L_3 \end{pmatrix}$, with

$$L_1 = -L_1^\dagger, \qquad L_3 = -L_3^\dagger \tag{3.30}$$

then conditions (3.29), (3.30) imply

$$L_3 = \bar{L}_1, \qquad L_2 = L_2^T;$$

therefore, any matrix A in $sp(k)$ has the form $A := \begin{pmatrix} L_1 & L_2 \\ -\bar{L}_2 & \bar{L}_1 \end{pmatrix}$, with L_1 a skew-Hermitian matrix but otherwise arbitrary, and $L_2 = L_2^T$. In L_1 we have k^2 free real parameters while, for L_2, we can choose $\frac{k(k+1)}{2}$ complex numbers and therefore $k(k+1)$ real parameters since L_2 is symmetric. The total number of free parameters, that is, the *dimension* of $sp(k)$, is

$$\dim sp(k) = k^2 + k(k+1) = k(2k+1).$$

Therefore, the dimension of $Sp(k)$ is $k(2k+1)$ as well.

3.3.1.7 Transitivity of the symplectic group

For our purposes, an important property of the symplectic group $Sp(k)$ is that it is transitive on the complex unit sphere in the $2k$-dimensional complex space, $S_{\mathbb{C}}^{2k-1}$.

Consider the point $\vec{\psi}_0 := [1, 0, ..., 0]^T$ in $S_{\mathbb{C}}^{2k-1}$ and let $\vec{\psi}_f : [v_1^T v_2^T]^T$ be a target point in $S_{\mathbb{C}}^{2k-1}$, where v_1 and v_2 are k-dimensional complex vectors. To construct a matrix $X \in Sp(k)$ such that $\vec{\psi}_f = X\vec{\psi}_0$, we choose the first column of X in (3.26), $[x_1^T y_1^T]^T$, equal to $\vec{\psi}_f := [v_1^T v_2^T]^T$. The $(k+1)$-th column is chosen according to (3.27). The second column is chosen as a unit vector which is orthogonal to the first and $(k+1)$-th columns. The $(k+2)$-th column is chosen according to (3.27). Proceeding this way for the following columns, one obtains the desired matrix.

The isotropy group in $Sp(k)$ for $\vec{\psi}_0 = [1, 0, ..., 0]$ is given by the special unitary matrices of the form

$$X = \begin{pmatrix} 1 & 0 & 0 & 0 \\ 0 & A & 0 & -\bar{B} \\ 0 & 0 & 1 & 0 \\ 0 & B & 0 & \bar{A} \end{pmatrix},$$

where A and B are $(k-1) \times (k-1)$ blocks and

$$\begin{pmatrix} A & -\bar{B} \\ B & \bar{A} \end{pmatrix} \in Sp(k-1).$$

The coset space $Sp(k)/Sp(k-1)$ has dimension $k(2k+1)-(k-1)(2(k-1)+1) = 4k-1$ which is in fact the real dimension of the sphere $S_{\mathbb{C}}^{2k-1}$.

3.3.1.8   Further properties of the symplectic Lie algebra

The symplectic Lie algebra $sp(k)$ is one of the most studied Lie algebras, being one of the so-called *classical Lie algebras* (along with $su(n)$ and $so(n)$, the Lie algebra of skew-symmetric matrices of dimension n). It has several properties. In the next three propositions, we list three properties which will be used in the following treatment. Proofs can be found in [7], [54].

Proposition 3.3.5 The Lie algebra $sp(\frac{n}{2})$ is a **maximal Lie subalgebra** of $su(n)$, i.e., there is no Lie subalgebra of $su(n)$ properly containing $sp(\frac{n}{2})$ other than $su(n)$ itself. Equivalently, the Lie algebra generated by $sp(\frac{n}{2})$ and any other element F of $su(n)$, $F \notin sp(\frac{n}{2})$, is equal to $su(n)$.

Another example of a maximal subalgebra of $su(n)$ is $so(n)$, [7], [54].

The main consequence for us of the property in Proposition 3.3.5 is that there is no connected proper Lie subgroup of $SU(n)$ transitive on the sphere that contains $Sp(\frac{n}{2})$ properly. If this were the case, the corresponding Lie algebra would contain $sp(\frac{n}{2})$ properly (see Exercise 3.4), which is not possible. Therefore, if we are trying to obtain transitive Lie groups as Lie groups which contain $Sp(\frac{n}{2})$ properly we will be able to find only $SU(n)$.

In subsection 3.1.1.3 we defined isomorphisms between two Lie algebras. If we have a matrix Lie group $e^{\mathcal{G}}$ corresponding to a Lie algebra $\mathcal{G}$, it is possible to define an isomorphism ϕ_U between two Lie subalgebras of $\mathcal{G}$, $\mathcal{L}_1$ and $\mathcal{L}_2$, i.e., $\phi_U : \mathcal{L}_1 \to \mathcal{L}_2$, as follows

$$\phi_U(A) := UAU^{-1}, \qquad \forall A \in \mathcal{L}_1,$$

where U is an element of $e^{\mathcal{G}}$. If this is the case, $\mathcal{L}_1$ and $\mathcal{L}_2$ are said to be **conjugate in** $\mathcal{G}$ and we write $\mathcal{L}_2 = U\mathcal{L}_1 U^{-1}$. For the corresponding Lie groups we have

$$U e^{\mathcal{L}_1} U^{-1} = e^{\mathcal{L}_2},$$

and the map $X \to UXU^{-1}$ is an isomorphism of the two Lie groups $e^{\mathcal{L}_1}$ and $e^{\mathcal{L}_2}$.

Proposition 3.3.6 Every Lie subalgebra of $su(n)$ isomorphic to $sp(\frac{n}{2})$ is conjugate to $sp(\frac{n}{2})$ in $su(n)$.

Since we know that $sp(\frac{n}{2})$ gives rise to a transitive Lie group, it is natural to investigate what happens for subalgebras of $su(n)$ which are isomorphic to $sp(\frac{n}{2})$. The property in Proposition 3.3.6 says that we only have to investigate Lie algebras *conjugate* to $sp(\frac{n}{2})$. It is clear that these subalgebras, $\mathcal{L}$, give rise to transitive Lie groups as well. In fact if $e^{\mathcal{L}} = U Sp(\frac{n}{2}) U^{\dagger}$, we can find $X \in e^{\mathcal{L}}$ such that $\vec{\psi}_f = X\vec{\psi}_0$, for any $\vec{\psi}_0$ and $\vec{\psi}_f$. We can choose X as $X = U\tilde{X}U^{\dagger}$ where $\tilde{X}$ is a matrix in $Sp(\frac{n}{2})$ such that $U^{\dagger}\vec{\psi}_f = \tilde{X}U^{\dagger}\vec{\psi}_0$.

Another property of the Lie algebra $sp(\frac{n}{2})$ which we will also use is the following one.

Proposition 3.3.7 The Lie algebras $su(n)$ and $sp(\frac{n}{2})$ are simple. The Lie algebra $so(n)$ is simple for $n \neq 4$, and semisimple for $n = 4$.[8]

3.3.2 Test for pure state controllability

From what we have seen so far, if the dynamical Lie algebra $\mathcal{L}$ is either $su(n)$ or a Lie algebra conjugate to $sp(\frac{n}{2})$, then the quantum control system is PSC. The following theorem which was proved in [7] says that these are essentially the only cases.

Theorem 3.3.8 *The quantum system is PSC if and only if the corresponding dynamical Lie algebra $\mathcal{L}$ satisfies one of the following*

1. $\mathcal{L} = su(n)$.

2. $\mathcal{L}$ is conjugate to $sp(\frac{n}{2})$.

3. $\mathcal{L} = u(n)$.

4. $\mathcal{L} = \mathrm{span}\{i\mathbf{1}_{n\times n}\} \oplus \tilde{\mathcal{L}}$, where $\tilde{\mathcal{L}}$ is a Lie algebra conjugate to $sp(\frac{n}{2})$.

---

[8]In this case, it is the direct sum of two copies of $so(3)$.

The cases 1 and 3 above can be checked by checking the dimension of $\mathcal{L}$. If the dimension is $n^2 - 1$ or n^2, then we have cases 1 and 3, respectively. To have cases 2 and 4, n must be even and the dimension of $\mathcal{L}$ must be equal to $\frac{n}{2}(n+1)$ or $\frac{n}{2}(n+1)+1$, respectively. However, this is not enough to conclude that we are in case 2 or 4. Let us consider case 2 first. We need to check that $\mathcal{L}$ is conjugate to $sp(\frac{n}{2})$ and therefore that there exists a unitary matrix U such that

$$UAU^\dagger J + J(UAU^\dagger)^T = 0, \tag{3.31}$$

for every $A \in \mathcal{L}$. In fact it is enough to check (3.31) for a set of generators $\{A_1, ..., A_s\}$ of $\mathcal{L}$, namely $UA_kU^\dagger J + J(UA_kU^\dagger)^T = 0$, $k = 1, ..., s$, which, defining

$$\tilde{J} = U^\dagger J \bar{U}, \tag{3.32}$$

means

$$A_k \tilde{J} + \tilde{J} A_k^T = 0,   k = 1, ..., s. \tag{3.33}$$

System (3.33) is a linear system of equations in the unknown $\tilde{J}$. If there exists a solution which is related to J as in (3.32), then $\mathcal{L}$ is conjugate to $sp(\frac{n}{2})$. This method was discussed in [251]. Another method to check the given isomorphism (conjugacy) is to use the structure theory of Lie algebras (see, e.g., [130], [144]). Another, much more direct, method will be described in section 3.3.4 below (cf. Theorem 3.3.12) after we develop some more theory. To check possibility 4, one writes $\mathcal{L}$ as $\mathcal{L} = \{\text{span}\{i\mathbf{1}_{n\times n}\}\} \oplus (\text{span}\{i\mathbf{1}_{n\times n}\})^\perp$, and check the above conjugacy for $(\text{span}\{i\mathbf{1}_{n\times n}\})^\perp$.

3.3.3   Equivalent state controllability

From a physics point of view, having equivalent state controllability (ESC) is equivalent to having pure state controllability. We expect that the mathematical conditions to have ESC would be the same as the ones for PSC. This is indeed the case as we shall see now.

An equivalent definition of ESC is as follows.

Definition 3.3.9 The system is ESC if and only if, for any pair of points on the complex sphere, $\vec{\psi}_0$ and $\vec{\psi}_1$, there exists an X in the reachable set $\mathcal{R}$ for (3.1) and $\phi \in \mathbf{R}$ such that

$$\vec{\psi}_1 = e^{-i\phi} X \vec{\psi}_0.$$

It is clear that PSC implies ESC. Conversely, assume that ESC is verified. Consider the (augmented) Lie algebra

$$\bar{\mathcal{L}} := \text{span}\{i\mathbf{1}_{n\times n}\} + \mathcal{L}.$$

The Lie group $e^{\bar{\mathcal{L}}}$ clearly contains all the elements $e^{-i\phi} X$, with $\phi \in \mathbf{R}$ and $X \in \mathcal{R} \subseteq e^{\mathcal{L}}$. Therefore, because of the ESC property, $e^{\bar{\mathcal{L}}}$ is transitive on the complex sphere. Since $\bar{\mathcal{L}}$ contains multiples of the identity, according to Theorem 3.3.8, it has to be equal to $u(n)$ or $\mathrm{span}\{i\mathbf{1}_{n\times n}\} \oplus \mathcal{D}$, with $\mathcal{D}$ any Lie algebra conjugate to $sp(\frac{n}{2})$. This implies that

$$[\bar{\mathcal{L}}, \bar{\mathcal{L}}] = [\mathcal{L}, \mathcal{L}] = [h, h] \subseteq \mathcal{L}, \tag{3.34}$$

where h stands for $su(n)$ or $\mathcal{D}$ (conjugate to $sp(\frac{n}{2})$). However, since both these Lie algebras are simple $[h, h] = h$ (otherwise, since h is not Abelian, $[h, h]$ would be a nontrivial ideal). Therefore, (3.34) implies

$$su(n) \subseteq \mathcal{L},$$

or

$$\mathcal{D} \subseteq \mathcal{L},$$

which implies PSC.

We summarize this discussion in the following proposition.

Proposition 3.3.10 Equivalent state controllability (ESC) and pure state controllability (PSC) are equivalent properties.

3.3.4 Equality of orbits and practical tests

Let $\mathcal{L}$ be the dynamical Lie algebra associated with the quantum control system at hand. If ρ_0 is the initial condition of an ensemble, after the evolution $X \in \mathcal{R} \subseteq e^{\mathcal{L}}$, the state of the ensemble is equal to $X\rho_0 X^\dagger$. Therefor, the set of states reachable from ρ_0 is, according to Theorem 3.2.1, dense in the **orbit**

$$\mathcal{O}_{\mathcal{L}}(\rho_0) := \{X\rho_0 X^\dagger | X \in e^{\mathcal{L}}\}. \tag{3.35}$$

The question we shall deal with in this section is to find conditions so that this set is equal to the largest possible set, namely the orbit[9]

$$\mathcal{O}_{u(n)}(\rho_0) := \{X\rho_0 X^\dagger | X \in U(n)\}. \tag{3.36}$$

Notice this depends in a critical way on the initial density matrix ρ_0. If $\rho_0 = \frac{1}{n}\mathbf{1}_{n\times n}$, a perfectly mixed ensemble of systems, then $\mathcal{O}_{\mathcal{L}} = \mathcal{O}_{u(n)} = \{\rho_0\}$, no matter what $\mathcal{L}$ is. In the special case where ρ_0 is a pure state, namely a matrix of rank equal to one, the equality $\mathcal{O}_{\mathcal{L}} = \mathcal{O}_{u(n)}$ is equivalent to ESC, and therefore to PSC, and it is satisfied only in the cases described in Theorem 3.3.8.

In the general case, we can proceed as follows. The isotropy group of ρ_0 in $U(n)$ is a closed Lie subgroup of $U(n)$ whose Lie algebra is the space of

---

[9]In Chapter 9, section 9.3, in the context of entanglement theory, we shall consider the question of testing whether given a Lie algebra $\mathcal{L}$ and two density matrices ρ_1 and ρ_2, $\mathcal{O}_{\mathcal{L}}(\rho_1) = \mathcal{O}_{\mathcal{L}}(\rho_2)$. In entanglement theory, the Lie algebra of interest $\mathcal{L}$ is the one of local Hamiltonians of a multipartite system.

matrices in $u(n)$ commuting with $i\rho_0$. This Lie algebra is called the *centralizer* of $i\rho_0$ in $u(n)$ and it is denoted by $\mathcal{C}_{\rho_0}$. Therefore, the isotropy group is denoted here by $e^{\mathcal{C}_{\rho_0}}$. The isotropy group of ρ_0 in $e^{\mathcal{L}}$ is a closed (in the topology of $e^{\mathcal{L}}$) Lie subgroup of $e^{\mathcal{L}}$ whose Lie algebra is $\mathcal{C}_{\rho_0} \cap \mathcal{L}$. This isotropy group is denoted here by $e^{\mathcal{C}_{\rho_0} \cap \mathcal{L}}$. Now, it follows from the general facts on coset spaces and homogeneous spaces discussed in subsection 3.3.1.2 (applied to the manifolds $\mathcal{O}_{\mathcal{L}}$ and $\mathcal{O}_{u(n)}$) that $\mathcal{O}_{\mathcal{L}}$ and $\mathcal{O}_{u(n)}$ are diffeomorphic to $e^{\mathcal{L}}/e^{\mathcal{C}_{\rho_0} \cap \mathcal{L}}$ and $U(n)/e^{\mathcal{C}_{\rho_0}}$ respectively. If they are equal

$$\dim \mathcal{O}_{u(n)} = \dim U(n)/e^{\mathcal{C}_{\rho_0}}$$

and

$$\dim \mathcal{O}_{\mathcal{L}} = \dim e^{\mathcal{L}}/e^{\mathcal{C}_{\rho_0} \cap \mathcal{L}}$$

are also equal. Therefore, we must have

$$\dim u(n) - \dim \mathcal{C}_{\rho_0} = \dim \mathcal{L} - \dim(\mathcal{C}_{\rho_0} \cap \mathcal{L}). \qquad (3.37)$$

This condition is also sufficient to have $\mathcal{O}_{\mathcal{L}} = \mathcal{O}_{u(n)}$. The proof of this fact is presented in [7].

Theorem 3.3.11 *The orbits (3.35) and (3.36) are equal if and only if condition (3.37) is verified.*

We can rewrite condition (3.37) in a different, more compact, form. Consider the linear map $ad_{i\rho_0} : u(n) \to u(n)$, which associates to an element $A \in u(n)$ the element $[i\rho_0, A]$. The Lie algebra $\mathcal{C}_{\rho_0}$ is the kernel of this map and, by standard results of linear algebra, the number on the left-hand side of (3.37) is the dimension of the range of $ad_{i\rho_0}$, $[i\rho_0, u(n)]$. Analogously, the number on the right-hand side is the dimension of $[i\rho_0, \mathcal{L}]$. Since $[i\rho_0, \mathcal{L}] \subseteq [i\rho_0, u(n)]$, condition (3.37) can be rewritten as

$$[i\rho_0, \mathcal{L}] = [i\rho_0, u(n)],$$

or

$$\dim[i\rho_0, \mathcal{L}] = \dim[i\rho_0, u(n)]. \qquad (3.38)$$

Condition (3.38) is always satisfied if $i\rho_0$ is a multiple of the identity. If ρ_0 is any matrix of rank one it gives an alternate and practical way to check pure state controllability and therefore equivalent state controllability and it will be satisfied only in the cases listed in Theorem 3.3.8. We state this test formally in the following theorem.

Theorem 3.3.12 *A quantum control system with dynamical Lie algebra $\mathcal{L}$ is pure state controllable and equivalent state controllable if and only if condition (3.38) is satisfied for a rank 1 matrix ρ_0, for example $\rho_0 = \mathrm{diag}(1, 0, \ldots, 0)$.*

Test (3.38) also gives a practical way to check the isomorphism between $\mathcal{L}$ and $sp(\frac{n}{2})$.

The dimension of $[i\rho_0, u(n)]$ can be calculated in terms of the eigenvalues of ρ_0. It is zero if ρ_0 has only one eigenvalue. In all the other cases it is equal to

$$\dim[i\rho_0, u(n)] = 2\sum_{j<k} n_j n_k, \tag{3.39}$$

where $n_{j,k}$ are the multiplicities of the eigenvalues of ρ_0 (cf. Exercise 3.10). In the special case where $\rho_0 = \operatorname{diag}(1, 0, \ldots, 0)$, the number on the right-hand side of (3.39) is $2(n-1)$. Therefore, we have this corollary to Theorem 3.3.12.

Corollary 3.3.13 The quantum control system with dynamical Lie algebra $\mathcal{L}$ is PSC if and only if

$$\dim[i\rho_0, \mathcal{L}] = 2(n-1),$$

with $\rho_0 = \operatorname{diag}(1, 0, \ldots, 0)$.

Example 3.3.14 (cf. [7]) Consider the dynamical Lie algebra $\mathcal{L}$ spanned by skew Hermitian matrices of the form

$$F := \begin{pmatrix} L+Z & T+C \\ -\bar{T}+\bar{C} & -L+Z^T \end{pmatrix}.$$

All the sub-matrices are 2×2, with L and T diagonal and Z having all entries on the main diagonal equal to zero, and C such that $C^T = -C$. The Lie algebra $\mathcal{L}$ so defined is isomorphic to $sp(2)$. However, instead of verifying this isomorphism we follow Corollary 3.3.13 and calculate $[i\rho_0, \mathcal{L}]$ with $\rho_0 := \operatorname{diag}(1, 0, \ldots, 0)$. This is given by matrices of the form

$$G := \begin{pmatrix} 0 & a+ib & -c+id & -e+if \\ -a+ib & 0 & 0 & 0 \\ c+id & 0 & 0 & 0 \\ e+if & 0 & 0 & 0 \end{pmatrix},$$

with a, b, c, d, e and f free parameters. Therefore, $\dim[i\rho_0, \mathcal{L}] = 6$ which is equal to $2(n-1)$ since $n = 4$. Therefore, this system is PSC.

Consider now the orbit associated to $\rho_0 := \operatorname{diag}(\frac{1}{2}, \frac{1}{2}, 0, 0)$ under the dynamical Lie algebra $\mathcal{L}$. To verify the conditions of Theorem 3.3.11, we use formula (3.39) and calculate

$$\dim[i\rho_0, u(n)] = 8.$$

However, $[i\rho_0, \mathcal{L}]$ is spanned by matrices of the type

$$G := \begin{pmatrix} 0 & 0 & a+ib & c+id \\ 0 & 0 & -c-id & e+if \\ -a+ib & c-id & 0 & 0 \\ -c+id & -e+if & 0 & 0 \end{pmatrix},$$

and it is therefore, six-dimensional. Therefore, the two orbits are not equal.

3.3.5   Density matrix controllability

The definition of density matrix controllability implies that, for every ρ_0, $\{X\rho_0 X^\dagger \mid X \in \mathcal{R}\} = \mathcal{O}_{u(n)}(\rho_0)$, and, therefore, since $\{X\rho_0 X^\dagger \mid X \in \mathcal{R}\} \subseteq \mathcal{O}_{\mathcal{L}}(\rho_0)$, $\mathcal{O}_{\mathcal{L}}(\rho_0)$ and $\mathcal{O}_{u(n)}(\rho_0)$ must coincide. This is a very strong requirement and it is possible if and only if $\mathcal{L}$ is equal to $su(n)$ or $u(n)$. In fact, since this has to be true in particular for ρ_0 of rank one, the only other possibility would be n even and the dynamical Lie algebra $\mathcal{L}$ (mod multiples of the identity) conjugate to the symplectic Lie algebra $sp\left(\frac{n}{2}\right)$. However, taking for example

$$\rho_0 = \begin{pmatrix} L & \mathbf{0}_{n \times n} \\ \mathbf{0}_{n \times n} & L \end{pmatrix},$$

with $L = \text{diag}(\frac{1}{2}, 0, \ldots, 0)$, one sees that equality (3.38) is not verified. Obviously, if $\mathcal{L} = su(n)$ or $\mathcal{L} = u(n)$, since from Theorem 3.2.1, $\mathcal{R} = e^{\mathcal{L}} = SU(n)$ or $\mathcal{R} = e^{\mathcal{L}} = U(n)$, $\{X\rho_0 X^\dagger \mid X \in \mathcal{R}\} = \{X\rho_0 X^\dagger \mid X \in U(n)\}$ for every density matrix ρ_0. Therefore, we have the following condition for density matrix controllability of Definition 3.3.3.

Theorem 3.3.15 *A quantum control system with dynamical Lie algebra $\mathcal{L}$ is DMC if and only if $\mathcal{L} = su(n)$ or $\mathcal{L} = u(n)$, i.e., it is operator controllable.*

The diagram in Figure 3.2 summarizes the relations among the various notions of controllability and the conditions under which they are verified.

3.4   Notes and References

3.4.1   Alternate tests of controllability

There have been several attempts in the literature to give tests of controllability which avoid the direct computation of the dynamical Lie algebra. This is particularly valuable in high-dimensional situations and when the physics of the problem itself can be used to check controllability. Most of the alternate methods are concerned with special classes of systems of interest in applications. In many cases, the results concern the use of conditions on a graph associated with the system to determine controllability (see, e.g., [6], [21]).

To give an example of the type of results in this area, we briefly review the main result of [276]. In that paper, a system of the type,

$$\dot{\psi} = (A + Bu)\psi, \tag{3.40}$$

is considered, with A diagonal and B real and therefore skew-symmetric. A graph is associated with the system. The n vertices of the graph represent the

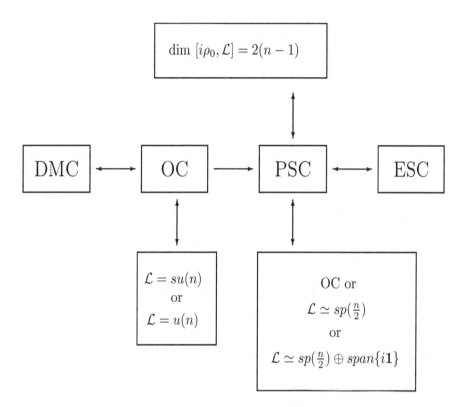

FIGURE 3.2: Relation among the various notions of controllability. DMC (density matrix controllability) is equivalent to OC (operator controllability) which implies PSC (pure state controllability) which is in turn equivalent to ESC (equivalent state controllability). OC is verified if and only if the dynamical Lie algebra $\mathcal{L}$ is equal to $su(n)$ or $u(n)$. PSC is verified if and only if the system is OC or $\mathcal{L}$ is isomorphic ($\simeq$) and therefore conjugate (cf. Proposition 3.3.6) to $sp(\frac{n}{2})$ or $sp(\frac{n}{2}) \oplus 1_{n \times n}$. PSC is equivalent to the condition $\dim[i\rho_0, \mathcal{L}] = 2(n-1)$, for $\rho_0 := \text{diag}(1, 0, ..., 0)$.

n eigenstates of A, which are the energy eigenstates when the control is equal to zero. An edge connects vertices j and k if the element j,k (and therefore the element k,j) of the matrix B is different from zero. This means that there is *coupling* between the eigenstates corresponding to the j-th and k-th eigenvalue. If the remaining entries of B were zero (and, for simplicity, A were zero) the solution of (3.40) would be $\psi(t) = X(t)\psi(0)$ where $X(t)$ is a matrix equal to the identity everywhere except for the elements at the intersection of the j-th and k-th rows and columns which form the matrix (assuming the corresponding nonzero element of B is equal to one)

$$U_{j,k}(t) = \begin{pmatrix} \cos(t) & \sin(t) \\ -\sin(t) & \cos(t) \end{pmatrix}.$$

Therefore, the dynamics transfer magnitude between the j-th and k-th eigenstates. The above graph is called in [276] a *connectivity graph*. The edges connecting the vertices corresponding to the eigenvalues $-i\lambda_j$ and $-i\lambda_k$ of A are labeled by $\omega_{jk} := |\lambda_j - \lambda_k|$. Since the λ_j's are the eigenvalues of the Hamiltonian without control, these are the energy differences for the various eigenstates. In Figure 3.3 we report, as an example, the connectivity graph of the pair

$$A := \begin{pmatrix} 3i & 0 & 0 & 0 \\ 0 & 2i & 0 & 0 \\ 0 & 0 & i & 0 \\ 0 & 0 & 0 & 10i \end{pmatrix}, \qquad B := \begin{pmatrix} 0 & 1 & 1 & 0 \\ -1 & 0 & 1 & 1 \\ -1 & -1 & 0 & 2 \\ 0 & -1 & -2 & 0 \end{pmatrix}. \tag{3.41}$$

The main result of [276] is as follows.

Theorem 3.4.1 *If the connectivity graph remains connected even after removing the edges which have equal labels, then system (3.40) is pure state controllable.*

3.4.2   Pure state controllability and existence of constants of motion

It is intuitive that the existence of a constant of motion for the quantum control system (3.16) would prevent pure state controllability. A **constant of motion** is an observable that gives the same value, with certainty, when measured along a trajectory. In finite-dimensional quantum mechanics, we associate Hermitian matrices with observables. To a constant of motion there will then correspond a matrix C which commutes with $H(u)$ in (3.16) for every value of u, and therefore with the whole dynamical Lie algebra $\mathcal{L}$, i.e.,

$$[C, \mathcal{L}] = 0. \tag{3.42}$$

In order to explain this correspondence first assume that (3.42) is true. If we measure C, we obtain the eigenvalue λ. According to the measurement

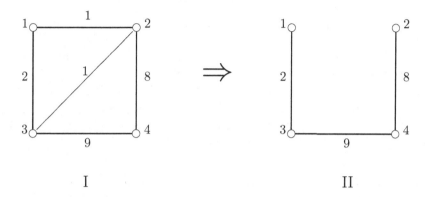

FIGURE 3.3: The connectivity graph for the pair $\{A, B\}$ in (3.41), Part I, and the graph obtained by removing equal labels, Part II. According to Theorem 3.4.1, the system is pure state controllable in this case.

postulate (see subsection 1.2.2) the state will collapse into an eigenstate $\vec{\psi}_0$ of C, i.e.,

$$C\vec{\psi}_0 = \lambda \vec{\psi}_0. \tag{3.43}$$

After an evolution $X \in e^{\mathcal{L}}$, the state will be $X\vec{\psi}_0$ which is again an eigenvector of C with eigenvalue λ since, from (3.43) and the fact that $CX = XC$, we obtain

$$CX\vec{\psi}_0 = XC\vec{\psi}_0 = \lambda X\vec{\psi}_0. \tag{3.44}$$

Therefore, a measurement of C will give the value λ again, with certainty. This shows that if (3.42) is true, C is a constant of motion. Conversely, if C is a constant of motion, for every eigenvector $\vec{\psi}_0$ and eigenvalue λ we have $C\vec{\psi}_0 = \lambda \vec{\psi}_0$ and $CX\vec{\psi}_0 = \lambda X\vec{\psi}_0$, for every $X \in e^{\mathcal{L}}$. This gives, for every eigenvector $\vec{\psi}_0$ and for every $X \in e^{\mathcal{L}}$,

$$C\vec{\psi}_0 = X^\dagger CX\vec{\psi}_0.$$

Being this true for every eigenvector $\vec{\psi}_0$, we have $C = X^\dagger CX$, for every $X \in e^{\mathcal{L}}$ which, in turn, implies (3.42).

We now wish to investigate the relation between existence of a nontrivial constant of motion, namely one that is not a multiple of the identity, and the pure state controllability of the system. We want to give a mathematical explanation of the intuition that the existence of such a constant of motion prevents controllability. In the next chapter, we will dig deeper into this issue by showing that the existence of constant of motion actually imply a decomposition of the underlying Hilbert space into invariant subspaces. Assume there exists a constant of motion C and write iC as $i\frac{\text{Tr}(C)}{n}\mathbf{1} + \tilde{C}$, with $\tilde{C} \in su(n)$, and assume by contradiction that the system is PSC. Then, if the system is PSC because $\mathcal{L} = u(n)$ or $\mathcal{L} = su(n)$, we have, from (3.42), $[\tilde{C}, su(n)] = 0$. However, this is possible only if $\tilde{C} = 0$ because otherwise span$\{\tilde{C}\}$ would be a nontrivial ideal of $su(n)$ and $su(n)$, being simple, has no nontrivial ideals. If the system is PSC because $\mathcal{L}$ is (conjugate to) span$\{i\mathbf{1}_{n\times n}\} \oplus sp(\frac{n}{2})$, then we must have (up to a conjugacy), $[\tilde{C}, sp(\frac{n}{2})] = 0$, but this is not possible. If $\tilde{C} \in sp(\frac{n}{2})$ it would contradict the fact that $sp(\frac{n}{2})$ is simple. If $\tilde{C} \notin sp(\frac{n}{2})$ it would contradict the fact that $sp(\frac{n}{2})$ is a maximal subalgebra in $su(n)$ (cf. Proposition 3.3.5). Therefore, we can conclude with the following proposition.

Proposition 3.4.2 If a nontrivial constant of motion exists then the system is not PSC.

The converse of Proposition 3.4.2 is in general not true, namely it is possible that the system is not PSC and there is no nontrivial constant of motion. For example, if the dynamical Lie algebra $\mathcal{L}$ is equal to $so(n)$, then the system is not PSC. However, since $so(n)$, for $n \neq 4$ is also a simple maximal subalgebra in $su(n)$, reasoning as above for $sp(\frac{n}{2})$, we conclude that there is no nontrivial constant of motion.

3.4.3   Bibliographical notes

Much of the treatment presented here is based on [7], in particular for what concerns controllability of the state $|\psi\rangle$ or ρ. This work was based on a previous paper [198] on the transitive action of transformation groups on spheres. A related work is [251]. The books [130], [157], [199], [290] are very good references for the basic material concerning Lie transformation groups. Various concepts on Lie algebras introduced can be found in some of the standard books on Lie algebras and Lie groups such as [127], [130] and [144]. The result on maximal subalgebras of a Lie algebra in Proposition 3.3.5 follows from the work of Dynkin [101] and was summarized in [54]. The discussion on the relation between constants of motion and controllability in the previous subsection 3.4.2 was motivated by a conjecture in [276] and was elaborated by the author in collaboration with Francesca Albertini.

The study of controllability of quantum systems is a large sub-field of quantum control theory which would deserve more space than the one devoted here. Many of the contributions deal with special classes of systems, such as spin networks. However, some others are of general and fundamental type. For instance, in [71] [82] the problem is treated of the state that can be achieved for a subsystem of a larger system.

3.4.4 Some open problems

There are several open problems concerning the study of controllability for quantum systems. In particular, the tests described in this chapter do not allow us in general to decide whether given two density matrices it is possible to go from one to the other in the case where the system is not controllable. A general test is needed to decide whether a density matrix is in an orbit of the form (3.35). An instance of this problem in the context of entanglement theory is described in section 9.3. Another area of interest is the characterization of the reachable sets. This problem is related to the minimum time optimal control in Chapter 7 (see in particular subsection 7.4.2). The extension of the Lie algebraic approach presented here to infinite-dimensional systems presents many interesting mathematical challenges (see, e.g., [36] and references therein). A problem of practical interest is the study of controllability for classes of control functions which do not include piecewise constant functions. This is motivated by the fact that in practice a jump in the control corresponds to very high-frequency Fourier components. High-frequency components correspond to large differences between the energy levels of the system (cf. section 8.1) and controls with high frequency may induce population transfer between levels that have been ignored in the modeling procedure. So, an interesting question is that of controllability with limited bandwidth. Some work in this direction was done in [133].

3.5 Exercises

Exercise 3.1 The Lie algebra $\{x_1, ..., x_n\}_{\mathcal{L}}$ contains not only the repeated Lie brackets of $\{x_1, ..., x_n\}$ but also their Lie brackets and Lie brackets of their Lie brackets. As an example it contains $[[x_1, x_2], [x_3, x_4]]$ which is *not* a repeated Lie bracket as defined in subsection 3.2.2. However, these can be expressed as linear combinations of repeated Lie brackets of $\{x_1, ..., x_n\}$ and this justifies the algorithm presented in subsection 3.2.2. This exercise concerns the proof of this fact.

To justify the algorithm of subsection 3.2.2, it is enough to show that given a Lie bracket $[P, Q]$ where P is a repeated Lie bracket of the elements $\{x_1, ..., x_n\}$ then $[P, Q]$ is a linear combination of elements of the type[10]

$$[x_{j_1}, [x_{j_2}, \cdots [x_{j_l}, Q] \cdots]] := ad_{x_{j_1}} ad_{x_{j_2}} \cdots ad_{x_{j_l}} Q, \qquad (3.45)$$

where $j_1, ..., j_l$ vary in the set of indices $\{1, ..., n\}$. This can be shown by induction on the depth of P and using Jacobi identity. By doing the same

---

[10]The operation ad_x on elements of a Lie algebra $\mathcal{L}$ is defined as $ad_x Q := [x, Q]$.

thing for $[x_{j_i}, Q]$ in (3.45) one shows that $[P, Q]$ is a linear combination of vectors of the type $ad_{x_{j_1}} ad_{x_{j_2}} \cdots ad_{x_{j_k}} x_h$, with $j_1, j_2, ..., j_k, h$ in $\{1, ..., n\}$, namely of repeated Lie brackets.

Provide the details and fill the gaps in this proof.

Exercise 3.2 As we have seen in Chapter 2, many quantum control systems have a bilinear structure

$$\dot{X} = AX + \sum_{k=1}^{m} B_k X u_k.$$

Assume that the set of the possible values for the controls contains a neighborhood of the origin in $\mathbf{R}^k$. Show that the dynamical Lie algebra coincides with the one generated by $A, B_1, ..., B_m$.

Exercise 3.3 Write a MATLAB program to test the operator controllability of finite-dimensional quantum systems based on the algorithm of subsection 3.2.2. The program should accept a set of matrices and return an answer *YES* or *NO* according to whether the system is controllable or not. Use this algorithm to verify whether or not the following systems are operator controllable:

1. The system of two interacting spin $\frac{1}{2}$ particles with nonzero Ising interaction and all the components of the magnetic field possibly used for control (simultaneously for both spins).

2. The system of two interacting spin $\frac{1}{2}$ particles with interaction equal to zero and all the components of the magnetic field possibly used for control (simultaneously for both spins).

3. The system of two interacting spin $\frac{1}{2}$ particles with nonzero Ising interaction and only the x component of the magnetic field possibly used for control (simultaneously for both spins).

Exercise 3.4 Use the correspondence between Lie algebras and connected Lie groups discussed in subsection 3.1.2.3 to show that if $\mathcal{L}_1$ and $\mathcal{L}_2$ are two Lie algebras and $e^{\mathcal{L}_1}$ and $e^{\mathcal{L}_2}$ the corresponding Lie groups, $\mathcal{L}_1 \subseteq \mathcal{L}_2$ if and only if $e^{\mathcal{L}_1} \subseteq e^{\mathcal{L}_2}$. Show that proper inclusion for the Lie algebras is equivalent to proper inclusion for the Lie groups.

Exercise 3.5 Prove that the bilinearity condition (3.3) along with condition (3.4) implies, for a general Lie algebra $\mathcal{L}$, that $[y, x]$ is the inverse element of $[x, y]$ in the vector space.

Exercise 3.6 Prove that a unitary matrix of the form (3.28) is automatically special unitary namely its determinant is equal to one.

Exercise 3.7 Give a basis of $su(n)$ and a basis of $sp(\frac{n}{2})$.

Exercise 3.8 Find a transformation X in $Sp(2)$ which performs the state transfer $\psi_1 = X\psi_0$, where $\psi_0 = [\frac{1}{\sqrt{2}}, 0, 0, \frac{1}{\sqrt{2}}]^T$ and $\psi_1 = [0, 0, \frac{1}{\sqrt{2}}, i\frac{1}{\sqrt{2}}]^T$.

Exercise 3.9 Display a basis of a subalgebra of $su(4)$ which is isomorphic to $sp(2)$ but it is not $sp(2)$. Use the test in subsection 3.3.4 to show that the corresponding group is transitive on the complex sphere.

Exercise 3.10 Prove formula (3.39).

Exercise 3.11 Study the controllability of the system in Example 3.2.5 in the case where the gyromagnetic ratios of the two spins are different.

Exercise 3.12 Assume that the dynamical Lie algebra $\mathcal{L} \subseteq u(4)$ is ten-dimensional and spanned by

$$\{i\sigma_x \otimes \mathbf{1}, i\sigma_y \otimes \sigma_x, i\sigma_y \otimes \sigma_y, i\sigma_y \otimes \sigma_z, i\sigma_z \otimes \sigma_x,$$
$$i\sigma_z \otimes \sigma_y, i\sigma_z \otimes \sigma_z, i\mathbf{1} \otimes \sigma_x, i\mathbf{1} \otimes \sigma_y, i\mathbf{1} \otimes \sigma_z\},$$

where the Pauli matrices $\sigma_{x,y,z}$ were defined in (1.20). Apply the test of Corollary 3.3.13 to show that the associated system is PSC and $\mathcal{L}$ is conjugate and therefore isomorphic to $sp(2)$.

Chapter 4

Uncontrollable Systems and Dynamical Decomposition

As discussed in section 3.2.4, operator controllability is a generic property for (closed) finite-dimensional quantum systems,

$$\dot{X} = -iH(u)X, \qquad X(0) = \mathbf{1}, \tag{4.1}$$

as long as the Hamiltonian $H(u)$ can attain at least two different values by varying the control u. Nevertheless, in real-life experiments, the presence of symmetries often prevents full controllability. In these cases, one would like to understand the structure of the dynamics and how it can be decomposed. In some sense, the situation is similar to what happens in linear systems theory [155] where one has the celebrated *Kalman decomposition* in controllable-uncontrollable (observable-unobservable) dynamics.

Consider, as an elementary example, the case of two identical spin $\frac{1}{2}$ particles, possibly interacting with each other and subject to a common electromagnetic field. Neglect all the degrees of freedom besides spin. Since the two spins are identical and they feel the same action of the control, unitary transformations which perform different unitaries on the two spins are forbidden and the system is not controllable. From a mathematical point of view, the Hamiltonian $H = H(u)$ in (4.1) will remain unchanged if we permute the two positions in the tensor products, i.e., it will present a *symmetry*, in fact, a *group of symmetries* given, in this case, by the group of permutations of two objects.

From a practical point of view, who analyzes the control system (4.1) has, as starting data, either a basis of the dynamical Lie algebra, $\mathcal{L}$, calculated using the algorithm of subsection 3.2.2 and-or a group of symmetries for the system, namely a group of matrices G such that $CH(u)C^{-1} = H(u)$, for every $C \in G$ and every control value u. Each such matrix C, if Hermitian, will represent a constant of motion as discussed in subsection 3.4.2. We shall show in this chapter, how to obtain a *decomposition of the dynamics* for quantum control systems, starting from these data. We shall see that the quantum control system (4.1) splits into the parallel of several control systems. In fact, in the presence of a finite group of symmetries such splitting reveals, in appropriate coordinates, a block diagonal form for system (4.1) so that there are several invariant subspaces for the quantum state. One then might ask the question of whether one has operator controllability on each of such invariant subspaces. Such property is called *subspace controllability*.

We shall give algorithms to obtain the dynamical decomposition of quantum systems which are not controllable. In the process, we shall introduce some more notions in the theory of Lie algebras, Lie groups and their representations. Our starting point is Theorem 3.1.8 of the previous chapter which gives a decomposition of any Lie subalgebra $\mathcal{L}$ of $u(n)$ into the direct sum of an Abelian and a semisimple part. In section 4.1, we show how to obtain, starting from a basis of the dynamical Lie algebra $\mathcal{L}$, bases of its Abelian part and of the simple ideals of its semisimple part. This gives the decomposition of uncontrollable dynamics we are looking for. In section 4.2, we study such a decomposition more in depth and, with the help of some notions of representation theory, and in particular of Schur lemma, we show that in a special basis, the dynamical Lie algebra takes the form of a tensor product. In section 4.3, we assume that we do not have knowledge of the dynamical Lie algebra $\mathcal{L}$ but only of a group of symmetries G which commutes with $\mathcal{L}$. The knowledge of a group of symmetries G tells us that $\mathcal{L}$ must be a subalgebra of $u^G(n)$, the Lie subalgebra of $u(n)$ commuting with G. We obtain a decomposition of $u^G(n)$, and therefore of $\mathcal{L}$, in block diagonal form. This decomposition in invariant subspaces justifies the notion of *subspace controllability*.

4.1 Dynamical Decomposition Starting from a Basis of the Dynamical Lie Algebra

From Theorem 3.1.8 and the definitions of reductive and semisimple Lie algebras, we know that the dynamical Lie algebra $\mathcal{L}$ is the direct sum of an Abelian Lie algebra, $\mathcal{Z}$, and a number of simple Lie algebras $\mathcal{S}_1, ..., \mathcal{S}_s$, i.e.,

$$\mathcal{L} = \mathcal{S}_1 \bar\oplus \mathcal{S}_2 \bar\oplus \cdots \bar\oplus \mathcal{S}_s \bar\oplus \mathcal{Z} = \mathcal{S} \oplus \mathcal{Z}, \qquad (4.2)$$

with $\mathcal{S} := \mathcal{S}_1 \bar\oplus \mathcal{S}_2 \bar\oplus \cdots \bar\oplus \mathcal{S}_s$. Therefore, for every control value u, $-iH(u)$ in (4.1) will decompose into commuting components belonging to the various subspaces $\mathcal{S}_1, ..., \mathcal{S}_s, \mathcal{Z}$. Our goal here is to find bases of these subspaces starting from a basis $L_1, ..., L_m$ of $\mathcal{L}$, obtained, for instance, with the algorithm of subsection 3.2.2. We assume that $m < n^2 - 1$, otherwise we would have $\mathcal{L} = su(n)$ or $\mathcal{L} = u(n)$. We shall also assume $\mathcal{L} \subseteq su(n)$. This is not a genuine restriction because if $-iH(u)$ in (4.1) has a nonzero component along the identity, it can be written as $-iH(u) = -i\frac{Tr(H(u))}{n}\mathbf{1}_n - i\tilde{H}(u)$ and we can ignore the element $-i\frac{Tr(H(u))}{n}\mathbf{1}_n$ which only gives a physically unimportant phase factor and does not contribute to the dynamical Lie algebra (except for itself) since it commutes with every matrix.

In the decomposition (4.2), $\mathcal{Z}$ is the center of the Lie algebra $\mathcal{L}$, i.e., the subspace of elements which commute with all of $\mathcal{L}$. Therefore, a basis of $\mathcal{Z}$ can be found by solving the linear system

$$\left[\sum_{j=1}^{m} x_j L_j, L_k \right] = \sum_{j=1}^{m} x_j \left[L_j, L_k \right] = 0, \qquad k = 1, 2, ..., m. \qquad (4.3)$$

Let $\{Z_1, ..., Z_{m-r}\}$ denote a basis of $\mathcal{Z}$.

We use the fact that

$$[\mathcal{L}, \mathcal{L}] = [\mathcal{S} \bar{\oplus} \mathcal{Z}, \mathcal{S} \bar{\oplus} \mathcal{Z}] = [\mathcal{S}, \mathcal{S}] = \mathcal{S},$$

where the last equality is due to the fact that $\mathcal{S}$ is semisimple. From this, we find that a basis of $\mathcal{S}$ can be found by selecting r linearly independent elements in the set $\{[L_j, L_k], j, k = 1, ..., m\}$. We denote such a basis by $\{S_1, ..., S_r\}$. An alternative method is given by the following Proposition 4.1.1 (cf. Exercise 4.3). We recall that $su(n)$ is equipped with the inner product $\langle A, B \rangle := Tr(AB^\dagger) = -Tr(AB)$, which has the property

$$\langle [A, B], C \rangle = \langle [B, C], A \rangle. \qquad (4.4)$$

The dynamical Lie algebra $\mathcal{L}$ inherits this inner product from $su(n)$, since it is a subalgebra of $su(n)$.

Proposition 4.1.1 The decomposition (4.2) is an orthogonal decomposition with respect to the above inner product. Therefore, $\mathcal{S}$ can be found as the orthogonal complement of $\mathcal{Z}$ in $\mathcal{L}$.

4.1.1 Finding the simple ideals

To refine the decomposition of $\mathcal{L}$ we need to find bases of the simple ideals, $\mathcal{S}_j$, $j = 1, .., s$, which make up $\mathcal{S}$, starting from a basis $\{S_1, ..., S_r\}$ of $\mathcal{S}$. This is obtained from the *primary decomposition* of the Lie algebra $\mathcal{S}$. In order to define the primary decomposition we need to introduce the concept of *Cartan subalgebra* (CSA).

4.1.1.1 Cartan subalgebras

Definition 4.1.2 A **Cartan subalgebra** (CSA) of a semisimple Lie algebra $\mathcal{S}$ is an Abelian subalgebra $\mathcal{A}$ of $\mathcal{S}$ such that if $[S, \mathcal{A}] \subseteq \mathcal{A}$, then $S \in \mathcal{A}$.

The space $\mathcal{N}_\mathcal{A} := \{S \in \mathcal{S} \mid [S, \mathcal{A}] \subseteq \mathcal{A}\}$ is called the *normalizer* of the Lie subalgebra $\mathcal{A}$. It is itself a Lie subalgebra as it is easily verified using the Jacobi identity. It is the largest Lie subalgebra of $\mathcal{S}$ for which $\mathcal{A}$ is an ideal. The definition of Cartan subalgebra can be restated by saying that $\mathcal{A}$ is an Abelian subalgebra which equal to its own normalizer. It also has the following characterization.

Proposition 4.1.3 A Lie subalgebra $\mathcal{A}$ of $\mathcal{S}$ is a CSA in $\mathcal{S}$ if and only if it is a *maximal* Abelian subalgebra.

Proof. Assume that $\mathcal{A}$ is a CSA and there is an Abelian subalgebra $\mathcal{A}'$ properly containing $\mathcal{A}$. Then we have $\mathcal{A} \subsetneq \mathcal{A}' \subseteq \mathcal{N}_\mathcal{A}$ which contradicts $\mathcal{A} = \mathcal{N}_\mathcal{A}$. Vice versa, assume that $\mathcal{A}$ is a maximal Abelian subalgebra but there is an element $S \in \mathcal{N}_\mathcal{A}$ such that $S \notin \mathcal{A}$. Since $\mathcal{A}$ is Abelian, all the elements in $\mathcal{A}$ can be simultaneously diagonalized. In this basis, consider an element $A \in \mathcal{A}$ which is now diagonal of the form $\mathtt{diag}(i\lambda_1 \mathbf{1}_{q_1}, i\lambda_2 \mathbf{1}_{q_2}, ..., i\lambda_b \mathbf{1}_{q_b})$, for some $b \geq 2$ and $\lambda_j \neq \lambda_k$.[1] Since $[S, A]$ is diagonal, by writing $S \in su(n)$ in blocks, we see that S has to be block diagonal with blocks of dimensions $q_1, q_2, ..., q_b$. Moreover, this form implies that $[S, A] = 0$. We can repeat this argument for any A in a basis of $\mathcal{A}$, and therefore we have $[S, \mathcal{A}] = 0$. The Lie algebra $\mathcal{A} \oplus \mathtt{span}\{S\}$ is therefore Abelian and properly contains $\mathcal{A}$ which contradicts the fact that $\mathcal{A}$ is maximal Abelian. $\square$

A practical algorithm to obtain a CSA was given in [80] (Section III A). We report it here referring to [80] for the proof that the resulting Lie algebra is, in fact, a CSA of $\mathcal{S}$.

Algorithm I

1. Set $\mathcal{A} = \{0\}$.

2. Select an element $A \in \mathcal{S}$, with $A \neq 0$.

3. Calculate the *centralizer* $\mathcal{D}$ of A in $\mathcal{S}$, that is, the Lie algebra of elements in $\mathcal{S}$ commuting with A. This is obtained as solution of a linear system of equations (similar to (4.3)). Since $\mathcal{D} \subseteq su(n)$, we can apply Theorem 3.1.8 to say that $\mathcal{D} = [\mathcal{D}, \mathcal{D}] \oplus \mathcal{Z}(\mathcal{D})$, where $\mathcal{Z}(\mathcal{D})$ is the center of $\mathcal{D}$ and $[\mathcal{D}, \mathcal{D}]$ is semisimple. These can be calculated adapting the previously described algorithms.

4. Update $\mathcal{A}$ as $\mathcal{A} = \mathcal{A} + \mathcal{Z}(\mathcal{D})$.

5. If $[\mathcal{D}, \mathcal{D}] = \{0\}$ STOP and return $\mathcal{A}$ as the CSA. Otherwise, replace $\mathcal{S}$ with $[\mathcal{D}, \mathcal{D}]$ and go back to Step 2 above.

The algorithm terminates because at every iteration $\mathcal{D}$ is a proper subalgebra of $\mathcal{S}$, otherwise there would be an element of $\mathcal{S}$ commuting with the whole $\mathcal{S}$ which would contradict the fact that $\mathcal{S}$ is semisimple. Moreover, $\mathcal{S}$ is finite-dimensional.

Example 4.1.4 In the special case where $\mathcal{S} = su(n)$, the previous algorithm gives (in an appropriate basis) all possible diagonal elements. Consider for

---

[1] If b was equal to 1, this would be a multiple of the identity which would commute with all of $\mathcal{S}$ (and $\mathcal{L}$) which we have excluded (it would be in the center of $\mathcal{L}$).

instance $su(3)$ and start the algorithm with $A = \begin{pmatrix} i & 0 & 0 \\ 0 & 0 & 0 \\ 0 & 0 & -i \end{pmatrix}$. The central-izer of A in $su(3)$ is the space of all diagonal elements of $su(3)$ which is its own center. Therefore, the algorithm ends in one iteration.

4.1.1.2 Primary decomposition

It is convenient to consider the *adjoint representation* of the Lie algebra S which associates to $X \in S$ the linear operator ad_X on S which acts as $ad_X(S) := [X, S]$ (cf. Exercise 4.1). It can be shown using Cartan's semisimplicity criterion (cf. Exercise 4.4) that the eigenvalues of ad_X are 0 and pairs of conjugate purely imaginary eigenvalues of the form $\pm ia$, with $a \neq 0$. Consider now a basis of the CSA $\mathcal{A}$, $\{A_1, ..., A_{\bar{s}}\}$, and an element $A := \sum_{j=1}^{\bar{s}} c_j A_j$ such that $ad_A = \sum_{j=1}^{\bar{s}} c_j ad_{A_j}$ has, besides 0, all *different* pairs of nonzero imaginary conjugate eigenvalues. Such an element is called a *splitting element* and it exists according to Corollary 4.11.3 of [92]. Given A, a splitting element of S, a CSA $\mathcal{A}$ is the eigenspace associated with the eigenvalue 0. Let us denote by $\mathcal{V}_j$, $j = 1, ..., \frac{\dim(S)-\dim(\mathcal{A})}{2}$, the two-dimensional subspaces associated with the pair of eigenvalues $\pm ia_j$. We have

$$S = \mathcal{A} \oplus \mathcal{V}_1 \oplus \mathcal{V}_2 \oplus \cdots \oplus \mathcal{V}_{\frac{\dim(S)-\dim(\mathcal{A})}{2}}. \tag{4.5}$$

The decomposition (4.5) is called the *primary decomposition* of the semisimple Lie algebra S. Each space $\mathcal{V}_j$ is the kernel of $ad_A^2 + a_j^2 \mathbf{1}$ for each pair of eigenvalues of A, $\pm ia_j$.

Example 4.1.5 Consider the dynamical Lie algebra

$$\mathcal{L} := \text{span}\{\bar{\sigma}_x \otimes \mathbf{1}, \mathbf{1} \otimes \bar{\sigma}_x, i\bar{\sigma}_z \otimes \bar{\sigma}_z, i\bar{\sigma}_y \otimes \bar{\sigma}_y, i\bar{\sigma}_z \otimes \bar{\sigma}_y, i\bar{\sigma}_y \otimes \bar{\sigma}_z\},$$

where $\bar{\sigma}_{x,y,z}$ are the Pauli matrices defined in (3.6). The Lie algebra $\mathcal{L}$ is semisimple since it has zero center. To find a CSA use the Algorithm I starting with $S = \bar{\sigma}_x \otimes \mathbf{1}$. The centralizer is the span of $\{\bar{\sigma}_x \otimes \mathbf{1}, \mathbf{1} \otimes \bar{\sigma}_x\}$ which does not have a semisimple component. Therefore, the algorithm ends at the first iteration. Now given $A_1 := \bar{\sigma}_x \otimes \mathbf{1}$ and $A_2 = \mathbf{1} \otimes \bar{\sigma}_x$, we calculate ad_{A_1} and ad_{A_2} in the given basis, with the help of the commutation relations (3.7). We have

$$ad_{A_1} = \begin{pmatrix} 0 & 0 & 0 & 0 & 0 & 0 \\ 0 & 0 & 0 & 0 & 0 & 0 \\ \hline 0 & 0 & 0 & 0 & 0 & 1 \\ 0 & 0 & 0 & 0 & -1 & 0 \\ 0 & 0 & 0 & 1 & 0 & 0 \\ 0 & 0 & -1 & 0 & 0 & 0 \end{pmatrix} \qquad ad_{A_2} = \begin{pmatrix} 0 & 0 & 0 & 0 & 0 & 0 \\ 0 & 0 & 0 & 0 & 0 & 0 \\ \hline 0 & 0 & 0 & 0 & 1 & 0 \\ 0 & 0 & 0 & 0 & 0 & -1 \\ 0 & 0 & -1 & 0 & 0 & 0 \\ 0 & 0 & 0 & 1 & 0 & 0 \end{pmatrix}. \tag{4.6}$$

For an element $A_1 + cA_2$, the matrix $ad_{A_1+cA_2}$ has eigenvalues $\lambda = 0$, with multiplicity 2, and $\lambda = \pm(1 + c)i$ and $\lambda = \pm(1 - c)i$.[2] Therefore, as long as $c \neq 0, 1, -1$, $A_1 + cA_2$ is a splitting element. Choose $c = 2$, so that the eigenvalues of $ad_{A_1+2A_2} = ad_{A_1} + 2ad_{A_2}$ are 0 (with multiplicity 2), $\pm 3i$, $\pm i$. The eigenspace associated with $\pm 3i$ is $\mathcal{V}_{\pm 3i} := \mathbf{span}\{i\bar{\sigma}_z \otimes \bar{\sigma}_y + i\bar{\sigma}_y \otimes \bar{\sigma}_z, i\bar{\sigma}_z \otimes \bar{\sigma}_z - i\bar{\sigma}_y \otimes \bar{\sigma}_y\}$, where the two elements in the indicated basis correspond to the vectors $[0,0,0,0,1,1]^T$ and $[0,0,1,-1,0,0]^T$ in $\mathbf{R}^6$, respectively. The eigenspace associated with $\pm i$ is $\mathcal{V}_{\pm i} := \mathbf{span}\{i\bar{\sigma}_z \otimes \bar{\sigma}_y - i\bar{\sigma}_y \otimes \bar{\sigma}_z, i\bar{\sigma}_z \otimes \bar{\sigma}_z + i\bar{\sigma}_y \otimes \bar{\sigma}_y\}$, where the two elements in the indicated basis correspond to the vectors $[0,0,0,0,1,-1]^T$ and $[0,0,1,1,0,0]^T$ in $\mathbf{R}^6$, respectively. The primary decomposition is given by $\mathcal{S} = \mathcal{A} \oplus \mathcal{V}_{\pm 3i} \oplus \mathcal{V}_{\pm i}$.

4.1.1.3   Decomposition into simple ideals

The primary decomposition is directly related to the decomposition into simple ideals we are looking for. We have the following result proved in [92] (Section IV.12).

Theorem 4.1.6 *Each simple ideal of a semisimple Lie algebra $\mathcal{S}$ is the ideal generated by one of the subspaces $\mathcal{V}_j$ in the primary decomposition (4.5).*

The ideal generated by a subspace $\mathcal{V}$ of a Lie algebra $\mathcal{S}$, is, by definition, the smallest ideal of $\mathcal{S}$ containing $\mathcal{V}$, it can be obtained as

$$\mathcal{I} := +_{k=0}^{\infty} ad_{\mathcal{S}}^k \mathcal{V}, \tag{4.7}$$

where $ad_{\mathcal{S}}^k \mathcal{V}$ is defined inductively as $ad_{\mathcal{S}}^0 \mathcal{V} := \mathcal{V}$, $ad_{\mathcal{S}}^k \mathcal{V} := [\mathcal{S}, ad_{\mathcal{S}}^{k-1}\mathcal{V}]$ when $k \geq 1$ (cf. Exercise 4.2). It can be calculated with an algorithm similar to the one described in subsection 3.2.2. Starting from a basis of $\mathcal{V}$ and a basis of $\mathcal{S}$, calculate $[\mathcal{S}, \mathcal{V}]$ and then $[\mathcal{S}, [\mathcal{S}, \mathcal{V}]]$, and so on until the dimension does not increase anymore. Notice that, because of finite dimensionality of $\mathcal{S}$, this algorithm will stop after a finite number of steps and therefore only a finite number of terms have to be considered in (4.7).

Example 4.1.7 (Continuation of Example 4.1.5) Let us consider $\mathcal{V}_{\pm 3i}$. From Exercise 4.5 we do not need to compute $[\mathcal{A}, \mathcal{V}_j]$, for any $\mathcal{V}_j$. By calculating $[\mathcal{V}_{\pm 3i}, \mathcal{V}_{\pm 3i}]$, we obtain $\mathbf{span}\{1 \otimes \bar{\sigma}_x + \bar{\sigma}_x \otimes 1\}$ while $[\mathcal{V}_{\pm 3i}, \mathcal{V}_{\pm i}] = 0$. At the next iteration the dimension does not grow. Therefore, the ideal generated by $\mathcal{V}_{\pm 3i}$ is

$$\mathcal{S}_1 := \mathbf{span}\{i\bar{\sigma}_z \otimes \bar{\sigma}_y + i\bar{\sigma}_y \otimes \bar{\sigma}_z, i\bar{\sigma}_z \otimes \bar{\sigma}_z - i\bar{\sigma}_y \otimes \bar{\sigma}_y, 1 \otimes \bar{\sigma}_x + \bar{\sigma}_x \otimes 1\}. \tag{4.8}$$

Analogously we find that the ideal generated by $\mathcal{V}_{\pm i}$ is

$$\mathcal{S}_2 := \mathbf{span}\{i\bar{\sigma}_z \otimes \bar{\sigma}_y - i\bar{\sigma}_y \otimes \bar{\sigma}_z, i\bar{\sigma}_z \otimes \bar{\sigma}_z + i\bar{\sigma}_y \otimes \bar{\sigma}_y, 1 \otimes \bar{\sigma}_x - \bar{\sigma}_x \otimes 1_2\}, \tag{4.9}$$

---

[2]To simplify this calculation use the formula for the determinant of block matrices $\det \begin{pmatrix} A & B \\ C & D \end{pmatrix} = \det(AD - BC)$ if C and D commute [265].

which is, as expected from Proposition 4.1.1, orthogonal to $\mathcal{S}_1$ in (4.8). So $\mathcal{L} = \mathcal{S}_1 \oplus \mathcal{S}_2$, with $\mathcal{S}_1$ and $\mathcal{S}_2$ defined in (4.8) and (4.9), is a decomposition into simple ideals of the semisimple Lie algebra $\mathcal{L}$ of Example 4.1.5.

4.1.2 Decomposition of the dynamics

Summarizing the above treatment, if system (4.1) is not controllable, the dynamical Lie algebra $\mathcal{L}$ splits into mutually orthogonal and mutually commuting Lie subalgebras all of them simple except one, $\mathcal{Z}$, which is Abelian. This is the decomposition (4.2). Therefore, for every u, the Hamiltonian $-iH(u)$ will be written as the sum of mutually orthogonal and mutually commuting terms $-iH(u) = -iH_{\mathcal{Z}}(u) + \sum_{j=1}^{s} -iH_{\mathcal{S}_j}(u)$ where $-iH_{\mathcal{S}_j}(u)$ is the component of $-iH(u)$ along the simple ideal $\mathcal{S}_j$ and $-iH_{\mathcal{Z}}(u)$ is the component of $-iH(u)$ along $\mathcal{Z}$. Since all these components commute, the solution X of (4.1) takes the form of the commuting product

$$ X = X_{\mathcal{Z}} \prod_{j=1}^{s} X_j, $$

where $X_{\mathcal{Z}}$ is the solution of $\dot{X}_{\mathcal{Z}} = -iH_{\mathcal{Z}}(u)X_{\mathcal{Z}}$, $X_{\mathcal{Z}}(0) = 1$, and, for $j = 1, 2, ..., s$, X_j is the solution of $\dot{X}_j = -iH_{\mathcal{S}_j}(u)X_j$, $X_j(0) = 1$. Thus the system (4.1) behaves effectively as the parallel of $s+1$ systems each on a Lie group associated with the Lie algebras $\mathcal{S}_j$, $j = 1, ..., s$, $\mathcal{Z}$. The situation is summarized in Figure 4.1.

4.2 Tensor Product Structure of the Dynamical Lie Algebra

We now examine more in depth the structure of the decomposition (4.2) of the dynamical Lie algebra $\mathcal{L}$. We will show that, in appropriate coordinates, $\mathcal{L}$ can be written as the tensor product of Lie algebras isomorphic to $\mathcal{S}_1,...,\mathcal{S}_s$, and $\mathcal{Z}$, i.e., $\mathcal{L}$ has a basis that in appropriate coordinates can be written as $\{A_1 \otimes 1 \otimes \cdots \otimes 1, 1 \otimes A_2 \otimes 1 \otimes \cdots \otimes 1, ..., 1 \otimes 1 \otimes \cdots \otimes 1 \otimes A_s \otimes 1, 1 \otimes \cdots \otimes 1 \otimes A_{\mathcal{Z}}\}$ where A_j gives a basis of a Lie algebra isomorphic to $\mathcal{S}_j$ and $A_{\mathcal{Z}}$ gives a basis of a Lie algebra isomorphic to $\mathcal{Z}$.

We need some notions of representation theory summarized in the following subsection (see, e.g., [112], [252], [273], [297]).

4.2.1 Some representation theory and the Schur Lemma

In Definition 3.1.3, we defined a representation of a Lie algebra. Analogously, we can define the representation of a group or an algebra.

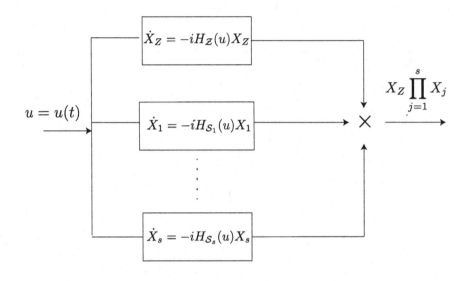

FIGURE 4.1:   Dynamical structure of an uncontrollable quantum system.

Definition 4.2.1 Given an algebra L (a group G (which could be a finite group or a Lie group)) a representation is a pair (π, V), where V is a complex inner product space, and π is an algebra (group) homomorphism from L (G) to the algebra (group) of (invertible) endomorphisms of V, $End(V)$, i.e., a map $\pi : L(G) \to End(V)$ which preserves the algebra (group) operations. A group representation of called unitary if $\pi(g)$ is unitary for every $g \in G$.[3]

Definition 4.2.2 A representation (π, V) (of an algebra L, group G, or Lie algebra $\mathcal{L}$) is called *irreducible* if there exists no, nonzero, subspace, $V_1 \subseteq V$ which is invariant for the representation, i.e., $\pi(g)V_1 \subseteq V_1$, $\forall g \in (L, G, \mathcal{L})$. If there exists such a subspace, the pair (π_1, V_1), where π_1 is the restriction to

---

[3]We mainly consider in the following $V = \mathbb{C}^n$. In a more abstract sense where we deal with general inner product spaces V, unitary representation means that the adjoint of $\pi(g)$ is equal to its inverse in $End(V)$.

V_1 of π, is also a representation. In this case, (π, V) is said to be *reducible* and (π_1, V_1) is called a *sub-representation*. A representation (π, V) is called *completely reducible* if $V = V_1 \oplus V_2 \oplus \cdots \oplus V_m$ for some subspaces $V_1, ..., V_m$ of V and each V_j, $j = 1, ..., m$, is invariant under π, so that (π_j, V_j) is an irreducible representation, where π_j is the restriction to V_j of π.

In the case of a completely reducible representation (π, V), $V = V_1 \oplus \cdots \oplus V_m$, in a basis obtained by putting together bases of $V_1, ... V_m$ in the given order, the matrices of $\pi(g)$, for every $g \in (L, G, \mathcal{L})$ take a *block diagonal form*.

Lemma 4.2.3 Every unitary representation of a group G, (π, V), is completely reducible, i.e., $V = V_1 \oplus V_2 \oplus \cdots \oplus V_m$, with $\{V_1, ..., V_m\}$ mutually orthogonal.

Proof. If (π, V) is irreducible the lemma is trivially true. Assume that it is reducible, so that there exists a nonzero invariant subspace $V_1 \subset V$. Then, $V_1^\perp$ is also invariant because, for $\vec{v} \in V_1^\perp$ and any $\vec{w}$ in V_1, we have (indicating by $(\cdot, \cdot)$ the inner product in V), for every $g \in G$,

$$(\vec{w}, \pi(g)\vec{v}) = (\pi^\dagger(g)\vec{w}, \vec{v}) = (\pi^{-1}(g)\vec{w}, \vec{v}) = (\pi(g^{-1})\vec{w}, \vec{v}) = 0,$$

since $\pi(g^{-1})\vec{w} \in V_1$. If V_1 and $V_1^\perp$ are irreducible the lemma is proved. Otherwise one continues the decomposition until one obtains a decomposition into irreducible representations. This always happens because of the finite dimension. By construction, the subspaces one obtains are mutually orthogonal.
$\square$

4.2.1.1   Schur lemma and some of its consequences

The next result, **Schur lemma**, is one of the most useful tools in representation theory. For simplicity of exposition we shall refer to the case of a representation of a group but the result and its consequences hold, mutatis mutandis, for the case of algebras and Lie algebras as well (cf. Exercise 4.10). We shall present the lemma as a series of facts concerning the situation where there are two representations (π_1, V_1) and (π_2, V_2) of a group G and a linear map $A : V_1 \to V_2$, which satisfies

$$\pi_2(g)A = A\pi_1(g), \qquad \forall g \in G. \tag{4.10}$$

If this happens, A is called a G-map.

Theorem 4.2.4 *(Schur lemma) Consider two representations of the group G, (π_1, V_1) and (π_2, V_2) and a linear, nonzero, map $A : V_1 \to V_2$, which satisfies (4.10). Then we have*

1. *$Ker(A)$ is a sub-representation of (π_1, V_1).*

2. $Im(A)$ *is a sub-representation of* (π_2, V_2).

3. *If* (π_1, V_1) *is irreducible, then* A *is injective.*

4. *If* (π_2, V_2) *is irreducible, then* A *is surjective.*

5. *If* $(\pi_1, V_1) = (\pi_2, V_2) := (\pi, V)$ *is irreducible then* $A = \lambda \mathbf{1}$*, i.e., it is a multiple of the identity map.*

The last statement is the statement usually referred to as 'Schur lemma'. *Proof.* Let $\vec{v} \in Ker(A)$. Then for every $g \in G$, $A\pi_1(g)\vec{v} = \pi_2(g)A\vec{v} = \pi_2(g)\vec{0} = \vec{0}$. Therefore, $\pi_1(g)\vec{v} \in Ker(A)$ and therefore $(\pi_1, Ker(A))$ is a sub-representation of (π_1, V_1). This proves statement 1. The proof of statement 2 is similar. From statement 1, if (π_1, V_1) is irreducible, we must have $Ker(A) = 0$ and therefore A must be injective, which proves 3. Analogously from 2 irreducibility implies that $Im(A)$ must be equal to V_2 and therefore implies surjectivity, which is 4. Let λ be an eigenvalue of A so that $Ker(A - \lambda \mathbf{1}) \neq 0$. The map $A - \lambda \mathbf{1}$ commutes with π_1. If $A - \lambda \mathbf{1}$ was different from zero, then from part 3 above we would have $Ker(A - \lambda \mathbf{1}) = 0$ which we have excluded. Therefore, necessarily $A = \lambda \mathbf{1}$. $\qquad \square$

In the following corollary of Schur lemma, we consider two *irreducible* representations of a group G, (π_1, V_1) and (π_2, V_2), and the space $Hom_G(V_1, V_2)$, defined as the space of homomorphisms $A\colon V_1 \to V_2$, such that $A\pi_1 = \pi_2 A$, that is, the space of G-maps. We say that two representations (π_1, V_1) and (π_2, V_2) are isomorphic if there exists an isomorphism in $Hom_G(V_1, V_2)$.

Corollary 4.2.5 Consider a group G and two irreducible representations of G, (π_1, V_1) and (π_2, V_2). The space $Hom_G(V_1, V_2)$ can only be 1 or 0− dimensional, according to whether or not (π_1, V_1) and (π_2, V_2) are isomorphic representations.

Therefore, according to the above corollary, all the G-maps are multiples of a fixed one in the case of irreducible representations.

Proof. Consider an $A \in Hom_G(V_1, V_2)$. If A is not an isomorphism then A is not one to one or it is not onto. If it is not one to one, $Ker(A) \neq \{0\}$. However, because of the irreducibility of (π_1, V_1) this implies from Schur lemma that $Ker(A) = V_1$ and therefore $A = 0$. Analogously, if A is not onto, since the Image of A is a sub-representation of V_2 and by irreducibility can only be $\{0\}$, this again implies that A is zero. Therefore, every element in $Hom_G(V_1, V_2)$ which is not zero must be an isomorphism. Consider now two such isomorphisms A and B. From $B\pi_1 = \pi_2 B$ we obtain $\pi_1 B^{-1} = B^{-1}\pi_2$. Using this, after multiplying $A\pi_1 = \pi_2 A$ by B^{-1} to the left, we have $B^{-1}A\pi_1 = \pi_1 B^{-1}A$. By Schur lemma then $B^{-1}A$ is a multiple of the identity, which implies that A is a multiple of B. Therefore, $Hom_G(V_1, V_2)$ is one-dimensional. $\qquad \square$

We can state some sort of converse of Schur lemma as follows.

Proposition 4.2.6 Let (π, V) be a unitary representation of a group G on $V \simeq \mathbb{C}^n$, which is reducible. Then there exists a skew-Hermitian matrix A which is not a multiple of the identity such that $A\pi(g) = \pi(g)A$, for every $g \in G$.

Proof. Recall, from Lemma 4.2.3, that every unitary representation is completely reducible. Since our representation is reducible, in appropriate coordinates all matrices $\pi(g)$ are in block diagonal form, with at least two blocks. In these coordinates, the desired diagonal matrix A is the one divided in the corresponding blocks and where the j-th block (of dimension n_j) is $i\mathbf{1}_{n_j}$. $\square$

The matrix A in Proposition 4.2.6 has the function of subdividing a representation in its irreducible components.

4.2.1.2 Completely reducible representation of a group; uniqueness of the decomposition

We have seen from Lemma 4.2.3 that unitary representations can be completely reduced to the sum of irreducible representations. One natural question is to what extent such a decomposition is unique. This is answered by the following theorem.

Theorem 4.2.7 *Let (π, V) be a unitary representation of a group G which is the sum of irreducible representations $V = \oplus_{j=1}^{n_1} S_j$, and consider another sum of irreducible representations of (π, V), $V = \oplus_{k=1}^{n_2} T_k$. Then each S_j is isomorphic to some T_k. Furthermore, let $S_R = \oplus_j S_j$ and $T_R := \oplus_k T_k$ be the sums of all the subspaces S_j and T_k isomorphic to a given irreducible subspace R. Then $S_R = T_R$. In particular the number of subspaces in S_R is equal to the number of subspaces in T_R, which implies $n_1 = n_2$.*

We reiterate that, in our context, isomorphism means G-isomorphism, that is, an isomorphism which is also a G-map.

Proof. Fix one S_j, say S_1, and one T_k, say T_1, and the map $\Pi_T i_S$ composition of Π_T, the orthogonal projection onto T_1 and i_S the inclusion map in S_1 (in other terms $\Pi_T i_S$ is the restriction of Π_T to S_1). We have, for every $g \in G$ and on S_1, using the fact that S_1 and T_1 are invariant under $\pi(g)$,[4]

$$\pi(g)\Pi_T i_S = \Pi_T \pi(g) i_S = \Pi_T i_S \pi(g).$$

Therefore, $\Pi_T i_S$ is a G-map of representations between S_1 and T_1, and according to Corollary 4.2.5 it is either zero or it is an isomorphism. Now if $\Pi_T i_S$ is an isomorphism then T_1 is isomorphic to S_1 and the first claim of the

---

[4]In appropriate coordinates Π_T has a form given by an identity in the first block and zero everywhere else, which commutes with the block diagonal form of $\pi(g)$. This explains the first equality. The second one follows from the fact that S_1 is invariant under $\pi(g)$.

theorem is proved. If it is zero, we can repeat the argument for T_2, T_3 and so on. If it is zero for all the subspaces, this would mean that $(\sum_j \Pi_{T_j})i_S = 0$ but this is impossible since $\sum_j \Pi_{T_j}$ is the identity map. This shows the first claim of the theorem.

Consider now an irreducible sub-representation (π, R) of (π, V) and, as in the statement of the theorem, let S_R and T_R the direct sum of all S_j and T_k, respectively, representations G-isomorphic to R. Call P_T^R the sum of the orthogonal projections onto the subspaces T_k in T_R and $P_T^{R\perp}$ the projection onto the orthogonal complements (which are not isomorphic to R). Since $P_T^R + P_T^{R\perp}$ is the identity map, the image of its restriction to S_R is S_R. However, such an image is also $P_T^R S_R$ since from Corollary 4.2.5 $P_T^{R\perp} S_R = 0$. From the above argument $S_R = P_T^R S_R \subseteq T_R$. Repeating the argument with T_R and S_R interchanged, we find that $T_R \subseteq S_R$ and therefore $T_R = S_R$. Equating the dimensions of S_R and T_R we see that S_R and T_R must have the same number of (isomorphic) subspaces (S_j and T_k) and repeating for every R, it follows that $n_1 = n_2$.

$\square$

4.2.1.3 Tensor products of representations of groups and Lie algebras

The direct sum $\oplus$ is a way to generate a new representation from two or more representations and, in the opposite direction, a way to decompose a representation into its (irreducible) components. Another way to build a representation out of two or more representations is the **tensor product** which we now define (cf. Exercise 4.6).

Definition 4.2.8 Consider two (possibly coinciding) groups G and H and their direct product $G \times H$. Consider representations (π_G, V_G) and (π_H, V_H) of G and H respectively. Then the tensor product representation $(\pi_G \otimes \pi_H, V_G \otimes V_H)$ is defined by
$$\pi_G \otimes \pi_H(g, h) = \pi_G(g) \otimes \pi_H(h),$$
where the symbol $\otimes$ on the right hand is the Kronecker product of matrices (assuming that V_G (V_H) is $\mathbb{C}^{n_G}$ ($\mathbb{C}^{n_H}$)).

The corresponding concept for Lie algebra representations is the following.

Definition 4.2.9 Consider two Lie algebras $\mathcal{L}_1$ and $\mathcal{L}_2$ and representations (π_1', V_1) and (π_2', V_2) of $\mathcal{L}_1$ and $\mathcal{L}_2$, respectively. Then the tensor product representation $(\pi_1' \otimes \pi_2', V_1 \otimes V_2)$ is a representation of the direct sum $\mathcal{L}_1 \oplus \mathcal{L}_2$, given for $X_1 \in \mathcal{L}_1$ and $X_2 \in \mathcal{L}_2$, by
$$\pi_1' \otimes \pi_2'(X_1 + X_2) = \pi_1'(X_1) \otimes 1 + 1 \otimes \pi_2'(X_2).$$

These definitions extend straightforwardly to the product or direct sum of an arbitrary number of groups and Lie algebras.

4.2.2 Tensor product structure for the irreducible representation of the product of two groups

We will now show that any irreducible representation of a product group has a tensor product structure. In the next subsection we will transfer this result to Lie algebras and then apply it to the dynamical Lie algebra of quantum control systems.

Theorem 4.2.10 *Consider an irreducible representation of the direct product of two (Lie) groups K and H, (π, V). Then (π, V) is the tensor product of a representation (π_K, V_K) of K and a representation (π_H, V_H) of H, respectively, with $V = V_H \otimes V_K$.*

Proof. (The proof is adapted from [97]) Consider the group H and the representation of H on V defined by $\pi(\mathbf{1}, h)$ for $h \in H$. Even though (π, V) is irreducible for $K \times H$, $\pi(\mathbf{1}, \cdot)$ is not necessarily irreducible for H. Therefore, there exists a minimal invariant subspace of V, W, which is invariant under H, so that $(\pi(\mathbf{1}, \cdot), W)$ is an irreducible representation of H. In the following for simplicity we shall denote by $h\vec{v}$ ($k\vec{v}$) multiplication in V by $\pi(\mathbf{1}, h)$ ($\pi(k, \mathbf{1})$), with $h \in H$ ($k \in K$). The space W above defined has the property that, for every $k \in K$, kW is also H-invariant because

$$hkW = khW \subseteq kW,$$

and also minimal because otherwise minimality of W would be violated. Consider now m subspaces of V, $W_1, ..., W_m$, defined recursively as follows: $W_1 = W$, and $W_j = k_j W$ for $k_j \in K$ chosen so that

$$k_j W \cap (W_1 + W_2 + \cdots + W_{j-1}) = 0, \qquad (4.11)$$

until such k_j exists. Notice that if $kW \cap (W_1 + W_2 + \cdots + W_{j-1}) \neq 0$, this would mean that $kW \cap (W_1 + W_2 + \cdots + W_{j-1}) = kW$ because otherwise we would have an invariant subspace for H, $kW \cap (W_1 + W_2 + \cdots + W_{j-1})$, smaller than kW which would contradict the minimality of W (and kW). By construction the sum $W_1 + W_2 + \cdots + W_m$ is a direct sum and, by finite dimensionality, the process will end after a finite number of steps giving $W_1 \oplus W_2 \oplus \cdots \oplus W_m \subseteq V$. Furthermore, this is the direct sum of minimal vector spaces invariant under H (as shown above) and the whole space is invariant under K. To see this last claim, assume that it is not true. Then there exists a $j \in \{1, ..., m\}$ and a $\vec{y} \in W_j$ as well as a $k \in K$ such that $k\vec{y} \notin W_1 \oplus \cdots \oplus W_m$. Writing $\vec{y}$ as $\vec{y} = k_j \hat{y}$, with $\hat{y} \in W$, we have $kk_j \hat{y} \notin W_1 \oplus \cdots \oplus W_m$. Now there are two possibilities: 1) $kk_j W \cap W_1 \oplus \cdots \oplus W_m \neq 0$ and 2) $kk_j W \cap W_1 \oplus \cdots \oplus W_m = 0$. The first case would be impossible because, as we have seen above, it would imply $kk_j W \subseteq W_1 \oplus \cdots \oplus W_m$ which we have excluded because $kk_j \hat{y} \notin W_1 \oplus \cdots \oplus W_m$. The case 2) is also impossible because we have assumed that we are not able to find any element $\hat{k} \in K$ such that the union with $\hat{k}W$ would enlarge such a subspace (cf. (4.11)).

Being invariant under both K and H, the subspace $W_1 \oplus \cdots \oplus W_m$, is in fact *equal* to V since (π, V) is irreducible. Therefore, we have a decomposition

$$V = W_1 \oplus W_2 \oplus \cdots \oplus W_m,$$

into (minimal) subspaces $W_1, ..., W_m$ invariant under H and such that $W_1 = W$ and $W_j = k_j W$.

Choose a basis in $W_1 = W$, $\{\vec{e}_1, ..., \vec{e}_s\}$, and bases $\{k_j \vec{e}_1, ..., k_j \vec{e}_s\}$ for W_j, so as to form a basis for V. Then every $h \in H$ acts in the same way on $\{\vec{e}_1, ..., \vec{e}_s\}$ and $\{k_j \vec{e}_1, ..., k_j \vec{e}_s\}$. Therefore, the matrix representative of $h \in H$ in this basis has a block diagonal form (because of invariance) and with all the blocks equal to each other, i.e., it is of the form $A(h) = \mathtt{diag}(S(h), S(h), ..., S(h))$. For $k \in K$, consider the action of k in the given basis and the associated matrix $B(k)$ partitioned in $m \times m$ blocks each of dimension $s \times s$, $B(k) := \{B_{r,l}(k)\}$. Here $r, l = 1, ..., m$ are indexes which label the blocks. Because of the commutativity of H and K we have for every $h \in H$ and $k \in K$, $A(h)B(k) = B(k)A(h)$ which, given the special form of $A(h)$, implies $B_{r,l}(k)S(h) = S(h)B_{r,l}(k)$. Thinking this for a fixed $k \in K$ and variable $h \in H$ and since S is an irreducible representation of H, it follows from Schür's lemma that $B_{r,l}(k)$ is a scalar matrix, i.e., $B_{r,l}(k) = \lambda_{r,l}(k)\mathbf{1}_{s \times s}$.

In conclusion, we have that, in the given basis, the matrix $A(h)$ has the form $A(h) = \mathbf{1}_{m \times m} \otimes S(h)$, while the matrix $B(k)$ has the form $\Lambda_{m \times m}(k) \otimes \mathbf{1}_{s \times s}$ for a matrix Λ. Therefore, the product $A(h)B(k)$ is the tensor product of two matrices one giving a representation of H and one giving a representation of K, i.e., $A(h)B(k) = \Lambda_{m \times m}(k) \otimes S(h)$ as desired.

$\square$

4.2.3   Tensor product structure of the dynamical Lie algebra

We shall now describe the consequences of Theorem 4.2.10 on the structure of the dynamical Lie algebra $\mathcal{L}$ (4.2) associated with a quantum control system. We first present the consequence of Theorem 4.2.10 for Lie algebra representations in general.

Theorem 4.2.11 *Let π' be an irreducible representation of a Lie algebra $\mathcal{L} := \mathcal{L}_1 \bar{\oplus} \mathcal{L}_2 \bar{\oplus} \cdots \bar{\oplus} \mathcal{L}_r$, direct sum of Lie algebras $\mathcal{L}_1, ... \mathcal{L}_r$, on a vector space V. Then V is a the tensor product of vector spaces $V_1, ..., V_r$ and π' is the tensor product representation of representations $(\pi'_1, V_1), ..., (\pi'_r, V_r)$ of $\mathcal{L}_1, ..., \mathcal{L}_r$, respectively.*

Proof. Consider the result of Theorem 4.2.10 naturally extended to the case of $r \geq 2$ groups and apply it to the Lie groups associated with the various Lie algebras $\mathcal{L}_1, \mathcal{L}_2, ..., \mathcal{L}_r$. This gives a decomposition of V as $V = V_1 \otimes V_2 \otimes \cdots \otimes V_r$ on which $e^{\mathcal{L}}$ acts as $X_1 \otimes X_2 \otimes \cdots \otimes X_m$, where X_j is in the image of the representation of $e^{\mathcal{L}_j}$. Taking a trajectory $X = X(t)$ in

$e^{\mathcal{L}}$, $X(t) = X_1(t) \otimes X_2(t) \otimes \cdots \otimes X_r(t)$, with $X(0)$ equal to the identity, and differentiating at $t = 0$, we have that an element of the Lie algebra $\mathcal{L}$ is written as $\dot{X}(0) = \dot{X}_1(0) \otimes \mathbf{1} \otimes \cdots \otimes \mathbf{1} + \mathbf{1} \otimes \dot{X}_2(0) \otimes \cdots \otimes \mathbf{1} + \cdots + \mathbf{1} \otimes \mathbf{1} \otimes \cdots \otimes \dot{X}_r(0)$ with $\dot{X}_j(0)$ giving a representation of $\mathcal{L}_j$ on V_j, for $j = 1, ..., r$.

$\square$

Theorem 4.2.11 now applies directly to the given representation of the dynamical Lie algebra of a quantum system (4.2) to tell us that, if it is irreducible, it has the form of a tensor product of Lie algebras isomorphic to $\mathcal{S}_1, ..., \mathcal{S}_s$, and $\mathcal{Z}$. An interesting consequence of this theorem is that if an irreducible representation of the dynamical Lie algebra (π', V) has dimension which is a prime number, then the associated Lie algebra has to be simple, not just reductive. We stress that this fact assumes however irreducibility.

In summary, there are **two mechanisms** that contribute to the decomposition (4.2): *Reduction* to the irreducible components and *decomposition* of the irreducible components as *tensor products*. Reducibility implies that there will be invariant subspaces and that the full space $V \simeq \mathbb{C}^n$ can be written as a sum of invariant subspaces which carry irreducible representations. To this purpose we also remark that representations of Lie algebras with values in $u(n)$, which are the ones we are dealing with, are also called unitary, and a proof similar to the one of Lemma 4.2.3 shows that unitary representations of Lie algebras are completely reducible (Exercise 4.7). On each irreducible component, a decomposition of the form (4.2) holds and, from the results of this section, it has the structure of a tensor product.

4.3 Dynamical Decomposition Starting from a Group of Symmetries; Subspace Controllability

We now assume that we have not calculated a basis of the dynamical Lie algebra $\mathcal{L}$. Perhaps, it is difficult to compute because the dimension of the system is too large. This is the case, for instance, when we have a network of N interacting spin $\frac{1}{2}$ particles. The dimension of the matrices involved grows like 2^N. It may be possible, however, to find a discrete group of transformations (which we shall call symmetries) G that leaves the Hamiltonians of the system unchanged. In the case of a network of interacting spins, the topology of the network itself may suggest a group G of symmetries for the Hamiltonians involved. For example, a network of three identical spin $\frac{1}{2}$ particles simultaneously controlled and with (Ising) interaction Hamiltonian $H_0 = \sigma_z \otimes \sigma_z \otimes \mathbf{1} + \sigma_z \otimes \mathbf{1} \otimes \sigma_z + \mathbf{1} \otimes \sigma_z \otimes \sigma_z$ can be represented by an equilateral triangle whose edges represent the mutual interactions among the spins which are represented by the nodes. The graph does not change if we interchange two or more nodes, which suggests a group of symmetries given by the symmetry group S_3. This system is further explored in the examples 4.3.8 and 4.3.9 below.

In the presence of a group of symmetries G, the dynamical Lie algebra $\mathcal{L}$ will be a Lie subalgebra of $u^G(n)$, the Lie subalgebra of $u(n)$ which is invariant under the action of G, i.e., $\mathcal{L} \subseteq u^G(n)$. In this section, we shall study how to describe $u^G(n)$. In order to do that, we first introduce some more notions of representation theory of finite groups.

4.3.1   Some more representation theory of finite groups: Group algebra, regular representation and Young Symmetrizers

Definition 4.3.1 Given a finite group G, the *group algebra* $\mathbb{C}[G]$ is defined as the algebra obtained by formal linear combinations of the elements in G with coefficients in $\mathbb{C}$ and extending bilinearly the product operation.

Example 4.3.2 The elements of the group of permutations of n objects, S_n, the *symmetric group*, are often denoted by the cycle notation. For example, in S_4 the element $(132)(4)$ denotes the permutation which sends position 1 to position 3, position 3 to position 2 and position 2 to position 1, while leaving position 4 unchanged. Thus, four objects in the order $ABCD$ are permuted by $(132)(4)$ to $BCAD$. The singleton cycles such as (4) are usually omitted in the notation. This notation allows us to perform products of permutations (from right to left) easily. For instance, again in S_4, the product $(132)(234)$ gives (134). This is calculated by starting with 1. The first permutation leaves 1 unchanged while the second one moves 1 to 3. So we have $(13 \cdots)$. Now consider 3. The first permutation moves 3 to 4 and the second one leaves 4 unchanged. So we have $(134 \cdots)$. Now consider 4. The first permutation moves 4 to 2. The second one moves 2 to 1, so that the cycle closes to become (134). Extending this multiplication to the algebra $\mathbb{C}[S_4]$ by bilinearity, we obtain, for example,

$$(12) \cdot (\lambda(12) + \mu(13)) = \lambda(1)(2)(3)(4) + \mu(132) - \lambda \mathbf{1} + \mu(132).$$

In this context $\mathbf{1}$ indicates the identity permutation.

The group G acts naturally on $\mathbb{C}[G]$ by extending by linearity the product in the group. That is, if $\{g_j\}$ are the elements of the group and λ_j arbitrary coefficients in $\mathbb{C}$, then $g\left(\sum_j \lambda_j g_j\right) := \sum_j \lambda_j g g_j$ for any $g \in G$. Therefore, $\mathbb{C}[G]$ gives a *representation* of G, which is called the *regular representation*. The regular representation contains several pieces of information about general representations of G. We summarize some of them in the following theorem whose proof can be found for instance in [112].

Theorem 4.3.3 *Every irreducible representation of a finite group G is G-isomorphic to a representation contained in the regular representation. Moreover, the number of times every irreducible representation appears (up to isomorphism) in the regular representation is equal to its dimension.*

Given a quantum dynamical control system, we naturally have a (unitary) representation of $\mathcal{L} \subseteq u^G(n)$ and $u(n)$ over a complex vector space $V \simeq \mathbb{C}^n$. At the same time, we have, over the same vector space, a representation of the symmetry group G, from which we can define a (algebra) representation for the group algebra $\mathbb{C}[G]$. The representation of G and the representation of $u^G(n)$ are intertwined in a way which is often referred to (sometimes in a more general context) as *Schur-Weyl duality* (see, e.g., [112]). The characterization of $u^G(n)$ we are going to describe is based on this fact and the regular representation is instrumental to this end.

The irreducible representations of G contained in the regular representation can be constructed starting from certain elements of $\mathbb{C}[G]$ called *Generalized Young Symmetrizers*. The terminology comes from the fact that in the case of the symmetric group, S_n, these are called *Young Symmetrizers* and 'generalized' refers to the fact that we are referring to a general finite group G.

Definition 4.3.4 A complete set of *Generalized Young Symmetrizers (GYS)* of a group G is a set of m elements $\{P_j\}$ in the group algebra $\mathbb{C}[G]$ satisfying the following conditions:

1. (Completeness)

$$\sum_{j=1}^{m} P_j = \mathbf{1}, \tag{4.12}$$

   where $\mathbf{1}$ is the identity of the group G.

2. (Orthogonality)

$$P_j P_k = \delta_{j,k} P_j, \qquad \forall j, k, \tag{4.13}$$

   where $\delta_{j,k}$ is the Kronecker delta.

3. (Primitivity) Given a GYS P_j, for every $g \in G$,

$$P_j g P_j = \lambda_g P_j, \tag{4.14}$$

   with λ_g a scalar that depends on g.

A complete set of GYS's always exists for any finite group G (see, e.g., Theorem III.2 in Appendix III of [273]). It is however not unique. In fact it is easy to verify that if $\{P_j\}$ is a complete set of GYS's, i.e., satisfying (4.12), (4.13) and (4.14), so is $\{gP_jg^{-1}\}$ for any $g \in G$. The left ideals generated by the GYS's, i.e., $\mathcal{C}_j := \mathbb{C}[G]P_j$, are the irreducible representations in the regular representation (cf. Theorem III.3 of the Appendix III of [273]). That is, $\mathbb{C}[G] = \bigoplus_{j=1}^{m} \mathcal{C}_j$, where $\mathcal{C}_j$ are the irreducible representations of G in $\mathbb{C}[G]$. Furthermore, we have the following (cf. Theorem III.4 in Appendix III of [273]) which gives us a test for two GYS's to give rise to isomorphic representations.

Proposition 4.3.5 Two left ideals $\mathcal{C}_1$ and $\mathcal{C}_2$, corresponding to symmetrizers P_1 and P_2, respectively, are G-isomorphic if and only if there exists an $r \in G$ such that $P_1 r P_2 \neq 0$.

An additional feature of GYS's, which is often required, is to be *Hermitian*. The concept of being Hermitian for an element of $\mathbb{C}[G]$ can be defined independently of the representation. Given an element $\sum_j \alpha_j g_j \in \mathbb{C}[G]$, with $g_j \in G$ and $\alpha_j \in \mathbb{C}$, we first define a † operation as

$$\left(\sum_j \alpha_j g_j \right)^\dagger := \sum_j \bar{\alpha}_j g_j^{-1}.$$

An element $a \in \mathbb{C}[G]$ is called *Hermitian* if and only if $a^\dagger = a$. Consider now the representation of G, (π, V) which is extended by linearity to a representation of $\mathbb{C}[G]$. If such a representation is unitary then, if a is Hermitian, $\pi(a)$ is Hermitian in the usual sense of matrices (Exercise 4.8). We may also assume the extra property:

4. The GYS are *Hermitian*, i.e., for every j,

$$P_j = P_j^\dagger. \tag{4.15}$$

4.3.2   Structure of the representations of $u^G(n)$ and G

Assume now that we have a complete set of Hermitian GYS's, $\{P_j\}$, and their corresponding images under a unitary representation (π, V) of G which we extend to a representation of the algebra $\mathbb{C}[G]$. From the completeness property, we have $V = \pi(P_1)V + \pi(P_2)V + \cdots + \pi(P_m)V$. Furthermore, for $j \geq 2$, assume $\vec{x} \in \pi(P_j)V \cap (\pi(P_1)V + \pi(P_2)V + \cdots \pi(P_{j-1})V) \neq \emptyset$. Then there exist $\vec{y}, \vec{y}_1, ..., \vec{y}_{j-1} \in V$ such that

$$\vec{x} = \pi(P_j)\vec{y} = \pi(P_1)\vec{y}_1 + \pi(P_2)\vec{y}_2 + \cdots \pi(P_{j-1})\vec{y}_{j-1}.$$

Applying $\pi(P_j)$ to this relation and using (4.13) we obtain $\vec{x} = 0$. Therefore, the sum $V = \pi(P_1)V + \pi(P_2)V + \cdots + \pi(P_m)V$ is a direct sum,

$$V = \pi(P_1)V \oplus \pi(P_2)V \oplus \cdots \oplus \pi(P_m)V. \tag{4.16}$$

We shall now write the elements of $u^G(n)$ in a basis chosen by putting together bases of the subspaces $\pi(P_j)V$, $j = 1, ..., m$. We shall see that, in this basis, $u^G(n)$ takes a special block diagonal form.

Remark 4.3.6 We remark that some $\pi(P_j)$ might be zero. However, if $\pi(P_j) = 0$ then $\pi(P_k) = 0$ for all P_k's corresponding to isomorphic representations as identified by Proposition 4.3.5. In order to see this, assume the isomorphism between $\mathcal{C}_j$ and $\mathcal{C}_k$ is verified and consider the G-homomorphism $\mathcal{C}_j \to \mathcal{C}_k$ given by right multiplication by $P_j r_1 P_k \neq 0$ for some $r_1 \in G$ of Proposition 4.3.5. Also consider the G-homomorphism $\mathcal{C}_k \to \mathcal{C}_j$ given by right multiplication by $P_k r P_j \neq 0$ for some $r \in G$. According to Corollary 4.2.5, up to a scalar λ, these maps must be inverse of each other and therefore applying them in sequence to P_k, we have (using (4.13))

$$P_k r P_j r_1 P_k = \lambda P_k. \tag{4.17}$$

Applying π to this relation and using the fact that π is a representation we have $\pi(P_k r)\pi(P_j)\pi(r_1 P_k) = \lambda \pi(P_k)$, which shows that $\pi(P_j) = 0$ implies $\pi(P_k) = 0$, and, exchanging the roles of P_k and P_j that the converse is also true. Moreover from formula (4.17) again applying π, we find that, in the case where $\pi(P_j)$ (and all $\pi(P_k)$'s) are different from zero $P_k r P_j \neq 0$ implies $\pi(P_k r P_j) \neq 0$. In Example 4.3.8 below various situations occur.

Let us group together the GYS's P_j corresponding to isomorphic representations $\mathcal{C}_j$. That is, we have subspaces $\pi(P_1)V,...,\pi(P_{m_A})V$ corresponding to isomorphic representations $\mathcal{C}_1,..., \mathcal{C}_{m_A}$, then $\pi(P_{m_A+1})V,...,\pi(P_{m_A+m_B})V$ corresponding to isomorphic representations $\mathcal{C}_{m_A+1},..., \mathcal{C}_{m_A+m_B}$, and so on, up to $\pi(P_{m_A+m_B+\cdots+1})V ,..., \pi(P_{m_A+m_B+\cdots+m_C})V$ corresponding to isomorphic representations $\mathcal{C}_{m_A+m_B+\cdots+1},..., \mathcal{C}_{m_A+m_B+\cdots+m_C}$, with $m_A+m_B+\cdots+m_C = m$. We only consider the P_j such that $\pi(P_j) \neq 0$.

Every subspace $\pi(P_j)V$ is invariant under $u^G(n)$ because, for any $F \in u^G(n)$, $F\pi(P_j)V = \pi(P_j)FV \subseteq \pi(P_j)V$. Therefore, in the given basis, every element F in $u^G(n)$ takes the form

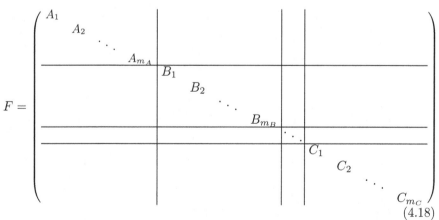

$$\tag{4.18}$$

We now want to obtain more information on the nature of the submatrices appearing in (4.18). This is achieved by better specifying the bases chosen for the various subspaces. In order to do that, we first introduce the space $\text{End}_G(V)$ as the space of endomorphisms on V commuting with G (in the given representation) of which $u^G(n)$ is a subset.[5] We have the following result adapted from [84] (Propositions 3.2 and 3.3).

Proposition 4.3.7 Consider two GYS, P_j and P_k corresponding to isomorphic irreducible sub-representations of the regular representation, $\mathcal{C}_j$, and $\mathcal{C}_k$. Then $\pi(P_j)V$ and $\pi(P_k)V$ are either both zero or they are irreducible representations of $\text{End}_G(V)$, which are isomorphic. Let $r \in G$ be chosen as in Proposition 4.3.5 so that $P_k r P_j \neq 0$. Then $\pi(P_k r P_j) \neq 0$ is an isomorphism $\pi(P_j)V \to \pi(P_k)V$ which commutes with $\text{End}_G(V)$. That is, the two representations of $\text{End}_G(V)$, $\pi(P_j)V$ and $\pi(P_k)V$, are $\text{End}_G(V)$-isomorphic.

One immediate consequence of this proposition is that $\pi(P_j)V$ and $\pi(P_k)V$ have the same dimension, therefore the matrices $A_1, ..., A_{m_A}$ in (4.18) have the same dimensions. The same is true for $B_1, ..., B_{m_B}$ and so on. Let us now restrict ourselves to the A_j matrices corresponding to isomorphic subspaces (the same reasoning will be valid for the other blocks). Also, for simplicity of notation, we shall omit the reference to the representation π. For instance, we shall denote $\pi(P_j)V$ by P_jV. However, all the expressions we discuss are meant to be in the given representation.

Consider a basis of P_1V $\{\vec{e}_1, ..., \vec{e}_{d_A}\}$. For P_2V, take the basis $\{\mu_2 P_2 r P_1 \vec{e}_1, ..., \mu_2 P_2 r P_1 \vec{e}_{d_A}\}$, where $P_2 r P_1$ is the chosen isomorphism from P_1V to P_2V, according to Proposition 4.3.7, μ_2 is a scalar parameter different from zero, which is otherwise free at this stage. Let $\vec{x} = \sum_{j=1}^{d_A} x_j \vec{e}_j \in P_1V$. Define, for any $F \in \text{End}_G(V)$, $F\vec{e}_j := \sum_{l=1}^{d_A} f_{jl} \vec{e}_l$. We have:

$$F\left(\sum_{j=1}^{d_A} x_j \vec{e}_j\right) = \sum_{l=1}^{d_A}\left(\sum_{j=1}^{d_A} x_j f_{jl}\right) \vec{e}_l,$$

and

$$F\left(\sum_{j=1}^{d_A} x_j \mu_2 P_2 r P_1 \vec{e}_j\right) = \mu_2 P_2 r P_1 F\left(\sum_{j=1}^{d_A} x_j \vec{e}_j\right) =$$

$$\mu_2 P_2 r P_1 \sum_{l=1}^{d_A}\left(\sum_{j=1}^{d_A} x_j f_{jl}\right) \vec{e}_l = \sum_{l=1}^{d_A}\left(\sum_{j=1}^{d_A} x_j f_{jl}\right) \mu_2 P_2 r P_1 \vec{e}_l.$$

Therefore, in the given basis, F transforms the coefficients in the basis $\{\vec{e}_j\}$ in the same way as the coefficients in the basis $\{\mu_2 P_2 r P_1 \vec{e}_j\}$. This shows that The matrices A_1 and A_2 in (4.18) are equal. The argument can be repeated

---

[5]We do not say a 'subspace' because $\text{End}_G(V)$ is meant to be a *complex* vector space while $u^G(n)$ is a *real* one.

with A_1 and A_3 (choosing the isomorphism $\mu_3 P_3 r P_1{}^6$ and so on. Therefore, we have $A_1 = A_2 = \cdots = A_{m_A} := A$, and analogously $B_1 = B_2 = \cdots = B_{m_B} := B$, and so on, up to $C_1 = C_2 = \cdots = C_{m_C} := C$. In the given basis, the matrix F in (4.18) takes the form

$$
F = \left(
\begin{array}{c|c|c|c}
\mathbf{1}_{m_A} \otimes A & & & \\ \hline
& \mathbf{1}_{m_B} \otimes B & & \\ \hline
& & \ddots & \\ \hline
& & & \mathbf{1}_{m_C} \otimes C
\end{array}
\right). \tag{4.19}
$$

The form (4.19) is valid for any matrix in $\mathrm{End}_G(V)$. We now consider in particular the elements $F \in u^G(n)$, that is $F \in u(n) \cap \mathrm{End}_G(V)$. The operators in $u(n)$ can be defined as the elements of $\mathrm{End}(V)$ whose representative in an orthonormal basis (and therefore in all orthonormal bases) is a skew-Hermitian matrix. We can further specify the basis above so that we obtain an orthonormal basis. It is here that the assumption that the GYS's are Hermitian is used. Pick the basis of $P_1 V$, $\{\vec{e}_1, ..., \vec{e}_{d_A}\}$ orthonormal, and consider the basis of $P_2 V$, $\{\mu_2 P_2 r P_1 \vec{e}_1, ..., \mu_2 P_2 r P_1 \vec{e}_{d_A}\}$, choosing μ_2 so that $\|\mu_2 P_2 r P_1 \vec{e}_1\| = 1$, i.e., using μ_2 to normalize the first element of the basis. Now consider the inner product of an element of the basis of $P_1 V$, say $\vec{e}_j$, with an element of the basis of $P_2 V$, say $\mu_2 P_2 r P_1 \vec{e}_k$. This is given by

$$
\mu_2 \vec{e}_j^{\dagger} P_2 r P_1 \vec{e}_k = \mu_2 \vec{e}_j^{\dagger} P_1^{\dagger} P_2 r P_1 \vec{e}_k = \mu_2 \vec{e}_j^{\dagger} P_1 P_2 r P_1 \vec{e}_k = 0,
$$

where we use the fact that P_1 is Hermitian and the orthogonality property $P_1 P_2 = 0$. Therefore, the set $\{\mu_2 P_2 r P_1 \vec{e}_1, ..., \mu_2 P_2 r P_1 \vec{e}_{d_A}\}$ is orthogonal to $\{\vec{e}_1, ..., \vec{e}_{d_A}\}$. The same proof works if $P_2 V$ is not isomorphic to $P_1 V$. To see that $\{\mu_2 P_2 r P_1 \vec{e}_1, ..., \mu_2 P_2 r P_1 \vec{e}_{d_A}\}$ is actually an orthonormal set, consider two elements in it $\mu_2 P_2 r P_1 \vec{e}_j$ and $\mu_2 P_2 r P_1 \vec{e}_k$ and take the inner product

$$
(\mu_2 P_2 r P_1 \vec{e}_j, \mu_2 P_2 r P_1 \vec{e}_k) = |\mu_2|^2 \vec{e}_j^{\dagger} P_1^{\dagger} r^{\dagger} P_2^{\dagger} P_2 r P_1 \vec{e}_k = |\mu_2|^2 \vec{e}_j^{\dagger} P_1 r^{\dagger} P_2 P_2 r P_1 \vec{e}_k.
$$

Since $P_2 r P_1$ is nonzero and it is an $\mathrm{End}_G(V)$-isomorphism $P_1 V \to P_2 V$, $P_1 r^{\dagger} P_2 = (P_2 r P_1)^{\dagger}$ is also nonzero and it is an $\mathrm{End}_G(V)$-isomorphism $P_2 V \to P_1 V$. Corollary 4.2.5, applied to the algebra $\mathrm{End}_G(V)$, tells us that $P_1 r^{\dagger} P_2$ must be a multiple of $(P_2 r P_1)^{-1}$. Therefore, we have for some $\lambda \neq 0$ (necessarily real)

$$
(\mu_2 P_2 r P_1 \vec{e}_j, \mu_2 P_2 r P_1 \vec{e}_k) = |\mu_2|^2 \lambda \vec{e}_j^{\dagger} (P_2 r P_1)^{-1} P_2 r P_1 \vec{e}_k = \lambda |\mu_2|^2 \vec{e}_j \vec{e}_k,
$$

which is zero if $k \neq j$, and it is equal to $\lambda |\mu_2|^2$ if $j = k$, independently of the choice of j. Since μ_2 was chosen to normalize the first vector in the basis, we necessarily have $\lambda = \frac{1}{|\mu_2|^2}$ so that all vectors in the basis have norm one.

$^6 r$ here is not necessarily equal to r in the previous case, $\mu_3 \neq 0$ a scalar parameter.

Example 4.3.8 Consider the symmetric group $G = S_3$ and its representation on $V = \mathbb{C}^2 \otimes \mathbb{C}^2 \otimes \mathbb{C}^2$, with $\{\vec{e}_1, \vec{e}_2\}$ the standard basis of $\mathbb{C}^2$, so that the basis of V is $\{\vec{e}_j \otimes \vec{e}_k \otimes \vec{e}_l\}$, $j, k, l = 1, 2$ and the action of $G = S_3$ on V is defined by permuting the elements of this basis. There are 6 elements in S_3, which we list in the cycle notation, denoting by **1** the identity permutation.

$$\mathbf{1}, \quad (12), \quad (13), \quad (23), \quad (123), \quad (213).$$

Symmetry with respect to this group occurs, for instance, when we have a controlled network of 3 spin $\frac{1}{2}$ particles, with identical interaction with each other. The topology of the network does not change if we permute two of the spins.

The following four elements of the group algebra satisfy the conditions (4.12), (4.13), (4.14) (4.15), that is, they form a complete set of Hermitian GYS's, as it can be verified by applying the definitions (cf. Exercise 4.9).

$$P_1 := \frac{1}{6}\left(\mathbf{1} + (12) + (13) + (23) + (123) + (213)\right),$$

$$P_2 := \frac{1}{6}\left(2 \cdot \mathbf{1} + 2 \cdot (12) - (13) - (23) - (213) - (123)\right),$$

$$P_3 := \frac{1}{6}\left(2 \cdot \mathbf{1} - 2 \cdot (12) + (13) + (23) - (123) - (213)\right),$$

$$P_4 := \frac{1}{6}\left(\mathbf{1} - (12) - (13) - (23) + (213) + (123)\right).$$

Furthermore, using the test of Proposition 4.3.5 we can check that P_2 and P_3 correspond to isomorphic ideals (we can for example use $r = (13)$ in the test) while all other pairs of ideals are not isomorphic. We now consider the images of P_1, P_2, P_3, P_4 under the given representation on $\mathbb{C}^2 \otimes \mathbb{C}^2 \otimes \mathbb{C}^2$. This gives $\pi(P_4) = 0$ and

$$\pi(P_1) = \frac{1}{3}\begin{pmatrix} 3 & 0 & 0 & 0 & 0 & 0 & 0 & 0 \\ 0 & 1 & 1 & 0 & 1 & 0 & 0 & 0 \\ 0 & 1 & 1 & 0 & 1 & 0 & 0 & 0 \\ 0 & 0 & 0 & 1 & 0 & 1 & 1 & 0 \\ 0 & 1 & 1 & 0 & 1 & 0 & 0 & 0 \\ 0 & 0 & 0 & 1 & 0 & 1 & 1 & 0 \\ 0 & 0 & 0 & 1 & 0 & 1 & 1 & 0 \\ 0 & 0 & 0 & 0 & 0 & 0 & 0 & 3 \end{pmatrix},$$

$$\pi(P_2) = \frac{1}{3}\begin{pmatrix} 0 & 0 & 0 & 0 & 0 & 0 & 0 & 0 \\ 0 & 2 & -1 & 0 & -1 & 0 & 0 & 0 \\ 0 & -1 & \frac{1}{2} & 0 & \frac{1}{2} & 0 & 0 & 0 \\ 0 & 0 & 0 & \frac{1}{2} & 0 & \frac{1}{2} & -1 & 0 \\ 0 & -1 & \frac{1}{2} & 0 & \frac{1}{2} & 0 & 0 & 0 \\ 0 & 0 & 0 & \frac{1}{2} & 0 & \frac{1}{2} & -1 & 0 \\ 0 & 0 & 0 & -1 & 0 & -1 & 2 & 0 \\ 0 & 0 & 0 & 0 & 0 & 0 & 0 & 0 \end{pmatrix},$$

$$\pi(P_3) = \frac{1}{2} \begin{pmatrix} 0 & 0 & 0 & 0 & 0 & 0 & 0 & 0 \\ 0 & 0 & 0 & 0 & 0 & 0 & 0 & 0 \\ 0 & 0 & 1 & 0 & -1 & 0 & 0 & 0 \\ 0 & 0 & 0 & 1 & 0 & -1 & 0 & 0 \\ 0 & 0 & -1 & 0 & 1 & 0 & 0 & 0 \\ 0 & 0 & 0 & -1 & 0 & 1 & 0 & 0 \\ 0 & 0 & 0 & 0 & 0 & 0 & 0 & 0 \\ 0 & 0 & 0 & 0 & 0 & 0 & 0 & 0 \end{pmatrix}.$$

Using these, we get that the dimensions of $\pi(P_1)V$ is $\mathbf{rank}(\pi(P_1)) = 4$, the dimensions of $\pi(P_2)V$ and $\pi(P_3)V$ are $\mathbf{rank}(\pi(P_2)) = \mathbf{rank}(\pi(P_3)) = 2$. More specifically, from $\pi(P_1)$, we can get an orthonormal basis of $\pi(P_1)V$ to be $\{\vec{b}_1, \vec{b}_2, \vec{b}_3, \vec{b}_4\}$ with $\vec{b}_1 := [1,0,0,0,0,0,0,0]^T$, $\vec{b}_2 := \frac{1}{\sqrt{3}}[0,1,1,0,1,0,0,0]^T$, $\vec{b}_3 := \frac{1}{\sqrt{3}}[0,0,0,1,0,1,1,0]^T$, $\vec{b}_4 = [0,0,0,0,0,0,0,1]^T$, and, from $\pi(P_2)$, an orthonormal basis of $\pi(P_2)V$ to be $\{\vec{g}_1, \vec{g}_2\}$, with $\vec{g}_1 := \frac{1}{\sqrt{6}}[0,2,-1,0,-1,0,0,0]^T$, $\vec{g}_2 := \frac{1}{\sqrt{6}}[0,0,0,1,0,1,-2,0]^T$. In order to obtain the appropriate orthonormal basis for $\pi(P_3)V$, following Proposition 4.3.5, we choose an $r \in S_3$ so that $\pi(P_3 r P_2) = \pi(P_3)\pi(r)\pi(P_2) \neq 0$ and choose an orthonormal basis by normalizing $\{\pi(P_3 r P_2)\vec{e}_1, \pi(P_3 r P_2)\vec{e}_2\}$. A possible choice for r is $r = (13)$. With this choice we obtain the basis of $\pi(P_3)V$ $\{\vec{f}_1, \vec{f}_2\}$ given by $\vec{f}_1 = \frac{1}{\sqrt{2}}[0,0,-1,0,1,0,0,0]^T$, $\vec{f}_2 = \frac{1}{\sqrt{2}}[0,0,0,-1,0,1,0,0]^T$. Details are left to the reader (cf. Exercise 4.9). The dynamical Lie algebra $\mathcal{L}$ of this quantum system is a subalgebra of $u^G(8)$, with $G = S_3$. In the above basis, every element of $\mathcal{L}$ takes a block diagonal form with one block of dimension 4×4 and two equal blocks of dimension 2×2.

4.3.3 Subspace controllability

In the presence of a group of symmetries G, the form (4.19) of an element F, in the dynamical Lie algebra $\mathcal{L}$, suggests that the state space V of the quantum systems splits into invariant subspaces given by the nonzero $\pi(P_j)V$. We remark that such a decomposition is not unique since the GYS's are not unique (cf. Exercise 4.11). A natural question to ask is whether the submatrices matrices $A, B, ..., C$ can attain all the possible values in $su(n_d)$ where n_d is the dimension of the block at hand. In this case, every special unitary operation will be possible on the single subspace. This property is called **subspace controllability**. We illustrate this with an example.

Example 4.3.9 (Example 4.3.8 continued)

Consider a network of 3 identical spin $\frac{1}{2}$ particles subject to a simultaneous control electromagnetic field, $u_{x,y,z}$, in the x, y and z direction. The Hamiltonian can be written (after scaling) as $H = H_0 + H_x u_x + H_y u_y + H_z u_z$, where the Hamiltonian H_0 modeling the interaction may be given, assuming Ising interaction, as

$$H_0 = \sigma_z \otimes \sigma_z \otimes 1 + \sigma_z \otimes 1 \otimes \sigma_z + 1 \otimes \sigma_z \otimes \sigma_z.$$

Here, $\sigma_{x,y,z}$ are the Pauli matrices defined in (1.20). The operators $H_{x,y,z}$ are defined as

$$H_j = \sigma_j \otimes 1 \otimes 1 + 1 \otimes \sigma_j \otimes 1 + 1 \otimes 1 \otimes \sigma_j, \qquad j = x, y, z.$$

Both the Hamiltonian H_0 and the Hamiltonians $H_{x,y,z}$ are invariant under the action of the permutation group S_3, an action which is defined by conjugation $H \to gHg^{-1}$, $g \in S_3$ in the given representation. Therefore, the dynamical Lie algebra generated by $iH_0, iH_{x,y,z}$ must be a subalgebra of $u^{S_3}(8)$ and, in the appropriate coordinates, it takes the form of the matrices in (4.19). The change of coordinates was described in Example 4.3.8. It gives a single block A of dimension 4×4 and two blocks B of dimensions 2×2. Let us obtain the expressions of H_0 and $H_{x,y,z}$ in the basis described in Example 4.3.8. It is convenient to write such a basis in terms of tensor products of the basis vectors in $\mathbb{C}^2$ in the Dirac notation $|0\rangle, |1\rangle$. The basis of $\pi(P_1)V$ is $\{\vec{b}_1, \vec{b}_2, \vec{b}_3, \vec{b}_4\}$, with $\vec{b}_1 = |000\rangle$, $\vec{b}_2 = \frac{1}{\sqrt{3}}(|001\rangle + |010\rangle + |100\rangle)$, $\vec{b}_3 = \frac{1}{\sqrt{3}}(|011\rangle + |101\rangle + |110\rangle)$, $\vec{b}_4 = |111\rangle$. The basis of $\pi(P_2)V$, $\{\vec{g}_1, \vec{g}_2\}$ is $\vec{g}_1 = \frac{1}{\sqrt{6}}(2|001\rangle - |010\rangle - |100\rangle)$, $\vec{g}_2 = \frac{1}{\sqrt{6}}(|011\rangle + |101\rangle - 2|110\rangle)$. The basis of $\pi(P_3)V$ is $\{\vec{f}_1, \vec{f}_2\}$, with $\vec{f}_1 = \frac{1}{\sqrt{2}}(-|010\rangle + |100\rangle)$, $\vec{f}_2 = \frac{1}{\sqrt{2}}(-|011\rangle + |101\rangle)$. Using the action of $\sigma_{x,y,z}$ on $|0\rangle$ and $|1\rangle$, i.e., $\sigma_z|0\rangle = |0\rangle$, $\sigma_z|1\rangle = -|1\rangle$, $\sigma_x|0\rangle = |1\rangle$, and so on, we obtain $H_0\vec{b}_1 = 3\vec{b}_1$, $H_0\vec{b}_2 = -\vec{b}_2$, and so on. Therefore, the expression of H_0 in the given coordinates is $H_0 = \mathtt{diag}(3, -1, -1, 3, -1, -1, -1, -1)$. Analogously we obtain the expressions for $H_{x,y,z}$. In fact, in order to calculate the dynamical Lie algebra in the given coordinates, it is enough to compute the expressions for H_x and H_z since (up to a nonzero proportionality factor) the commutator of iH_x and iH_z gives iH_y. We verify $H_x\vec{b}_1 = \sqrt{3}\vec{b}_2$, $H_x\vec{b}_2 = \sqrt{3}\vec{b}_1 + 2\vec{b}_3$, $H_x\vec{b}_3 = \sqrt{3}\vec{b}_4 + 2\vec{b}_2$, $H_x\vec{b}_4 = \sqrt{3}\vec{b}_3$, $H_x\vec{g}_1 = \vec{g}_2$, $H_x\vec{g}_2 = \vec{g}_1$, $H_x\vec{f}_1 = \vec{f}_2$, $H_x\vec{f}_2 - \vec{f}_1$ and analogously for H_z. The resulting expressions for H_z and H_x are: $H_z = \mathtt{diag}(3, 1, -1, -3, 1, -1, 1, -1)$ and

$$H_x = \left(\begin{array}{cccc|cc|cc} 0 & \sqrt{3} & 0 & 0 & & & & \\ \sqrt{3} & 0 & 2 & 0 & & & & \\ 0 & 2 & 0 & \sqrt{3} & & & & \\ 0 & 0 & \sqrt{3} & 0 & & & & \\ \hline & & & & 0 & 1 & & \\ & & & & 1 & 0 & & \\ \hline & & & & & & 0 & 1 \\ & & & & & & 1 & 0 \end{array} \right).$$

In these coordinates the matrices iH_0, iH_z, iH_x have the block diagonal form which highlights the invariant subspaces. By calculating the dynamical Lie algebra on each invariant subspace, for example using the algorithm of sub-section 3.2.2 of the previous chapter, we see that each dynamical Lie algebra is equal to $su(n)$ for the appropriate n. This system is therefore subspace controllable.

4.3.4 Decomposition without knowing the generalized Young Symmetrizers

We have seen so far that, in appropriate coordinates, a matrix F in the Lie algebra $u^G(n)$ takes a block diagonal form (4.19). Therefore, in the same coordinates, the elements of each dynamical Lie algebra $\mathcal{L}$ which has G as symmetry group, i.e., $\mathcal{L} \subseteq u^G(n)$, also have the form (4.19). Finding this natural basis is relatively easy, once we have a set of GYS's for the group G. We have not, however, explained *how* to find such GYS's. For example, we have not explained how we found the GYS's in Example 4.3.8 but only invited the reader to verify that they satisfy the defining conditions (4.12), (4.13), (4.14) and (4.15). Explicit methods to find GYS's are known only for some groups, most notably, for the symmetric group S_n, for which the method is based on the use of *Young tableaux*[111]. To obtain the GYS's in the Example 4.3.8, we used in fact the method of [18] and [158] which modified the classical procedure to obtain operators that are also Hermitian. We shall not describe here methods to obtain Young symmetrizers which can be found in several detailed references on group theory in physics such as [18], [111], [158], [273] and references therein. However, we shall devote this subsection to explain how we can *circumvent* such an explicit calculation and still obtain the desired change of coordinates to put $u^G(n)$ in a block diagonal form.

A basis for $u^G(n)$ can be obtained from a basis $u(n)$ by imposing the condition $[u(n), G] = 0$. We can then consider the general space of matrices commuting with $u^G(n)$. We call this space the *commutant* and denote it by $\mathcal{C}$. The space $\mathcal{C}$ includes, as special cases, the representation of G itself and the matrices in $u(n)$ which commute with $u^G(n)$. Assume we are in the coordinates where any $F \in u^G(n)$ can be written as (4.19). Consider a general matrix $H \in \mathcal{C}$ and partition it according to the partition in (4.19) in blocks $\{H_{j,k}\}$. By imposing commutation with matrices F in (4.19) where A, B,...,or C is a nonzero multiple of the identity, we have that only the blocks corresponding to the big diagonal blocks of dimensions $m_A \times d_A$, $m_B \times d_B$,..., $m_C \times d_C$, may be different from zero. Let us now consider the big block corresponding to the submatrices A in (4.19). This block has dimension $m_A \times d_A$. For every $j, k = 1, ..., m_A$, we have $H_{j,k}A = AH_{j,k}$. Since A can take any value in $u(n_{d_A})$ and it is an irreducible representation of $u(n_{d_A})$, it follows from Schur lemma, Theorem 4.2.4, that $H_{j,k}$ can only be a multiple of the identity, i.e., $H_{j,k} = h_{j,k} 1_{d_A}$, for a scalar $h_{j,k}$. Collecting all such scalars in a matrix Λ of dimension $m_A \times m_A$, we have that the big sub-block corresponding to the A matrices has the form $\Lambda \otimes 1_{d_A}$. Repeating this for the other blocks we obtain that every matrix H in the commutant $\mathcal{C}$ has the form

$$
H = \left(\begin{array}{cccc}
\Lambda_{m_A} \otimes 1_{d_A} & & & \\
\hline
& \Upsilon_{m_B} \otimes 1_{d_B} & & \\
\hline
& & \ddots & \\
\hline
& & & \Gamma_{m_C} \otimes 1_{d_C}
\end{array}\right) \tag{4.20}
$$

for matrices Λ_{m_A}, Υ_{m_B}, ..., Γ_{m_C}, of dimensions $m_A \times m_A$, $m_B \times m_B$, ..., $m_C \times m_C$.

Let us focus on the case where the elements H in (4.20) are in $\mathcal{C} \cap u(n)$. In particular, given the fact that the matrices are written in an orthonormal basis, the matrices Λ_{m_A}, Υ_{m_B},...,Γ_{m_C} are arbitrary matrices in $u(m_A)$, $u(m_B)$,..., $u(m_C)$. Since $\mathcal{C} \cap u(n)$ is a subalgebra of $u(n)$, it is reductive (Theorem 3.1.8). It has a maximal Abelian (Cartan) subalgebra which we can calculate putting together a basis for the Cartan subalgebra of the semisimple part and a basis for its center. By inspection of (4.20), a basis of a maximal Abelian subalgebra for $\mathcal{C} \cap u(n)$ is obtained by taking in (4.20) $\Upsilon_{m_B} = 0$, ..., $\Gamma_{m_C} = 0$ and $\Lambda_{m_A} = \text{diag}(i, 0, ..., 0)$, then $\Lambda_{m_A} = \text{diag}(0, i, ..., 0)$, ..., until $\Lambda_{m_A} = \text{diag}(0, 0, ..., i)$, and then, $\Lambda_{m_A} = 0$, ..., $\Gamma_{m_C} = 0$ and $\Upsilon_{m_B} = \text{diag}(i, 0, ..., 0)$, then $\Upsilon_{m_B} = \text{diag}(0, i, ..., 0)$,..., until $\Upsilon_{m_B} = \text{diag}(0, 0, ..., i)$, and so on until $\Lambda_{m_A} = 0$, $\Upsilon_{m_B} = 0$,..., and $\Gamma_{m_C} = \text{diag}(i, 0, ..., 0)$, then $\Gamma_{m_C} = \text{diag}(0, i, ..., 0)$,..., until $\Gamma_{m_C} = \text{diag}(0, 0, ..., i)$. The CSA has dimension $m_A + m_B + \cdots + m_C$ and the given basis has a block diagonal form with identities of dimension d_A, d_B,...,d_C on the diagonal.

The crucial observation now is that these blocks identify the subspaces $\pi(P_j)V$ we have used to obtain formula (4.19). Orthogonal projections onto such subspaces are, in fact, the GYS's. The change of basis that puts the CSA of $\mathcal{C} \cap u(n)$ in the above form also puts $u^G(n)$ in the block diagonal form (4.19). We have seen in Algorithm I of section 4.1.1 how to find a maximal Abelian subalgebra for any semisimple (and therefore for every reductive) Lie subalgebra of $u(n)$ *in any coordinates*. This suggests an algorithm to find the bases of the spaces used in the expression (4.19) for matrices $F \in u^G(n)$.

Algorithm II

1. From a basis of $u(n)$, using the condition $[G, u(n)] = 0$, find a basis of $u^G(n)$.

2. From the condition $[\mathcal{C}, u^G(n)] = 0$, find a basis for $\mathcal{C} \cap u(n)$. Notice that $\mathcal{C} \cap u(n)$ being a subalgebra of $u(n)$, is reductive (Theorem 3.1.8). Thus it admits a maximal Abelian subalgebra obtained by putting together a basis of its center and a CSA of its semisimple part.

3. Find the center of $\mathcal{C} \cap u(n)$ and a basis for it, $\mathcal{B}_Z$.

4. Find the semisimple part of $\mathcal{C} \cap u(n)$ (section 4.1) and a basis $\mathcal{B}_A$ of its CSA using Algorithm I of subsection 4.1.1.

5. Consider the basis $\{\mathcal{B}_Z, \mathcal{B}_A\}$ of the maximal Abelian subalgebra of $\mathcal{C} \cap u(n)$ and simultaneously diagonalize these matrices.

6. Perform a Gaussian elimination procedure to a reduced row echelon form to obtain a basis for the CSA of the form $\text{diag}(i1_{d_A}, 0, ..., 0)$, $\text{diag}(0, i1_{d_A}, 0, ..., 0)$, ..., $\text{diag}(0, ..., i1_{d_B}, 0, ..., 0)$, ..., $\text{diag}(0, ..., 0, i1_{d_C})$ (see discussion below). This corresponds to taking linear combinations of the diagonalized matrices in $\{\mathcal{B}_Z, \mathcal{B}_A\}$. Taking the same linear combinations for the matrices $\{\mathcal{B}_Z, \mathcal{B}_A\}$ in the original coordinates we find a basis of the Cartan subalgebra of elements which have eigenvalues i and 0 with the desired multiplicity.

7. The orthonormal eigenvectors corresponding to the eigenvalues i of these matrices give the change of coordinates to put the matrices $u^G(n)$ in the block diagonal form (4.18).

Step 6 of the algorithm requires some extra explanation and justification. We start with a set of diagonal matrices giving a basis of the CSA of $\mathcal{C} \cap u(n)$, in new coordinates, and want to find a basis of the same space of the indicated form. Let us represent diagonal matrices with row vectors where we list their diagonal elements. Since we want to find a basis corresponding to the indicated form, we must have vectors of the type $i(1\,1\,\cdots\,1\,0\cdots 0)$, $i(0\,0\,\cdots\,0\,1\,\cdots\,1\,0\cdot 0)$ and so on up to $i(0\,\cdots\,0\,\cdots\,0\,1\cdots 1)$ where the ones are taken d_A times, then d_A times and so on, then d_B times and so on, up to d_C times. We want to obtain these by taking linear combinations of the original diagonalized matrices in the CSA of $\mathcal{C} \cap u(n)$. We notice that (omitting the un-important i factor) the row vectors thus found would form the (unique) reduced row echelon form of the original matrix of row vectors. We therefore perform a row reduction algorithm (see, e.g., [176]) to obtain the reduced row echelon form starting from the original matrix (omitting the i factor). At this point, however, there is no reason to believe that the reduced row echelon form of the initial matrix has the desired form and therefore corresponds to the CSA obtained from the general form of elements in $\mathcal{C} \cap u(n)$ in (4.20). We show that this is the case. To this goal, recall that all CSA are conjugate to each other.[7] Therefore, there exists a unitary matrix transforming the span of the basis obtained after the row reduction algorithm to the span of the basis described after formula (4.20). We claim that this unitary matrix must be the

---

[7]The conjugacy of CSA's is a standard question in Lie algebra theory (see, e.g., [168]) and it is valid for complex Lie algebras and not always valid for real ones (like the ones we are considering here). In our case of subalgebras of $u(n)$ it is however valid. Consider two CSA's $\mathcal{A}_1$ and $\mathcal{A}_2$. Since the elements of $\mathcal{A}_1$ can be simultaneously diagonalized, there exists a unitary U such that $U\mathcal{A}_1 U^\dagger$ are all diagonal. Analogously there exists a unitary F such that $F\mathcal{A}_2 F^\dagger$ are all diagonal. Pick an $A \in \mathcal{A}_2$. Since $FAF^\dagger$ commutes with $U\mathcal{A}_1 U^\dagger$, we have that $U^\dagger FAF^\dagger U$ commutes with $\mathcal{A}_1$. However since $\mathcal{A}_1$ is *maximal* Abelian, this implies $U^\dagger FAF^\dagger U \in \mathcal{A}_1$, and since A is general, $U^\dagger F\mathcal{A}_2 F^\dagger U \subseteq \mathcal{A}_1$. Analogously we obtain $F^\dagger U\mathcal{A}_1 U^\dagger F \subseteq \mathcal{A}_2$. This shows $U^\dagger F\mathcal{A}_2 F^\dagger U = \mathcal{A}_1$.

identity and the two bases coincide. First of all, the two sets have the same number of elements which is $m_A + m_B + \cdots + m_C$. Furthermore, conjugacy (by a unitary matrix) does not modify the set of possible eigenvalues of the linear combination of the two sets of matrices nor their multiplicities. Consider now the matrix in reduced row echelon form obtained with the above algorithm (omitting like usual the i coefficient). It has the form

$$
\begin{pmatrix}
1 & a_1 & \cdots & a_{r_1} & 0 & c_1 & \cdots & c_{r_2} & 0 & \cdots & 0 & d_1 & \cdots & d_{r_3} \\
0 & 0 & \cdots & 0 & 1 & b_1 & \cdots & b_{r_2} & 0 & \cdots & 0 & e_1 & \cdots & e_{r_3} \\
\cdots & \cdot & \cdots & 0 & 0 & \cdot & \cdots & \cdots & 0 & \cdots & 0 & \cdot & \cdots & \cdot \\
\cdots & \cdot & \cdots & \cdot & \cdot & \cdot & \cdots & \cdots & \cdot & \cdots & \cdot & \cdot & \cdots & \cdot \\
\cdots & \cdot & \cdots & \cdot & \cdot & \cdot & \cdots & \cdots & \cdot & \cdots & \cdot & \cdot & \cdots & \cdot \\
\cdots & \cdot & \cdots & \cdot & 0 & \cdot & \cdots & \cdots & \cdot & \cdots & 1 & f_1 & \cdots & f_{r_3}
\end{pmatrix},
$$

of a matrix of dimensions $(m_A + m_B + \cdots + m_C) \times n$. The numbers $a_1, ..., a_{r_1}$ must be equal to 1. Otherwise we would have a number of possible different eigenvalues larger than $m_A + m_B + \cdots + m_C$. For the same reason, the numbers $b_1, ..., b_{r_2},...,f_1, ..., f_{r_3}$ are also equal to 1. In order to avoid a number of different eigenvalues larger than $m_A + m_B + \cdots + m_C$, we also must have $c_1, ..., c_{r_2}$ all equal to zero as well as $d_1, ..., d_{r_3}$ and $e_1, ..., e_{r_3}$. Therefore, the reduced row echelon form is actually

$$
\begin{pmatrix}
1 & 1 & \cdots & 1 & 0 & 0 & \cdots & 0 & 0 & \cdots & 0 & 0 & \cdots & 0 \\
0 & 0 & \cdots & 0 & 1 & 1 & \cdots & 1 & 0 & \cdots & 0 & 0 & \cdots & 0 \\
\cdots & \cdot & \cdots & 0 & 0 & \cdot & \cdots & \cdots & 0 & \cdots & 0 & \cdot & \cdots & \cdot \\
\cdots & \cdot & \cdots & \cdot & \cdot & \cdot & \cdots & \cdots & \cdot & \cdots & \cdot & \cdot & \cdots & \cdot \\
\cdots & \cdot & \cdots & \cdot & \cdot & \cdot & \cdots & \cdots & \cdot & \cdots & \cdot & \cdot & \cdots & \cdot \\
\cdots & \cdot & \cdots & \cdot & 0 & \cdot & \cdots & \cdots & \cdot & \cdots & 1 & 1 & \cdots & 1
\end{pmatrix}.
$$

The number of 1's in each row indicates the possible multiplicities of the eigenvalues. Since these also have to coincide, we have that the two sets are the same. Example 4.3.12 below illustrates the full Algorithm II.

Remark 4.3.10 At the end of the algorithm, we obtain a block diagonal form of the elements of $u^G(n)$. The algorithm does not say, however, which blocks can be taken equal as in (4.19). Each block corresponds to an irreducible representation of $u^G(n)$.[8] Therefore, we have a map from $u^G(n)$ to each block. Blocks with the same dimension might correspond to isomorphic representations. We can check isomorphism directly. Let $A = A(p)$ be the function which gives the block A as a function of the parameters p of the matrices in $u^G(n)$, and let $Q = Q(p)$ the analogous function for the

---

[8]Notice if it was reducible according to Corollary 4.2.5 there would exist a non trivial (one that is not a multiple of the identity) block commuting with all the elements in the block. But this (together with all the other diagonal blocks being zero) would give another linearly independent element in the Cartan subalgebra which contradicts the fact that the Cartan subalgebra is a *maximal* Abelian subalgebra.

block Q. If they are isomorphic, there exists a unitary matrix U such that $A(p) = UQ(p)U^\dagger$ for every value of p. Once such a matrix U is known, it gives one more change of coordinates to put the matrices in $u^G(n)$ exactly in the format of F in (4.19), i.e., with all the blocks corresponding to isomorphic irreducible representations equal to each-other.

Remark 4.3.11 The calculation of a basis of $u^G(n)$ at the beginning of Algorithm II can be simplified in some cases when $n = 2^N$ (for example, for the case of a spin network of N spin $\frac{1}{2}$ particles considered in Example 4.3.9). In this case, we can consider a basis of $u(2^N)$ given by matrices of the form $i\sigma_1 \otimes \sigma_2 \otimes \cdots \otimes \sigma_N$, where σ_j is equal to the identity $\mathbf{1}_2$ or one of the Pauli matrices (1.20). Then assume that G acts on the 'words' $i\sigma_1 \otimes \sigma_2 \otimes \cdots \otimes \sigma_N$ by changing word into word like for example for the case of a subgroup of the symmetric group. Then each orbit of this action corresponds to a basis element of $u^G(2^N)$ given by the sum of all elements in the orbit (cf. Proposition 5.2 in [84]).

Example 4.3.12 Reconsider the case $G = S_3$ of Example 4.1.4. To calculate a basis for $u^G(8)$ we can use the method in Remark 4.3.11. Consider, the set of words $\sigma_1 \otimes \sigma_2 \otimes \sigma_3$ with $\sigma_{1,2,3}$ equal to the identity matrix or one of the Pauli matrices, $\sigma_{x,y,z}$, in (1.20), which form a basis of $iu(8)$. The *orbits* of S_3 on these sets of 'words' are given by words which have an equal number of occurrences of $\mathbf{1}$'s and $\sigma_{x,y,z}$'s. They can be labeled by triple (k_x, k_y, k_z) which indicate the occurrence of σ_x, σ_y and σ_z respectively with $0 \leq k_x + k_y + k_z \leq 3$. There are 20 of such triples and therefore the dimension of $u^{S_3}(8)$ is 20.[9] Denote by X_{k_x,k_y,k_z} the sum of all elements with $k_x \, \sigma_x$'s, $k_y \, \sigma_y$'s, $k_z \, \sigma_z$'s. They form a basis of $u^{S_3}(8)$. From the relation $[\mathcal{C}, u^{S_3}(8)] = 0$ we obtain a basis of the commutant $\mathcal{C} \cap u(8)$, which is given by E_1, E_2, E_3, E_4, E_5, with $iE_1 := \mathbf{1} \otimes \mathbf{1} \otimes \mathbf{1}$, $iE_2 := \sigma_x \otimes \mathbf{1} \otimes \sigma_x + \sigma_y \otimes \mathbf{1} \otimes \sigma_y + \sigma_z \otimes \mathbf{1} \otimes \sigma_z$, $iE_3 = \sigma_x \otimes \sigma_x \otimes \mathbf{1} + \sigma_y \otimes \sigma_y \otimes \mathbf{1} + \sigma_z \otimes \sigma_z \otimes \mathbf{1}$, $iE_4 = \mathbf{1} \otimes \sigma_x \otimes \sigma_x + \mathbf{1} \otimes \sigma_y \otimes \sigma_y + \mathbf{1} \otimes \sigma_z \otimes \sigma_z$, $iE_5 = \sigma_x \otimes (\sigma_y \otimes \sigma_z - \sigma_z \otimes \sigma_y) + \sigma_y \otimes (\sigma_z \otimes \sigma_x - \sigma_x \otimes \sigma_z) + \sigma_z \otimes (\sigma_x \otimes \sigma_y - \sigma_y \otimes \sigma_x)$. The following table gives the commutation relations between the elements $\{E_2, E_3, E_4, E_5\}$

$[\cdot, \cdot]$	E_2	E_3	E_4	E_5
E_2	0	$2E_5$	$-2E_5$	$4(E_4 - E_3)$
E_3	$-2E_5$	0	$2E_5$	$4(E_2 - E_4)$
E_4	$2E_5$	$-2E_5$	0	$4(E_3 - E_2)$
E_5	$4(E_3 - E_4)$	$4(E_4 - E_2)$	$4(E_2 - E_3)$	0

From this table, it follows that $\mathcal{E} = \{E_1, E_2 + E_3 + E_4, E_5\}$ is a basis for a maximal Abelian subalgebra of the commutant $\mathcal{C} \cap u(8)$ (such a Lie algbera

---

[9]The dimension of $u^{S_n}(2^n)$ in general is calculated in [12] to be $\binom{n+3}{n}$.

is Abelian and equal to its normalizer). Another CSA is spanned by the matrices $i\pi(P_1)$, $i\pi(P_2)$, $i\pi(P_3)$ found in Example 4.3.8 which give rise to the decomposition of the form (4.19). Let us focus on the basis $\mathcal{E} = \{E_1, E_2 + E_3 + E_4, E_5\}$. The matrices of $\mathcal{E}$ in the computational basis are given by $E_1 = i\mathbf{1}_8$,

$$
E_2 + E_3 + E_4 = -i \begin{pmatrix}
3 & 0 & 0 & 0 & 0 & 0 & 0 & 0 \\
0 & -1 & 2 & 0 & 2 & 0 & 0 & 0 \\
0 & 2 & -1 & 0 & 2 & 0 & 0 & 0 \\
0 & 0 & 0 & -1 & 0 & 2 & 2 & 0 \\
0 & 2 & 2 & 0 & -1 & 0 & 0 & 0 \\
0 & 0 & 0 & 2 & 0 & -1 & 2 & 0 \\
0 & 0 & 0 & 2 & 0 & 2 & -1 & 0 \\
0 & 0 & 0 & 0 & 0 & 0 & 0 & 3
\end{pmatrix},
$$

$$
E_5 = \begin{pmatrix}
0 & 0 & 0 & 0 & 0 & 0 & 0 & 0 \\
0 & 0 & -2 & 0 & 2 & 0 & 0 & 0 \\
0 & 2 & 0 & 0 & -2 & 0 & 0 & 0 \\
0 & 0 & 0 & 0 & 0 & 2 & -2 & 0 \\
0 & -2 & 2 & 0 & 0 & 0 & 0 & 0 \\
0 & 0 & 0 & -2 & 0 & 0 & 2 & 0 \\
0 & 0 & 0 & 2 & 0 & -2 & 0 & 0 \\
0 & 0 & 0 & 0 & 0 & 0 & 0 & 0
\end{pmatrix}.
$$

Let us find the eigenvalues and eigenvectors of $F_2 := i(E_2 + E_3 + E_4)$ and $F_3 := iE_5$ (set F_1 to be the 8×8 identity).[10] The matrix F_2 has eigenvalue 3 with eigenspace spanned by the orthogonal vectors

$$
\vec{v}_1 = \begin{pmatrix} 1 \\ 0 \\ 0 \\ 0 \\ 0 \\ 0 \\ 0 \\ 0 \end{pmatrix}, \quad
\vec{v}_2 = \begin{pmatrix} 0 \\ 0 \\ 0 \\ 0 \\ 0 \\ 0 \\ 0 \\ 1 \end{pmatrix}, \quad
\vec{v}_3 = \begin{pmatrix} 0 \\ 0 \\ 0 \\ 1 \\ 1 \\ 0 \\ 1 \\ 0 \end{pmatrix}, \quad
\vec{v}_4 = \begin{pmatrix} 0 \\ 0 \\ 0 \\ 1 \\ 0 \\ 1 \\ 1 \\ 0 \end{pmatrix},
$$

and eigenvalue -3 with eigenspace spanned by the orthogonal vectors

$$
\vec{w}_1 = \begin{pmatrix} 0 \\ -1 \\ 1 \\ 0 \\ 0 \\ 0 \\ 0 \\ 0 \end{pmatrix}, \quad
\vec{w}_2 = \begin{pmatrix} 0 \\ -1 \\ -1 \\ 0 \\ 2 \\ 0 \\ 0 \\ 0 \end{pmatrix}, \quad
\vec{w}_3 = \begin{pmatrix} 0 \\ 0 \\ 0 \\ -1 \\ 0 \\ 1 \\ 0 \\ 0 \end{pmatrix}, \quad
\vec{w}_4 = \begin{pmatrix} 0 \\ 0 \\ 0 \\ -1 \\ 0 \\ -1 \\ 2 \\ 0 \end{pmatrix}.
$$

---

[10] We made the matrices Hermitian so as to have real eigenvalues.

The matrix F_3 has the eigenvalue 0, with eigenspace spanned by $\{\vec{v}_1, \vec{v}_2, \vec{v}_3, \vec{v}_4\}$ above, the eigenvalue $\sqrt{12}$ with eigenspace spanned by $\mathcal{B}_1 := \{-i\sqrt{3}\vec{w}_1 + \vec{w}_2, i\sqrt{3}\vec{w}_3 + \vec{w}_4\}$ and the eigenvalue $-\sqrt{12}$ with eigenspace spanned by $\mathcal{B}_2 = \{i\sqrt{3}\vec{w}_1 + \vec{w}_2, -\sqrt{3}i\vec{w}_3 + \vec{w}_4\}$. Therefore, in the basis obtained by putting together $\{\vec{v}_1, \vec{v}_2, \vec{v}_3, \vec{v}_4\}$ and $\mathcal{B}_1$ and $\mathcal{B}_2$, the matrix corresponding to F_2 is given by $\mathtt{diag}(3, 3, 3, 3, -3, -3, -3, -3)$ and the matrix corresponding to F_3 is $\mathtt{diag}(0, 0, 0, 0, \sqrt{12}, \sqrt{12}, -\sqrt{12}, -\sqrt{12})$. We now carry out the Gaussian row reduction algorithm of Step 6 starting from the matrix

$$M = \begin{pmatrix} 1 & 1 & 1 & 1 & 1 & 1 & 1 & 1 \\ 3 & 3 & 3 & 3 & -3 & -3 & -3 & -3 \\ 0 & 0 & 0 & 0 & \sqrt{12} & \sqrt{12} & -\sqrt{12} & -\sqrt{12} \end{pmatrix},$$

the first row representing the identity F_1 and the second and third rows representing F_2 and F_3, respectively. The associated reduced row echelon form is

$$rref(M) = \begin{pmatrix} 1 & 1 & 1 & 1 & 0 & 0 & 0 & 0 \\ 0 & 0 & 0 & 0 & 1 & 1 & 0 & 0 \\ 0 & 0 & 0 & 0 & 0 & 0 & 1 & 1 \end{pmatrix},$$

with $EM = rref(M)$ where $E = \begin{pmatrix} \frac{1}{2} & \frac{1}{6} & 0 \\ \frac{1}{4} & -\frac{1}{12} & \frac{\sqrt{12}}{24} \\ \frac{1}{4} & -\frac{1}{12} & -\frac{\sqrt{12}}{24} \end{pmatrix}$. From this we find that the spaces giving the block diagonal form for $u^G(8)$ are the eigenspaces corresponding to the eigenvalue $+1$ for the matrices $\frac{1}{2}F_1 + \frac{1}{6}F_2$, $\frac{1}{4}F_1 - \frac{1}{12}F_2 + \frac{\sqrt{12}}{24}F_3$, $\frac{1}{4}F_1 - \frac{1}{12}F_2 - \frac{\sqrt{12}}{24}F_3$, respectively.

4.4 Notes and References

We have given several methods to decompose the dynamics of an uncontrollable quantum system. The main references we have followed are [80], [84], [92], [217], which deal with the general methodologies to decompose Lie algebras and therefore dynamics of quantum systems. Related papers are [304] and [305]. There have recently been several papers dealing with the Lie algebra decomposition and subspace controllability for the case of certain systems of applied interest such as spin networks [13], [23], [61], [285], [286]. An important ingredient of the techniques developed in this chapter is the interplay between (matrix type) methods and algorithms for Lie algebras such as the ones described in section 4.1 and group theoretic methods such as the ones described in section 4.3. It is interesting to note how, in some cases, both points of view can be used leading to the same solution of a given problem. The treatment of subsection 4.3.4 is an example of both these points of view leading to the same result.

4.5   Exercises

Exercise 4.1 Prove that the adjoint representation is, in fact, a representation.

Exercise 4.2 Show that the ideal generated by an element of the primary decomposition $\mathcal{V}$ as in (4.7) is a simple Lie algebra.

Exercise 4.3 Give the proof of Proposition 4.1.1 and observe that it can be used to find $\mathcal{S}$ from $\mathcal{Z}$ and $\mathcal{L}$ in the decomposition (4.2) and to show that all simple ideals are orthogonal to each other.

Exercise 4.4 Prove that with the given inner product on $su(n)$, $\langle A, B \rangle = Tr(AB^\dagger)$, $ad_X^T = -ad_X$ for any X in a semisimple Lie algebra $\mathcal{S} \subseteq su(n)$. Recall the definition of ad_X^T, $\langle ad_X^T A, B \rangle := \langle A, ad_X B \rangle$, for every $A, B \in \mathcal{S}$. *Hint:* Use Cartan's semisimplicity criterion in Appendix C. Show that in an orthonormal basis of $\mathcal{S}$, ad_X is a skew-symmetric matrix, which has therefore only purely imaginary eigenvalues, which if they are nonzero, come in pair $\pm ia$.

Exercise 4.5 Consider the primary decomposition of $\mathcal{S}$ (4.5). Show that the two-dimensional spaces $\mathcal{V}_j$ are invariant under Lie bracket with any element of the Cartan subalgebra $\mathcal{A}$. Observe that this can be used to slightly simplify the calculation of the ideals in (4.7).

Exercise 4.6 Use the properties of the Kronecker product of matrices (cf. Chapter 1) to show that the definitions of tensor product representations, Definitions 4.2.8 and 4.2.9, give representations of groups and Lie algebras, respectively.

Exercise 4.7 A representation of a Lie algebra is called unitary if it has values in $u(n)$. Adapt the proof of Lemma 4.2.3 to prove that every unitary representation of a Lie algebra is completely reducible.

Exercise 4.8 Prove that if the representation of the group G, π is unitary, then, for any element a of the group algebra, $a = a^\dagger$ implies $\pi(a) = [\pi(a)]^\dagger$, where the second † operation is the standard conjugate transpose of a matrix.

Exercise 4.9 The goal of this exercise is to give more details on the calculations of Example 4.3.8. All the symbols refer to that example. You may want to use MATLAB for some of the manipulations with 8×8 matrices. 1) Verify that P_2 is Hermitian (similarly one can verify $P_{1,3,4}$). 2) Verify that P_2 and P_3 are orthogonal (similarly one can verify the other pairs), 3) Verify that the completeness relation (4.12) holds in this case. 4) Calculate $\pi(P_3)\pi((12))\pi(P_2)$. 5) Obtain the orthonormal basis $\{\vec{f_1}, \vec{f_2}\}$ for $\pi(P_3)V$ from the basis $\{\vec{e_1}, \vec{e_2}\}$ of $\pi(P_2)V$.

Exercise 4.10 Verify the proofs of Theorem 4.2.4, Corollary 4.2.5 and Proposition 4.2.6 for the case of representations of algebras and Lie algebras.

Exercise 4.11 We have mentioned that there are several ways to decompose V as a representation of $\text{End}_G(V)$ which give the form (4.19). Investigate to which extent such a decomposition is unique up to $\text{End}_G(V)$-isomorphisms by adapting Theorem 4.2.7 and its proof by replacing a group G with the algebra $\text{End}_G(V)$.

Chapter 5

Observability and State Determination

It is known from measurement theory (cf. section 1.2) that the measurement of an observable gives a certain result with a probability depending on the current state. This suggests that measurements of appropriate observables on several copies of the same system can be used to determine the state of the system with some level of confidence. In systems theory, the study of the *observability* of a system describes to what extent it is possible to determine the initial state by a combined action of dynamics and output measurement.

In this chapter, we shall first briefly summarize the main ideas of *quantum state tomography* which is the technique used to determine a quantum state. Then, we shall turn to a system theoretic treatment of the problem of state determination namely to the study of the observability of quantum systems. We shall partition the space of density matrices into equivalence classes of states that cannot be distinguished by appropriate control and measurement. This allows us to consider the equation that describes the dynamics (Liouville's equation in this case) on a lower-dimensional space.

5.1 Quantum State Tomography

Quantum state tomography (also called *quantum process tomography*) is a method to determine the state of a quantum system via a series of measurements. It also allows us to estimate the expectation value of an observable A for a system in a state ρ, that is, $\langle A \rangle_\rho$. The method uses the results of a large number N of *selective measurements* of a quorum of observables on N identical quantum systems.

5.1.1 Example: Quantum tomography of a spin-$\frac{1}{2}$ particle

Consider the case of a spin-$\frac{1}{2}$ particle. The underlying Hilbert space is two-dimensional. An orthonormal basis in the space of the Hermitian operators on this Hilbert space is given by the 2×2 multiple of the identity matrix

$\tilde{\sigma}_0 = \frac{1}{\sqrt{2}}\mathbb{1}$, along with the multiples of the Pauli matrices defined in (1.20), $\tilde{\sigma}_{x,y,z} := \frac{1}{\sqrt{2}}\sigma_{x,y,z}$. Therefore, we can expand any Hermitian operator A as

$$A = \sum_{j=0,x,y,z} \text{Tr}(A\tilde{\sigma}_j)\tilde{\sigma}_j. \tag{5.1}$$

In particular, this expansion is valid for the density matrix ρ. Therefore, recalling that the expectation value $\langle \tilde{\sigma}_j \rangle_\rho$ is defined as $\langle \tilde{\sigma}_j \rangle_\rho = \text{Tr}(\rho\tilde{\sigma}_j)$, we can write ρ as

$$\rho = \frac{1}{2}\mathbb{1} + \sum_{j=x,y,z} \langle \tilde{\sigma}_j \rangle_\rho \tilde{\sigma}_j = \frac{1}{2}\mathbb{1} + \sum_{j=x,y,z} \sum_{m=\pm\frac{1}{\sqrt{2}}} P_\rho(m,j)m\tilde{\sigma}_j. \tag{5.2}$$

In this formula, we have denoted by $P_\rho(m,j)$ the probability of obtaining the result $m = \pm\frac{1}{\sqrt{2}}$ when measuring $\tilde{\sigma}_j$, in the state ρ. From this formula, if one is able to measure the expectation value of the *quorum* of observables $\tilde{\sigma}_{x,y,z}$, one reconstructs the state ρ. The observables $\tilde{\sigma}_{x,y,z}$ correspond to the values of the spin angular momentum in the x,y,z directions respectively. Their eigenvalues are $\pm\frac{1}{\sqrt{2}}$.

It is convenient in the sum (5.2) to highlight the factors which depend only on the quorum of observables and the factors given by the probabilities which depend on the results of the measurements. One defines an *estimator* (or *kernel function*) as

$$\mathcal{R}(m,j) := \tilde{\sigma}_j m,$$

and writes (5.2) as

$$\rho = \frac{1}{2}\mathbb{1} + \sum_{j=x,y,z} \sum_{m=\pm\frac{1}{\sqrt{2}}} \mathcal{R}(m,j)P_\rho(m,j). \tag{5.3}$$

The expectation value of an arbitrary observable A can be obtained directly from (5.2) and (5.3). We have

$$\langle A \rangle_\rho = \frac{1}{2}\text{Tr}(A) + \sum_{j=x,y,z} \langle \tilde{\sigma}_j \rangle_\rho \text{Tr}(A\tilde{\sigma}_j) \tag{5.4}$$

$$= \frac{1}{2}\text{Tr}(A) + \sum_{j=x,y,z} \sum_{m=\pm\frac{1}{\sqrt{2}}} \mathcal{K}_A(m,j)P_\rho(m,j),$$

where we have used the *scalar kernel function* $\mathcal{K}_A(m,j) := m\,\text{Tr}(A\tilde{\sigma}_j)$.

In practice, the probabilities $P_\rho(m,j)$ are obtained by counting the results on a large number of measurements.

5.1.2 General quantum tomography

The scheme described in the above example can be greatly generalized to give a unifying scheme for quantum tomography procedures.[1] Consider a set of operators $\{C_k\}$, $k = 1, 2, ...$, not necessarily Hermitian, and a dual set $\{B_k\}$ such that every Hermitian operator A can be expanded as (cf. (5.1))

$$A = \sum_j \text{Tr}(AB_j^\dagger)C_j. \tag{5.5}$$

Proposition 5.1.1 The sets $\{B_k\}$ and $\{C_k\}$ give the expansion (5.5) for every observable A if the following two conditions are satisfied:

1. The set $\{C_k\}$ is *complete*, i.e., every Hermitian operator A can be expanded as (cf. (5.1))

$$A := \sum_k a_k C_k.$$

2.

$$\text{Tr}(C_j B_k^\dagger) = \delta_{jk}.$$

Proposition 5.1.2 The sets $\{B_k\}$ and $\{C_k\}$ give the expansion (5.5) for every observable A if the following condition is satisfied: For every orthonormal basis in the Hilbert space of the system $\{|j\rangle\}$,

$$\sum_\lambda \langle j|B_\lambda^\dagger|m\rangle\langle n|C_\lambda|k\rangle = \delta_{m,n}\delta_{j,k}. \tag{5.6}$$

Either one of the two propositions can be used to check whether the operators $\{B_k\}$ and $\{C_k\}$ satisfy the desired property. The proof of 5.1.1 is obvious while the proof of 5.1.2 is slightly more complicated[2] (also cf. Exercise 5.1). Given the sets $\{B_j\}$ and $\{C_j\}$, a **quorum** of observables is defined as a set of observables $\{Q_j\}$, $j = 1, 2, ...$, such that there exist functions f_j, with $f_j(Q_j) = B_j$. In this case, from the expansion (5.5) applied to ρ, we have

$$\rho = \sum_j \text{Tr}(\rho f_j(Q_j)^\dagger)C_j. \tag{5.7}$$

---

[1] We consider the case of Hilbert space with countable basis. There are natural generalizations to the case of uncountable basis.

[2] Write A as

$$A = \sum_{j,m} \langle m|A|j\rangle|m\rangle\langle j| = \sum_{j,m,k,n} \langle m|A|j\rangle|n\rangle\langle k|\delta_{m,n}\delta_{j,k}.$$

Using (5.6), we write

$$A = \sum_\lambda \sum_{j,m,k,n} \langle m|A|j\rangle\langle j|B_\lambda^\dagger|m\rangle\langle n|C_\lambda|k\rangle|n\rangle\langle k|,$$

which, using the completeness relation (cf. (1.9)) $\sum_j |j\rangle\langle j| = \mathbf{1}$, $\sum_m \langle m|AB_\lambda^\dagger|m\rangle = \text{Tr}(AB_\lambda^\dagger)$, and $C_\lambda = \sum_{n,k} \langle n|C_\lambda|k\rangle|n\rangle\langle k|$, gives (5.5).

By writing Q_j in terms of its eigenvalues and eigenvectors as

$$Q_j = \sum_{m_j} m_j |m_j\rangle\langle m_j|,$$

and $f(Q_j) = \sum_{m_j} f(m_j)|m_j\rangle\langle m_j|$ and recalling that $\langle m_j|\rho|m_j\rangle$ is the probability of having the result m_j when measuring Q_j, i.e., $P_\rho(m_j, j)$, we can rewrite (5.7) as

$$\rho = \sum_j \sum_{m_j} \mathcal{R}(m_j, j) P_\rho(m_j, j), \qquad (5.8)$$

where the estimator $\mathcal{R}(m_j, j)$ is defined by $\mathcal{R}(m_j, j) := C_j f_j(m_j)^*$, and it only depends on the quorum observables but not on the results of the measurements. Formula (5.8) can also be written in terms of the expectation values of the functions f_j, denoted here by $\langle f_j\rangle_\rho := \text{Tr}(\rho f_j(Q_j))$. We have

$$\rho = \sum_j \langle f_j\rangle_\rho^* C_j. \qquad (5.9)$$

The expectation value of an observable A can be found from (5.8), (5.9), as

$$\langle A\rangle_\rho = \sum_j \sum_{m_j} \mathcal{K}_A(m_j, j) P_\rho(m_j, j) = \sum_j \langle f_j\rangle_\rho^* \text{Tr}(AC_j), \qquad (5.10)$$

where the scalar kernel function is defined as $\mathcal{K}_A(m_j, j) := \text{Tr}(A\mathcal{R}(m_j, j))$. Once experiments are carried out to estimate the probabilities $P_\rho(m_j, j)$, formulas (5.8), (5.10) can be used to calculate the state ρ or the expectation value of any observable A.

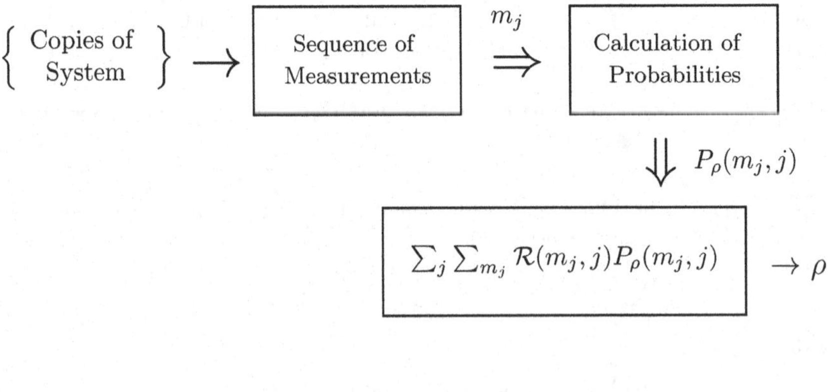

FIGURE 5.1: Scheme of the various stages of a quantum tomographic procedure.

In the scheme described in the previous subsection, $\tilde{\sigma}_{0,x,y,z}$ play the role of the operators C_j as well as the one of the corresponding dual operators B_j (i.e., in this case, every operator coincides with its dual). Moreover, they coincide with the quorum operators Q_j, i.e., the maps f_j above defined are all equal to the identity.

The above scheme naturally extends to the case of an uncountable number of operators C_λ and B_λ. In this case, we have a set of operators C_λ and B_λ, λ in a possibly uncountable but measurable set Λ, such that every observable A can be expressed as[3] (cf. (5.5))

$$A = \int_\Lambda \mathrm{Tr}(AB_\lambda^\dagger)C_\lambda d\lambda. \tag{5.11}$$

It is easy to show that this is the case if (cf. Proposition 5.1.2), for every orthonormal basis in the Hilbert space of the system $\{|j\rangle\}$,

$$\int_\Lambda \langle j|B_\lambda^\dagger|m\rangle\langle n|C_\lambda|k\rangle d\lambda = \delta_{m,n}\delta_{j,k}. \tag{5.12}$$

Given a quorum of observables Q_λ, the formula for the reconstruction of ρ corresponding to (5.8) reads as (cf. (5.8))

$$\rho = \int_\Lambda \sum_{m_\lambda} \mathcal{R}(m_\lambda, \lambda)P_\rho(m_\lambda, \lambda)d\lambda, \tag{5.13}$$

where the estimator $\mathcal{R}(m_\lambda, \lambda)$ is defined by $\mathcal{R}(m_\lambda, \lambda) := C_\lambda(f_\lambda(m_\lambda))^*$.

Quantum tomography schemes specialize the above approach to various situations.

5.1.3 Example: Quantum tomography of a spin-$\frac{1}{2}$ particle (ctd.)

We conclude this section by reconsidering the problem of state determination for a spin $\frac{1}{2}$ treated in subsection 5.1.1, but with a different quorum of observables parametrized by a continuous set Λ. In particular, let Λ be the set of angles θ, ϕ, ψ,

$$\Lambda := \{\theta, \phi, \psi | 0 \leq \theta \leq 2\pi, 0 \leq \phi \leq \pi, -\pi \leq \psi \leq \pi\}.$$

Consider the quorum of observables given by multiples of the spin operators in all the possible directions. If the direction $\vec{n}$ is given by

$$\vec{n} = (\cos(\theta)\sin(\phi), \sin(\theta)\sin(\phi), \cos(\phi))^T$$

---

[3] For an example see the following subsection, 5.1.3.

and the spin operator $\vec{\sigma}$ is represented by the Pauli matrices (1.20) $\vec{\sigma} := \{\sigma_x, \sigma_y, \sigma_z\}$, then the spin operator in the direction $\vec{n}$ can be written as $\vec{\sigma} \cdot \vec{n}$. The quorum of observables $\{Q_\lambda\}$ with $\lambda \in \Lambda$ is given by

$$Q_\lambda = \vec{\sigma} \cdot \vec{n}(\theta, \phi)\psi.$$

The function f_λ which maps Q_λ to the operator C_λ is taken as the exponential $f_\lambda(Q_\lambda) := e^{-iQ_\lambda} = e^{-i\vec{\sigma}\cdot\vec{n}\psi} := C_\lambda$. Such a unitary operator physically can be seen as a rotation around the direction $\vec{n}$ by an angle 2ψ (cf., e.g., [245] section 3.2). Its matrix expression can be easily calculated and it is given by (with $\lambda = (\theta, \phi, \psi)$)

$$C_\lambda = \begin{pmatrix} \cos(\psi) - i\cos(\phi)\sin(\psi) & -i\sin(\phi)e^{-i\theta}\sin(\psi) \\ -i\sin(\phi)\sin(\psi)e^{i\theta} & \cos(\psi) + i\cos(\phi)\sin(\psi) \end{pmatrix}. \tag{5.14}$$

We choose, for every λ, $B_\lambda = C_\lambda$. With this choice, and taking as the volume element in the set Λ, $d\lambda = \frac{1}{2\pi^2}\sin^2(\psi)\sin(\phi)$, it is a routine calculation (using as a basis the set of eigenvectors of σ_z) to verify (5.12),[4] i.e.,

$$\frac{1}{2\pi^2}\int_0^\pi \int_0^{2\pi} \int_{-\pi}^\pi \langle j|B_\lambda^\dagger|m\rangle\langle n|C_\lambda|k\rangle \sin^2(\psi)\sin(\phi)d\psi d\theta d\phi = \delta_{m,n}\delta_{j,k}. \tag{5.15}$$

Therefore, we can apply the continuous version of Proposition 5.1.2 and write every operator A as (cf. (5.11))

$$A = \frac{1}{2\pi^2}\int_0^\pi \int_0^{2\pi} \int_{-\pi}^\pi \text{Tr}(Ae^{i\vec{\sigma}\cdot\vec{n}\psi})e^{-i\vec{\sigma}\cdot\vec{n}\psi}\sin^2(\psi)\sin(\phi)d\psi d\theta d\phi.$$

With the given quorum of observables Q_λ, the formula for the reconstruction of ρ reads as (cf. 5.13)

$$\rho = \frac{1}{2\pi^2}\int_\Lambda \int_{-1}^1 \mathcal{R}(m_{\theta,\phi,\psi}, \theta, \phi, \psi)P_\rho(m_{\theta,\phi,\psi})dm_{\theta,\phi,\psi}\sin^2(\psi)\sin(\phi)d\psi d\theta d\phi,$$

with the kernel $\mathcal{R}(m_{\theta,\phi,\psi}, \theta, \phi, \psi)$ given by

$$\mathcal{R}(m_{\theta,\phi,\psi}, \theta, \phi, \psi) := e^{-im_{\theta,\phi,\psi}}e^{i\vec{\sigma}\cdot\vec{n}\psi}.$$

---

[4]Formula (5.15) is a consequence of more general properties and in particular of the fact that by varying θ, ϕ and ψ, (5.14) gives an irreducible representation of $SU(2)$ and that the element of volume $d\lambda := \frac{1}{2\pi^2}\sin^2(\psi)\sin(\phi)$ is invariant under group translations. We have discussed some representation theory of groups in Chapter 4 and references particularly useful in our context here are [201], [203]. In particular the formula on pg. 216 of [201] corresponds to (5.12), while Chapters 5 and 6 of [203] give the definition of the volume element and a method to calculate it. In fact, group theory can be used to generalize the treatment here to more general systems which include, for example, spin different from $\frac{1}{2}$ and the electromagnetic radiation field, when considered as a quantum system (rather than classically as in Chapter 2 with the semiclassical approximation). This approach is known as *Group Tomography*. For this, we refer to [90] and the references therein.

5.2 Observability

As seen in the previous section, the formulas used in quantum tomography express the quantum state in terms of the probabilities and the expectation values of a quorum of observables. The evaluation of the probabilities has to be carried out by performing a large number of measurements of the quorum observables on exact copies of the same system. As the quorum observables consist typically of a large (sometimes uncountably infinite, see subsection 5.1.3) number of observables, this involves many types of measurement experiments. Often, as in the spin-$\frac{1}{2}$ example of subsections 5.1.1 and 5.1.3, the quorum of observables $\{Q_\lambda\}$ can be obtained by a set of unitary similarity transformations, U_λ, on a single observable Q, i.e.,

$$Q_\lambda = U_\lambda^\dagger Q U_\lambda, \qquad \lambda \in \Lambda.$$

In these cases, we have, for the expectation values $\langle Q_\lambda \rangle_\rho$,

$$\langle Q_\lambda \rangle_\rho = \mathrm{Tr}(\rho U_\lambda^\dagger Q U_\lambda) = \mathrm{Tr}(U_\lambda \rho U_\lambda^\dagger Q), \tag{5.16}$$

and for the probabilities $P_\rho(m, \lambda)$ in (5.8), (5.13), denoting by P_m the projection onto the eigenspace of Q corresponding to the eigenvalue m,

$$P_\rho(m, \lambda) = \mathrm{Tr}(\rho U_\lambda^\dagger P_m U_\lambda) = \mathrm{Tr}(U_\lambda \rho U_\lambda^\dagger P_m) \tag{5.17}$$

(cf. subsection 1.2.3). Formulas (5.16) and (5.17) suggest that, instead of changing the measurement apparatus, for any type of measurement, we could as well rely on the dynamics of the system $\rho \to U\rho U^\dagger$ and then measure always the same observable. This raises the question of characterizing, given a certain dynamics of the control system, the class of initial states that will give the same probabilities (expectation value) for any possible evolution.

To pose this problem in system theoretic terms as a study of observability, we consider the quantum control system (2.69) with output given by the probabilities $y(t) := \mathrm{Pr}\,(m) = \mathrm{Tr}(\rho(t)P_m)$, or with output given the expectation value $y(t) := \langle Q \rangle_\rho = \mathrm{Tr}(\rho(t)Q)$. In the last two formulas, $\rho(t)$ is the solution of the differential equation (2.69). Notice that, in both cases, the output y is a linear function of the system state. Therefore, we shall study the observability properties of system (2.69)

$$\dot{\rho} = [-iH(u(t)), \rho], \qquad \rho(0) = \rho_0, \tag{5.18}$$

with output

$$y(t) = \mathrm{Tr}(S\rho), \tag{5.19}$$

for some Hermitian operator S (which could be P_m or Q). Denote by $\rho(t, u, \tilde{\rho})$ the solution of (5.18) with initial condition $\rho_0 = \tilde{\rho}$ and control u at time t. We have the following definition.

Definition 5.2.1 Two states $\tilde{\rho}_1$ and $\tilde{\rho}_2$ are called **indistinguishable** if, for any control u, and any t (cf. (5.19)),

$$\text{Tr}(S\rho(t, u, \tilde{\rho}_1)) = \text{Tr}(S\rho(t, u, \tilde{\rho}_2)).$$

Definition 5.2.2 System (5.18), (5.19) is called **observable**, if two states $\tilde{\rho}_1$ and $\tilde{\rho}_2$ are indistinguishable if and only if $\tilde{\rho}_1 = \tilde{\rho}_2$.

5.2.1   Equivalence classes of indistinguishable states; partition of the state space

As seen in Chapter 3, given the dynamical Lie algebra $\mathcal{L}$ associated with the quantum system (5.18), there exists a control giving the unitary evolution X if and only if X belongs to the corresponding Lie group $e^{\mathcal{L}}$. This assumes that $e^{\mathcal{L}}$ is compact which, as we have seen in Chapter 3 can be assumed to be the case except for the center of the Lie group $e^{\mathcal{L}}$. We shall assume this to be the case in the following and identify the reachable set with $e^{\mathcal{L}}$.

Given the form (2.70) of the solution of (5.18), we easily obtain:

Proposition 5.2.3 Two states $\tilde{\rho}_1$ and $\tilde{\rho}_2$ are indistinguishable if and only if

$$\text{Tr}(X^\dagger S X \tilde{\rho}_1) = \text{Tr}(X^\dagger S X \tilde{\rho}_2), \qquad \forall X \in e^{\mathcal{L}}.$$

In order to state the condition of indistinguishability in linear algebraic terms, it is convenient to define the *observability spaces* as follows. Let

$$\tilde{S} := S - \frac{\text{Tr}(S)}{n} \mathbf{1}_{n \times n}, \tag{5.20}$$

where n is the dimension of the system. The observability space $\mathcal{V}$ is defined as

$$\mathcal{V} := \bigoplus_{j=0}^{\infty} ad_{\mathcal{L}}^j \, \text{span}\{i\tilde{S}\}, \tag{5.21}$$

where $ad_{\mathcal{L}}^j \mathcal{A}$ is defined as spanned by all the repeated Lie brackets

$$[R_j, [\cdots [R_2, [R_1, iA]] \cdots]],$$

with $R_1, \ldots, R_j \in \mathcal{L}$ and $iA \in \mathcal{A}$.

With these definitions, we have:

Theorem 5.2.4 *System (5.18) with output y in (5.19) is observable if and only if*

$$\mathcal{V} = su(n). \tag{5.22}$$

More generally, write $\rho = \rho_1 + \rho_2$ where ρ_1 is the component of ρ in[5] $i\mathcal{V}$ and ρ_2 is the component along $i\mathcal{V}^\perp$ where $\mathcal{V}^\perp$ is the orthogonal complement of $\mathcal{V}$ in $u(n)$. Then, the following decomposition of the dynamics holds true

$$\dot{\rho}_1 = -i[H(u), \rho_1],$$
$$\dot{\rho}_2 = -i[H(u), \rho_2], \tag{5.23}$$

and we have

$$y(t) := \mathrm{Tr}\big(S\rho(t)\big) = \mathrm{Tr}(S) + \mathrm{Tr}\big(S\rho_1(t)\big). \tag{5.24}$$

Initial states $\tilde{\rho}_1$ and $\tilde{\rho}_2$ are indistinguishable if and only if $(\tilde{\rho}_1 - \tilde{\rho}_2) \in i\mathcal{V}^\perp$.

Proof. Write (5.18) with $\rho = \rho_1 + \rho_2$, $\rho_1 \in i\mathcal{V}$, $\rho_2 \in i\mathcal{V}^\perp$,

$$\dot{\rho}_1 + \dot{\rho}_2 = [-iH(u(t)), \rho_1] + [-iH(u(t)), \rho_2], \tag{5.25}$$

and notice that, as $i\mathcal{V}$ is invariant under the commutator operation with $-iH(u(t))$, the first commutator on the right-hand side of (5.25) is always in $i\mathcal{V}$. The invariance of $i\mathcal{V}$ also implies the invariance of $i\mathcal{V}^\perp$, i.e.,[6]

$$[-iH(u), i\mathcal{V}^\perp] \subseteq i\mathcal{V}^\perp.$$

Separating the components in $i\mathcal{V}$ and $i\mathcal{V}^\perp$ in (5.25), we obtain (5.23). Moreover, using the fact that ρ_1 is traceless while $\mathrm{Tr}(\rho_2) = 1$, one obtains (5.24). By decomposing $\rho(t, u, \tilde{\rho}_1)$ and $\rho(t, u, \tilde{\rho}_2)$ according to their components in $i\mathcal{V}$ and $i\mathcal{V}^\perp$ and applying (5.23) and (5.24) one sees that the outputs coincide for every control if the components of $\tilde{\rho}_1$ and $\tilde{\rho}_2$ in $i\mathcal{V}$ coincide. This proves that if $\tilde{\rho}_1 - \tilde{\rho}_2 \in \mathcal{V}^\perp$, $\tilde{\rho}_1$ and $\tilde{\rho}_2$ are indistinguishable. Conversely, assume that $\tilde{\rho}_1$ and $\tilde{\rho}_2$ are indistinguishable. Then, from Proposition 5.2.3, for every $X \in e^{\mathcal{L}}$ we have

$$\mathrm{Tr}(X^\dagger \tilde{S} X(\tilde{\rho}_1 - \tilde{\rho}_2)) = 0. \tag{5.26}$$

In particular, if $R_1, ..., R_m$ are arbitrary elements of $\mathcal{L}$ and $t_1, .., t_m$ arbitrary real numbers, we have

$$e^{R_1 t_1} e^{R_2 t_2} \cdots e^{R_m t_m} \tilde{S} e^{R_1^\dagger t_1} e^{R_2^\dagger t_2} \cdots e^{R_m^\dagger t_m} (\tilde{\rho}_1 - \tilde{\rho}_2) \equiv 0.$$

Taking (possibly higher order) derivatives with respect to $t_1, t_2, ..., t_m$ at $t_1 = t_2 = ... = t_m = 0$ and taking into account that $R_1, ..., R_m$ are arbitrary, we see the $\mathrm{Tr}(A(\tilde{\rho}_1 - \tilde{\rho}_2)) = 0$ for A varying in a spanning set of $\mathcal{V}$. This shows that $\tilde{\rho}_1 - \tilde{\rho}_2 \in i\mathcal{V}^\perp$. In the special case where $\mathcal{V} = su(n)$, we have

---

[5] $i\mathcal{V}$ denotes the vector space of Hermitian matrices obtained by multiplying by i the skew-Hermitian matrices in $\mathcal{V}$.

[6] If $A \in i\mathcal{V}^\perp$, for every $B \in i\mathcal{V}$, we have $\mathrm{Tr}([-iH, A]B) = \mathrm{Tr}([B, -iH]A) = 0$ since $[B, -iH] \in i\mathcal{V}$.

$(\tilde{\rho}_1 - \tilde{\rho}_2) \in \mathrm{span}\{1\}$ and since $\mathrm{Tr}(\tilde{\rho}_1) = \mathrm{Tr}(\tilde{\rho}_2) = 1$, we have $\tilde{\rho}_1 = \tilde{\rho}_2$ and the system is observable.   □

The observability space $\mathcal{V}$ determines a partition of the space of density matrices into equivalence classes of indistinguishable states. Two states $\tilde{\rho}_1$ and $\tilde{\rho}_2$ belong to the same equivalence class if and only if they have the same component along $\mathcal{V}$. Moreover, such an equivalence class is an invariant set for the dynamics as expressed by (5.23). Consider as an example the two-level case where ρ in (1.19), (1.20) is represented by a point (x, y, z) in the Bloch sphere (cf. Figure 1.1). Assume $\mathcal{V} = \mathrm{span}\{i\sigma_z\}$. Then the situation is the one depicted in Figure 5.2. Two points $\tilde{\rho}_1$ and $\tilde{\rho}_2$ with the same value of z in (1.19) are indistinguishable. Moreover, assume $H(u) = \sigma_z$, for every control u. Then the dynamics is given by

$$\tilde{\rho}_1 \to e^{-i\sigma_z t} \tilde{\rho}_1 e^{i\sigma_z t}$$

$$= \frac{1}{2} 1 + z\sigma_z + (\cos(2t)x + \sin(2t)y)\sigma_x + (-\sin(2t)x + \cos(2t)y)\sigma_y,$$

that is, it is a circle in a surface $z = \mathrm{const}$ and therefore it is such that the component along $\mathcal{V}$ remains the same. This situation extends to higher dimensions according to Theorem 5.2.4.

The following remarks sketch a number of generalizations of the Lie algebraic characterization of observability of Theorem 5.2.4.

Remark 5.2.5 Theorem 5.2.4 extends easily to the case of several outputs. This is typically the case when we want to find the set of initial states which will give the same probabilities for any dynamics. One modifies the definition of observability spaces (5.21) by replacing $\mathrm{span}\{i\tilde{S}\}$ with

$$\mathrm{span}\{i\tilde{P}_1, i\tilde{P}_2, \ldots, i\tilde{P}_m\},$$

where $\tilde{P}_j = P_j - \mathrm{Tr}(P_j)1$, $j = 1, ..., m$, where the P_j's are the output operators (which might represent projections onto the eigenspaces of a given observable). The rest of the argument goes through with only formal modifications.

Remark 5.2.6 Another generalization consists of considering generalized measurement theory as described in Appendix A. In this case, all the treatment goes through by replacing the projections P_m with the effects F_m in subsection A.3.

Remark 5.2.7 One more generalization can be obtained by considering a sequence of measurements. As the state is modified at each measurement, according to measurement theory (cf. section 1.2), indistinguishability at the first measurement is not equivalent to indistinguishability at a following measurement. The treatment, however, can be extended to this case assuming we perform a nonselective measurement of the expectation value so that the modification of the state is assumed not to depend on the result (cf. formula

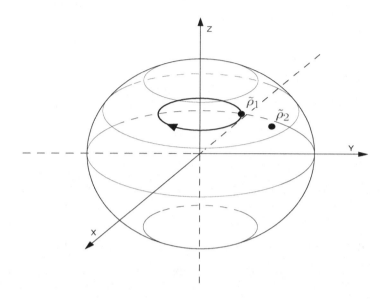

FIGURE 5.2: Partition of the Bloch sphere into equivalence classes of indistinguishable states. The states $\tilde{\rho}_1$ and $\tilde{\rho}_2$ which have the same value of z, are indistinguishable and every indistinguishable state evolves into indistinguishable states, as for example for the trajectory starting at $\tilde{\rho}_1$.

(1.40) in Chapter 1). Let $\mathcal{F}$ be the linear transformation of the state $\rho \to \mathcal{F}(\rho)$ due to the measurement and $\mathcal{F}^\dagger$ denote the dual map defined by $\mathrm{Tr}(A\mathcal{F}(\rho)) = \mathrm{Tr}(\mathcal{F}^\dagger(A)\rho)$, for every A and ρ. One defines indistinguishability in k steps for two states that cannot be distinguished at the k-th measurement. To give conditions for indistinguishability in k steps, one defines the *generalized observability spaces* recursively as

$$\mathcal{V}_0 := \mathrm{span}\{i\tilde{S}\}, \qquad \mathcal{V}_1 := \bigoplus_{j=0}^{\infty} ad_{\mathcal{L}}^j \mathcal{V}_0,$$

$$\mathcal{V}_k := \bigoplus_{j=0}^{\infty} ad_{\mathcal{L}}^j i\mathcal{F}^\dagger(i\mathcal{V}_{k-1}).$$

The criterion for indistinguishability (and observability) in k steps reads as Theorem 5.2.4 by replacing indistinguishability with indistinguishability in k

steps and $\mathcal{V}$ with $\mathcal{V}_k$. The proof is a generalization of the proof of 5.2.4 (see [75]).

Excluding the trivial case where S is a scalar matrix, if $\mathcal{L} = su(n)$ or $\mathcal{L} = u(n)$, i.e., the system is operator controllable, then the observability space $\mathcal{V}$ has to be (or to contain) the whole $su(n)$. This is a consequence of the fact that $su(n)$ is a simple Lie algebra (cf. Proposition 3.3.7). As $\mathcal{V}$ is an ideal of $u(n)$ it can only be spanned by a multiple of the identity (a case we have excluded) or contain $su(n)$. We state this in the following proposition.

Proposition 5.2.8 Operator controllability implies observability.

5.3   Observability and Methods for State Reconstruction

5.3.1   Observability conditions and tomographic methods

Consider now a quantum tomographic method where only a single observable is measured and the remaining information is obtained using the dynamics of the system. The practical scheme follows the same steps as in Figure 5.1 where in the sequence of measurements every measurement consists of an evolution followed by the measurement of the given observable. Such a scheme is characterized, in the discrete parameter set case, by a formula of the type (5.8) or, in the continuous case, by a formula (5.13) where the quorum $\{Q_\lambda\}$ is obtained from a single observable Q by unitary similarity transformations, i.e., $Q_\lambda = X_\lambda^\dagger Q X_\lambda$. Here $X_\lambda \in e^{\mathcal{L}}$, for λ in an appropriate discrete or continuous set. Therefore, if P_m denotes a projection associated with the observable Q and the result m, formulas (5.8) and (5.13) can be rewritten respectively as

$$\rho = \sum_\lambda \sum_m f_\lambda(m)^* C_\lambda \operatorname{Tr}(X_\lambda^\dagger P_m X_\lambda \rho) := \sum_\lambda \mathcal{M}_{X_\lambda, C_\lambda, f_\lambda}(\rho), \qquad (5.27)$$

or

$$\rho = \int_\Lambda \sum_m f_\lambda(m)^* C_\lambda \operatorname{Tr}(X_\lambda^\dagger P_m X_\lambda \rho) d\lambda := \int_\Lambda \mathcal{M}_{X_\lambda, C_\lambda, f_\lambda}(\rho) d\lambda.$$

Here we have introduced the notation of the function $\mathcal{M}_{X_\lambda, C_\lambda, f_\lambda}$ which depends on parameters λ in a set Λ. This function characterizes the tomographic scheme. Given the dynamics and the observable which has to be measured Q, the particular scheme we adopt depends on the choice of the operators C_λ, the functions f_λ, and the unitary evolutions X_λ in $e^{\mathcal{L}}$. It follows from the above discussion that, unless the system is fully observable, with respect to the outputs P_m (cf. Remark 5.2.5) associated with the observable Q, determination of the state with the quantum tomographic method $\mathcal{M}_{X_\lambda, C_\lambda, f_\lambda}(\rho)$ is possible only up to terms in the orthogonal of the observability space, namely $i\mathcal{V}^\perp$. We formalize this fact in the following proposition which we state for the case where Λ is a discrete set.

Proposition 5.3.1 Let the observability space $\mathcal{V}$ be calculated as in Remark 5.2.5 starting from the projections P_m associated with the observable Q. Then $\rho_1 - \rho_2 \in i\mathcal{V}^\perp$ if and only if, for *any* tomographic method $\mathcal{M}_{X_\lambda, C_\lambda, f_\lambda}$, $\mathcal{M}_{X_\lambda, C_\lambda, f_\lambda}(\rho_1) = \mathcal{M}_{X_\lambda, C_\lambda, f_\lambda}(\rho_2)$.

Proof. If $\rho_1 - \rho_2 \in i\mathcal{V}^\perp$, ρ_1 and ρ_2 are indistinguishable because of Theorem 5.2.4, and using (5.26) with the various projections P_m instead of $\tilde{S}$, we have

$$\mathcal{M}_{X_\lambda, C_\lambda, f_\lambda}(\rho_1) - \mathcal{M}_{X_\lambda, C_\lambda, f_\lambda}(\rho_2) = \sum_m f_\lambda(m)^* C_\lambda \operatorname{Tr}(X_\lambda^\dagger P_m X_\lambda(\rho_1 - \rho_2)) = 0.$$

Conversely, assume $\mathcal{M}_{X_\lambda, C_\lambda, f_\lambda}(\rho_1) = \mathcal{M}_{X_\lambda, C_\lambda, f_\lambda}(\rho_2)$ for every $\lambda \in \Lambda$ and any tomographic method $\mathcal{M}_{X_\lambda, C_\lambda, f_\lambda}$. Choose $X_1, ..., X_r$ and projections $P_1, ..., P_r$ so that $X_1 P_1 X_1^\dagger, ..., X_r P_r X_r^\dagger$ span a basis of $i\mathcal{V}$ (see Exercise 5.2). Choose $f_\lambda(m_\lambda)^* C_\lambda$, $\lambda = 1, ..., r$ (where m_λ is the eigenvalue corresponding to the projection P_λ) Hermitian with norm one, such that $f_\lambda(m_\lambda)^* C_\lambda \in i\mathcal{V}$ is orthogonal to $X_l P_l X_l^\dagger$, for $l \neq \lambda$, which implies that $f_\lambda(m_\lambda)^* C_\lambda$, $\lambda = 1, ..., r$, are an orthonormal basis of $i\mathcal{V}$. With these choices, $\mathcal{M}_{X_\lambda, C_\lambda, f_\lambda}(\rho_1) = \mathcal{M}_{X_\lambda, C_\lambda, f_\lambda}(\rho_2)$ using (5.27) implies $\operatorname{Tr}(X_\lambda P_\lambda X_\lambda^\dagger(\rho_1 - \rho_2)) = 0, \forall \lambda = 1, ..., r$, i.e., $\rho_1 - \rho_2 \in i\mathcal{V}^\perp$. $\square$

Remark 5.3.2 Notice that the quantum tomographic scheme described in the second part of the proof of the theorem allows the determination of the component of ρ in $i\mathcal{V}$.

5.3.2 System theoretic methods for quantum state reconstruction

The study of system theoretic properties, in particular observability, is not only a way to analyze the effectiveness of quantum tomographic methods, but it also gives rise to alternative procedures for the determination of the state. These can be obtained by applying the methods of system theory. The concept of observability plays a prominent role in this context as well. In the following we derive, an integral, system theoretic, formula for quantum state reconstruction based on the measurement of the expectation value (5.19) of a given observable.

Given the output (5.19) and the associated observability space $i\mathcal{V}$ (5.20), (5.21), we can choose a control $u = u(t)$, $t \in [0, T]$ so that the corresponding solution X_u of Schrödinger operator equation

$$\dot{X} = -iH(u)X, \quad X(0) = 1, \tag{5.28}$$

satisfies

$$\text{span}_{t \in [0,T]} \{X_u^\dagger(t) \tilde{S} X_u(t)\} = i\mathcal{V}. \tag{5.29}$$

One way to do this is to select $r := \dim(\mathcal{V})$ matrices $X_1, ..., X_r$ so that $X_1^\dagger S X_1, ..., X_r^\dagger S X_r$ are linearly independent and then concatenate the controls steering the matrix X in (5.28) to $X_1, X_2 X_1^\dagger, X_3 X_2^\dagger, ..., X_r X_r^\dagger$.

Now, consider a trajectory X_u of (5.28) having property (5.29). From (5.19), (5.20), we obtain

$$y(t) = \text{Tr}(X_u^\dagger \tilde{S} X_u (\rho_0 - \frac{1}{n}\mathbf{1})) + \text{Tr}(S),$$

which gives

$$\int_0^T X_u^\dagger(t) \tilde{S} X_u(t)(y(t) - \text{Tr}(S))dt = \tag{5.30}$$

$$\int_0^T X_u^\dagger(t) \tilde{S} X_u(t) \, \text{Tr}\left(X_u^\dagger(t) \tilde{S} X_u(t)(\rho_0 - \frac{1}{n}\mathbf{1})\right) dt.$$

Define the linear operator $\mathcal{W}$ to map $n \times n$ Hermitian matrices with zero trace into $n \times n$ Hermitian matrices with zero trace as follows

$$\mathcal{W}_u(\hat{\rho}_0) := \int_0^T X_u^\dagger(t) \tilde{S} X_u(t) \, \text{Tr}\left(X_u^\dagger(t) \tilde{S} X_u(t)\hat{\rho}_0\right) dt. \tag{5.31}$$

The range of the operator $\mathcal{W}_u$ is $i\mathcal{V}$. This follows from (5.29) and the fact that the kernel of $\mathcal{W}_u$ is the component of $i\mathcal{V}^\perp$ in $isu(n)$, i.e., the vector space of the traceless matrices in $i\mathcal{V}^\perp$. It is clear, in fact, that if $\hat{\rho}_0 \in i\mathcal{V}^\perp$, $\mathcal{W}_u(\hat{\rho}_0) = 0$. Conversely, assume $\mathcal{W}(\hat{\rho}_0) = 0$. Then, from (5.31), we have

$$0 = \text{Tr}(\hat{\rho}_0 \mathcal{W}_u(\hat{\rho}_0)) = \int_0^T (\text{Tr}(X_u^\dagger(t) \tilde{S} X_u(t)\hat{\rho}_0))^2 dt,$$

which implies

$$\text{Tr}(X_u^\dagger(t) \tilde{S} X_u(t)\hat{\rho}_0) = 0, \qquad a.e.$$

This, using (5.29), implies $\hat{\rho}_0 \in i\mathcal{V}$.

Formula (5.30) gives a linear system of equations whose solution allows one to calculate the initial state ρ_0 modulo elements in $i\mathcal{V}^\perp$. In particular, if the system is observable, then $\mathcal{W}_u$ is invertible and we have

$$\rho_0 = \frac{1}{n}\mathbf{1} + \mathcal{W}_u^{-1}\left(\int_0^T X_u^\dagger(t) \tilde{S} X_u(t)(y(t) - \text{Tr}(S))dt\right). \tag{5.32}$$

Formula (5.32) represents a system theoretic alternative to methods for quantum state tomography.

We summarize the discussion in the following theorem.

Theorem 5.3.3 *Consider system (5.18) with output (5.19). If the system is observable (in one step), then there exists a control such that formula (5.32) gives the initial state.*

Example 5.3.4 If the system is observable, in order to obtain the initial state from formula (5.32), according to Theorem 5.3.3, the control has to be chosen so that the corresponding evolution X_u satisfies (5.29). Consider as an example a two-level quantum system such that $\tilde{S} := \sigma_x$ in (1.20) and the control is chosen so that $H(u) := \sigma_z$ in $[0, 2\pi)$ and $H(u) := \sigma_y$ in $[2\pi, 4\pi)$. We calculate

$$
\mathcal{W}_u(\hat{\rho}_0) = \int_0^{2\pi} e^{i\sigma_z t} \sigma_x e^{-i\sigma_z t} \operatorname{Tr}\left(e^{i\sigma_z t} \sigma_x e^{-i\sigma_z t} \hat{\rho}_0\right) dt
$$

$$
+ \int_{2\pi}^{4\pi} e^{i\sigma_y t} \sigma_x e^{-i\sigma_y t} \operatorname{Tr}\left(e^{i\sigma_y t} \sigma_x e^{-i\sigma_y t} \hat{\rho}_0\right) dt.
$$

We obtain

$$
\mathcal{W}_u(\hat{\rho}_0) = \pi \left(2 \operatorname{Tr}(\sigma_x \hat{\rho}_0) \sigma_x + \operatorname{Tr}(\sigma_y \hat{\rho}_0) \sigma_y + \operatorname{Tr}(\sigma_z \hat{\rho}_0) \sigma_z\right),
$$

and

$$
\mathcal{W}_u^{-1}(a\sigma_x + b\sigma_y + c\sigma_z) = \frac{1}{\pi}\left(\frac{a}{4}\sigma_x + \frac{b}{2}\sigma_y + \frac{c}{2}\sigma_z\right).
$$

5.4   Notes and References

Our discussion of quantum tomography follows the general unifying scheme presented in [89], to which we also refer for a survey on the subject and relevant references. In particular, this paper presents some statistical aspects of quantum tomography as well as the connection with conventional tomographic methods in medical imaging. We have focused here on the fundamental aspects and the formulas for the reconstruction of the state emphasizing the role of the dynamics in the process of state reconstruction. Some, more recent, contributions on quantum state tomography are given in [38], [221], [270].

Paper [90] contains a more detailed description of spin tomography based on group theory, as well as a proposal for its practical implementation.

It must be mentioned that quantum state tomography is often one of the key steps in algorithms for identification of the quantum Hamiltonian determining the dynamics. One algorithm of this type is described in [289].

The treatment of observability for quantum systems is taken from [75] and [85], to which we refer for the generalizations mentioned in Remarks 5.2.5, 5.2.6, 5.2.7. This paper also presents a system theoretic treatment of observers for quantum systems related to formula (5.32). An alternative constructive algorithm which uses multiple measurements is given in [77]. A study of quantum observability which also presents some generalizations to the open systems case can be found in [149].

5.5 Exercises

Exercise 5.1 Prove that, if we relax the assumption that A is an Hermitian operator, the conditions of Propositions 5.1.1, 5.1.2 are also necessary for (5.5).

Exercise 5.2 Let $P_1,...,P_l$ be the orthogonal projections associated to an observable S as in the spectral decomposition (cf. 1.2.1.1). Let $\mathcal{L}$ be the dynamical Lie algebra and $e^{\mathcal{L}}$ the corresponding connected Lie group. Let the observability space $\mathcal{V}$ be calculated starting from $P_1,...,P_l$, as in Remark 5.2.5. Then prove that

$$i\mathcal{V} = \text{span}_{X_1,X_2,...,X_l \in e^{\mathcal{L}}} \{X_1^\dagger P_1 X_1, ..., X_l^\dagger P_l X_l\}.$$

Exercise 5.3 Consider a spin $\frac{1}{2}$ particle under Hamiltonian $H(u(t)) = \sigma_x u_x(t) + \sigma_y u_y(t)$ with measurement of the expectation value of the spin in the z direction, $\langle \sigma_z \rangle_\rho$. Write ρ in the Bloch representation (1.19) and consider the control $u_x(t) \equiv 1, u_y(t) \equiv 0, t \in [0,1]$, $u_x(t) \equiv 0, u_y(t) \equiv 1, t \in (1,2]$. Give a formula for the determination of x, y and z in (1.19) based on (5.32).

Exercise 5.4 Referring to Proposition 5.2.8, give an example of a system which is not operator controllable but is observable.

Chapter 6

Lie Group Decompositions and Control

Consider a Lie algebra $\mathcal{L}$ and the corresponding connected Lie group $e^{\mathcal{L}}$. A *decomposition* of $e^{\mathcal{L}}$ is a factorization of all the elements X of $e^{\mathcal{L}}$ as products

$$X = X_r X_{r-1} \cdots X_1, \tag{6.1}$$

with $X_j \in G_j$, $j = 1, ..., r$, and G_j are proper subsets of $e^{\mathcal{L}}$. These subsets typically have some further algebraic or geometric structure; for example, they are Lie subgroups of $e^{\mathcal{L}}$.

An example of a decomposition of a Lie group can be found in the proof of the controllability test of Chapter 3 which is given in Appendix D. The discussion in subsection 3.2.3 shows that, if the Lie group $e^{\mathcal{L}}$ is compact, as for example $SU(n)$, and $\{A_1, ..., A_s\}$ is a set of generators of the Lie algebra $\mathcal{L}$, then there exists a number r such that every X in $e^{\mathcal{L}}$ can be written as

$$X = e^{\tilde{A}_r t_r} e^{\tilde{A}_{r-1} t_{r-1}} \cdots e^{\tilde{A}_1 t_1}, \tag{6.2}$$

with $t_j \geq 0$ and $\tilde{A}_j$ given matrices in the set $\{A_1, ..., A_s\}$, $j = 1, ..., r$. Formula (6.2) is a Lie group decomposition of the type (6.1) where the role of the subsets G_j is played by the semigroups $\{e^{\tilde{A}_j t} | t \geq 0\}$.

A typical control problem for the Schrödinger operator equation

$$\dot{X} = -iH(u)X, \qquad X(0) = \mathbf{1}, \tag{6.3}$$

is to find the control function u which drives X from the identity $\mathbf{1}$ to a desired target $X_f \in e^{\mathcal{L}}$. Now, if we have a Lie group decomposition of $e^{\mathcal{L}}$, we can write X_f as a product of elements X_{jf} in subsets of $e^{\mathcal{L}}$, i.e.,

$$X_f = X_{rf} X_{r-1f} \cdots X_{1f}. \tag{6.4}$$

Assume now that we are able for every X_{jf} to find a control u_j which drives the identity to X_{jf}. Then, by the right invariance property of system (6.4),[1]

---

[1]This means that if $X(t)$ is the solution with initial condition equal to the identity, the solution with initial condition equal to S is $X(t)S$.

the concatenation of the controls u_1, u_2,...,u_r, in this order, drives the solution of (6.3) from the identity to the desired value X_f. So the problem of steering control is reduced to a sequence of subproblems. Lie group decompositions are of course useful if the subproblems are easier to solve than the original problem. This is usually the case if the subsets G_j, $j = 1,...,r$ are smaller subgroups or semigroups of $e^{\mathcal{L}}$. If they are one-dimensional semigroups $G_j :=$ $\{e^{\tilde{A}_j t}|t \geq 0\}$, $j = 1,...,r$, and there are values $\tilde{u}_j$ such that $-iH(\tilde{u}_j) = \tilde{A}_j$, the j-th control u_j will be equal to $\tilde{u}_j$ for an appropriate amount of time and the resulting overall control will be piecewise constant.

There are several Lie group decompositions known in the literature and developed independently of control theory. They are an important part of Lie group theory and linear algebra. For quantum systems, Lie group decompositions are not only a tool to design controls but also a way to analyze dynamics, as will be discussed in Chapter 9. The subject is very vast. For much of this chapter, we shall focus on decompositions deriving from Cartan classification of the symmetric spaces of the classical Lie groups also known as *Cartan decompositions*. We shall emphasize the computational aspects which are relevant for control purposes. Further investigations on how to generate new decompositions starting from Cartan decompositions and how to use these results for quantum dynamical analysis will be presented in Chapter 9.

This chapter is organized as follows. We shall start by discussing in detail the case of two-level systems and the associated Lie group $SU(2)$. In this case, the decomposition used is the classical Euler's resolution of a rotation. In section 6.2, we give a first example of a general decomposition of $SU(n)$, namely the decomposition into planar rotation, which is quite simple and can be obtained by straightforward computations. This decomposition gives a direct way to parametrize the special unitary group. In section 6.3 we present the basics of Cartan decompositions of semisimple Lie groups. Finally, section 6.4 gives some more examples of applications of Lie group decompositions to control which use the theory described in the previous sections. In the treatment, we shall introduce some more notions of Lie group and Lie algebra theory which complement the ones in Chapters 3 and 4 and will be further developed in Chapter 9.

6.1 Decompositions of $SU(2)$ and Control of Two-Level Systems

6.1.1 The Lie groups $SU(2)$ and $SO(3)$

6.1.1.1 $SU(2)$

The Lie algebra associated with $SU(2)$ (cf. subsection 3.3.1.3) is the Lie algebra $su(2)$ of 2×2, skew-Hermitian matrices which is spanned by the matrices (cf. the Pauli matrices in (1.20))

$$\bar{\sigma}_x := \frac{1}{2} \begin{pmatrix} 0 & i \\ i & 0 \end{pmatrix}, \qquad \bar{\sigma}_y := \frac{1}{2} \begin{pmatrix} 0 & -1 \\ 1 & 0 \end{pmatrix}, \qquad \bar{\sigma}_z := \frac{1}{2} \begin{pmatrix} i & 0 \\ 0 & -i \end{pmatrix}, \quad (6.5)$$

satisfying the commutation relations

$$[\bar{\sigma}_x, \bar{\sigma}_y] = \bar{\sigma}_z, \qquad [\bar{\sigma}_y, \bar{\sigma}_z] = \bar{\sigma}_x, \qquad [\bar{\sigma}_z, \bar{\sigma}_x] = \bar{\sigma}_y. \qquad (6.6)$$

The Lie algebra $su(2)$ is also an inner product space where the inner product between two elements A and B can be taken given by

$$\langle A, B \rangle = \mathrm{Tr}(AB^\dagger). \qquad (6.7)$$

With this inner product $\bar{\sigma}_{x,y,z}$ are orthogonal to each other.

6.1.1.2 $SO(3)$

$SO(3)$ is the Lie group of 3×3 real orthogonal matrices with determinant equal to one. The associated Lie algebra is $so(3)$, namely the Lie algebra of 3×3 skew-symmetric matrices. It is spanned by the matrices

$$S_x := \begin{pmatrix} 0 & 0 & 0 \\ 0 & 0 & 1 \\ 0 & -1 & 0 \end{pmatrix} \qquad S_y := \begin{pmatrix} 0 & 0 & 1 \\ 0 & 0 & 0 \\ -1 & 0 & 0 \end{pmatrix} \qquad S_z := \begin{pmatrix} 0 & 1 & 0 \\ -1 & 0 & 0 \\ 0 & 0 & 0 \end{pmatrix},$$

$$(6.8)$$

which satisfy the commutation relations

$$[S_x, S_y] = S_z, \qquad [S_y, S_z] = S_x, \qquad [S_z, S_x] = S_y. \qquad (6.9)$$

Every matrix X in $SO(3)$ acting on a vector applied at the origin $\vec{v} \in \mathbf{R}^3$ represents a rotation of $\vec{v}$ about an axis. Any $X \in SO(3)$ different from the identity has exactly one eigenvalue equal to one. The corresponding eigenvector $\vec{v}$ such that $X\vec{v} = \vec{v}$ is the axis of rotation, namely it gives the direction of the vectors which are not modified by X. The distance of every other point from the axis of rotation is unchanged by the operation X. The axis of rotation can also be determined by writing X as $X = e^A$, with $A \in so(3)$, which is always possible,[2] and then finding a vector $\vec{v} \in \mathbf{R}^3$ such that $A\vec{v} = 0$. Consider S_z in (6.8); $e^{S_z t}$ is given by

$$e^{S_z t} = \begin{pmatrix} \cos(t) & \sin(t) & 0 \\ -\sin(t) & \cos(t) & 0 \\ 0 & 0 & 1 \end{pmatrix}.$$

It rotates a vector clockwise about the z axis by an angle of t radians. Analogously, $e^{S_x t}$ and $e^{S_y t}$ rotate a vector about the x (clockwise) and y (counterclockwise) axes.

6.1.1.3 Relation between $SU(2)$ and $SO(3)$

It follows from the commutation relations (6.6), (6.9) that the map $\phi : su(2) \rightarrow so(3)$ defined as

$$\phi(\bar{\sigma}_x) = S_x, \qquad \phi(\bar{\sigma}_y) = S_y, \qquad \phi(\bar{\sigma}_z) = S_z \qquad (6.10)$$

is an isomorphism (cf. Definition 3.1.3).[3] This isomorphism induces a homomorphism (cf. Definition 3.1.10) $\Phi : SU(2) \rightarrow SO(3)$ which is defined as follows. For every matrix in $X \in SU(2)$ there exists a matrix $A \in su(2)$ such that $X = e^A$. If $X = e^A$, then

$$\Phi(X) = \Phi(e^A) := e^{\phi(A)}. \qquad (6.11)$$

This map does not depend on the choice of A and it is a homomorphism. Moreover, it is an onto and two-to-one map in that, for any $S \in SU(2)$, $\Phi(-S) = \Phi(S)$ and S and $-S$ are the only matrices giving the value $\Phi(S)$. For the proofs of these facts and more details see, e.g., [70], [231], [273], [297].

6.1.2 Euler decomposition of $SU(2)$ and $SO(3)$

It is a well-known fact of classical mechanics, and it is discussed in detail for example in [25], [245], that a rotation about an arbitrary axis can be decomposed into three rotations, the first one about the z axis, the second one about the y axis and the third one about the z axis, in that order. This means that every matrix X in $SO(3)$ can be written as

$$X = e^{S_z t_3} e^{S_y t_2} e^{S_z t_1}, \qquad (6.12)$$

where the parameters $t_{1,2,3}$ can be chosen nonnegative and represent the angles in radians of the corresponding rotations. The decomposition (6.12) for $SO(3)$

---

[2]The exponential map is surjective for compact connected Lie groups such as $SU(n)$ and $SO(n)$ (cf. [244]).

[3]Therefore, both $su(2)$ and $so(3)$ are isomorphic to $\mathbf{R}^3$ with the cross product operation, cf. Example 3.1.4.

is called **Euler's decomposition** and the parameters t_1, t_2 and t_3 are called *Euler angles*. If $\tilde{X} \in SU(2)$ is such that $\Phi(\tilde{X}) = X$, with Φ in (6.11) and X given in (6.12), then $\tilde{X}$ is given by

$$\tilde{X} = e^{\bar{\sigma}_z t_3} e^{\bar{\sigma}_y t_2} e^{\bar{\sigma}_z t_1}, \tag{6.13}$$

or

$$\tilde{X} = -e^{\bar{\sigma}_z t_3} e^{\bar{\sigma}_y t_2} e^{\bar{\sigma}_z t_1} = e^{2\sigma_z \pi} e^{\bar{\sigma}_z t_3} e^{\bar{\sigma}_y t_2} e^{\bar{\sigma}_z t_1} = e^{\bar{\sigma}_z (t_3 + 2\pi)} e^{\bar{\sigma}_y t_2} e^{\bar{\sigma}_z t_1}. \tag{6.14}$$

In order to see this, apply Φ to both sides of

$$L := e^{\bar{\sigma}_z t_3} e^{\bar{\sigma}_y t_2} e^{\bar{\sigma}_z t_1}.$$

Using the definition of Φ in (6.11) and (6.10) we obtain

$$\Phi(L) = X.$$

Therefore, from the properties of the homomorphism Φ, we have $\tilde{X} = \pm L$, which gives formulas (6.13) and (6.14). Formula (6.13) is the *Euler decomposition of $SU(2)$*. The parameters $t_{1,2,3}$ can be chosen nonnegative and are called *Euler's angles*, as well.

Remark 6.1.1 Notice that the Pauli matrices (6.5) can be transformed one into the other by a similarity transformation according to $\bar{\sigma}_x \to \bar{\sigma}_y \to \bar{\sigma}_z \to \bar{\sigma}_x$. Therefore, Euler decomposition could have been stated by using rotation about the z and x axes or x and y axes. In fact, every two rotations about two orthogonal axes could have been used (see also subsection 6.1.4 for detailed calculations).

6.1.3 Determination of the angles in the Euler decomposition of $SU(2)$

Given a matrix X in $SU(2)$, we show in this subsection how to determine the Euler's angles. Notice that, from the above discussion, this also gives Euler's angles for the image of X in $SO(3)$ under Φ, which can also be determined independently (see, e.g., [231] Chapter 8). An alternative method will follow from the general methods for determination of Cartan decompositions given in section 6.3 below. In fact we shall see that Euler's decomposition is a special case of Cartan decomposition.

An arbitrary matrix $X \in SU(2)$ is written as

$$X : \begin{pmatrix} Me^{i\phi} & Ne^{i\psi} \\ Ge^{i\eta} & Te^{i\gamma} \end{pmatrix}.$$

Since $M^2 + N^2 = 1$, $M^2 + G^2 = 1$, and $N^2 + T^2 = 1$, we can rewrite X as

$$X = \begin{pmatrix} Me^{i\phi} & \sqrt{1 - M^2} e^{i\psi} \\ -\sqrt{1 - M^2} e^{i\eta} & Me^{i\gamma} \end{pmatrix}. \tag{6.15}$$

The signs are chosen arbitrarily since we can always change them by modifying the phases in the exponentials in (6.15). In particular, we can set $M = \cos(\theta)$, $\sqrt{1 - M^2} = \sin(\theta)$, with $\theta \in [0, \frac{\pi}{2}]$. Therefore, we have

$$X = \begin{pmatrix} \cos(\theta)e^{i\phi} & \sin(\theta)e^{i\psi} \\ -\sin(\theta)e^{i\eta} & \cos(\theta)e^{i\gamma} \end{pmatrix}.$$

The condition $\det(X) = 1$ gives

$$\det(X) = \cos^2(\theta)e^{i(\phi+\gamma)} + \sin^2(\theta)e^{i(\psi+\eta)} = 1,$$

which implies $\phi + \gamma = 0$ and $\psi + \eta = 0$ if $\theta \notin \{0, \frac{\pi}{2}, \pi\}$. We can always choose $\gamma = -\phi$ and $\eta = -\psi$ because in the case $\theta = 0$ and $\theta = \pi$, the values of ψ and η do not matter (the phase factors are multiplied by zero) and analogously when $\theta = \frac{\pi}{2}$, the values of γ and ϕ do not matter. In conclusion, a general matrix $X \in SU(2)$ can be written as

$$X = \begin{pmatrix} \cos(\theta)e^{i\phi} & \sin(\theta)e^{i\psi} \\ -\sin(\theta)e^{-i\psi} & \cos(\theta)e^{-i\phi} \end{pmatrix}, \tag{6.16}$$

with $\theta \in [0, \frac{\pi}{2}]$ and $\phi, \psi \in [0, 2\pi)$. Now notice that

$$e^{\bar{\sigma}_y t} = \begin{pmatrix} \cos(\frac{t}{2}) & \sin(\frac{t}{2}) \\ -\sin(\frac{t}{2}) & \cos(\frac{t}{2}) \end{pmatrix}, , \qquad e^{\bar{\sigma}_z t} = \begin{pmatrix} e^{i(\frac{t}{2})} & 0 \\ 0 & e^{-i(\frac{t}{2})} \end{pmatrix}. \tag{6.17}$$

Using (6.17) and (6.16), and choosing

$$\frac{t_2}{2} = \theta, \qquad \frac{t_1}{2} = \frac{\phi - \psi}{2}, \qquad \frac{t_3}{2} = \frac{\phi + \psi}{2},$$

a straightforward computation shows that

$$e^{\bar{\sigma}_z t_3} e^{\bar{\sigma}_y t_2} e^{\bar{\sigma}_z t_1} = X, \tag{6.18}$$

with X in (6.16). Notice that it is always possible to choose $t_{1,2,3} \geq 0$, since we can always add to them multiples of 4π without modifying the product in (6.18).

6.1.4   Application to the control of two-level quantum systems

Consider a two-level quantum control system whose Schrödinger operator equation reads as

$$\dot{X} = (u_z \bar{\sigma}_z + u_y \bar{\sigma}_y)X, \tag{6.19}$$

with u_z and u_y two independent controls. This could represent, in appropriate units, the evolution of a spin $\frac{1}{2}$ particle where u_z and u_y are the z and y components of the electromagnetic field, and the x-component is set to be identically equal to zero (cf. subsection 2.4.2). Assume the problem of control is to drive the state from the identity to a target $X_f \in SU(2)$, and assume we have calculated Euler's decomposition of X_f,

$$X_f = e^{\bar{\sigma}_z t_3} e^{\bar{\sigma}_y t_2} e^{\bar{\sigma}_z t_1},$$

with $t_1, t_2, t_3 \geq 0$. Then a control law $u_z \equiv 1, u_y \equiv 0$, for time t_1, followed by $u_z \equiv 0, u_y \equiv 1$ for time t_2, followed by $u_z \equiv 1, u_y \equiv 0$, for time t_3 will drive the identity to X_f, in time $t_1 + t_2 + t_3$.

The above two-level system (6.19) has a very special structure which itself suggests the use of Euler's decomposition. However, the idea can be applied to much more general two-level systems

$$\dot{X} = -iH(u)X. \tag{6.20}$$

The only requirement is that there exist two values of the control say u_1 and u_2 such that $Z_1 := -iH(u_1)$, and $Z_2 := -iH(u_2)$ are *orthogonal*. If this is the case, let T_1 be a unitary matrix which diagonalizes Z_1 (which always exists since Z_1 is skew-Hermitian). We have

$$T_1 Z_1 T_1^\dagger = \lambda \bar{\sigma}_z,$$

for some $\lambda \in \mathbf{R}$, and

$$T_1 Z_2 T_1^\dagger = a \bar{\sigma}_y + b \bar{\sigma}_x,$$

for some real coefficients a and b. The matrix $T_1 Z_2 T_1^\dagger$ has zero component along $T_1 Z_1 T_1^\dagger$ and therefore zero coefficient for $\bar{\sigma}_z$. A further unitary transformation

$$T_2 := \begin{pmatrix} 1 & 0 \\ 0 & \frac{a+ib}{\sqrt{a^2+b^2}} \end{pmatrix}$$

is such that

$$T_2 T_1 Z_1 T_1^\dagger T_2^\dagger = \lambda \bar{\sigma}_z, \qquad T_2 T_1 Z_2 T_1^\dagger T_2^\dagger = \sqrt{a^2 + b^2} \bar{\sigma}_y.$$

Therefore, in appropriate coordinates, Z_2 and Z_1 are proportional to $\bar{\sigma}_z$ and $\bar{\sigma}_y$, respectively, and therefore an Euler factorization of any matrix exists in terms of Z_1 and Z_2, which is the same as the one in terms of $\bar{\sigma}_z$ and $\bar{\sigma}_y$, except that the Euler's angles $t_{1,2,3}$ have to be divided by λ or $\sqrt{a^2 + b^2}$. The algorithm for control can be applied in this case too. In practice, given the control system (6.20), one chooses two controls u_1 and u_2 to make $-iH(u_1)$ and $-iH(u_2)$ orthogonal, if possible. Then one determines the matrix $T := T_2 T_1$, where T_1 and T_2 are chosen as above, and the Euler decomposition of $T X_f T^\dagger$, where X_f is the target. From this, one finds the control algorithm.

In the most common case, two-level control systems have only one control. The corresponding Schrödinger operator equation is written as

$$\dot{X} = (A + Bu)X.$$

In this case the choice $u_1 = +k$, $u_2 = -k$, with

$$k := \sqrt{\frac{\text{Tr}(A^2)}{\text{Tr}(B^2)}},$$

is a possible choice to give two orthogonal matrices for $A + Bu$. Notice the quantity k is a measure of the 'control authority' of the given system.

6.2 Decomposition in Planar Rotations

The decomposition of the unitary group into planar rotations was described in [202]. It provides a simple and explicit way to parametrize the group $SU(n)$.

A **planar rotation** in $SU(n)$, $U_{j,k}(\theta, \beta)$, $j < k$, is an $n \times n$ matrix which is equal to the identity except for the elements at the intersection between the j-th and k-th rows and columns which are occupied by

$$\tilde{U}(\theta, \beta) := \begin{pmatrix} \cos(\theta) & -\sin(\theta)e^{-i\beta} \\ \sin(\theta)e^{i\beta} & \cos(\theta) \end{pmatrix}.$$

An important property of planar rotations is

$$[U_{j,k}(\theta, \beta)]^{-1} = U_{j,k}^\dagger(\theta, \beta) = U_{j,k}(-\theta, \beta). \tag{6.21}$$

We shall set $D(\alpha_1, \alpha_2, ..., \alpha_{n-1}) := \text{diag}(e^{i\alpha_1}, e^{i\alpha_2}, ..., e^{i\alpha_{n-1}}, e^{-i(\sum_{l=1}^{n-1} \alpha_l)})$. The following theorem gives a decomposition of any special unitary matrix into planar rotations and a matrix of the form $D(\alpha_1, \alpha_2, ..., \alpha_{n-1})$. The proof is constructive. It also show, how to determine the parameters involved in the decomposition.

Theorem 6.2.1 *For any matrix X in $SU(n)$ there exist $n - 1$ parameters, $\alpha_1, ..., \alpha_{n-1}$, $\frac{n(n-1)}{2}$ parameters, $\theta_1, ..., \theta_{\frac{n(n-1)}{2}}$ and $\frac{n(n-1)}{2}$ parameters, $\beta_1, ..., \beta_{\frac{n(n-1)}{2}}$, such that*

$$X = D(\alpha_1, ..., \alpha_{n-1})U_{1,2}(\theta_1, \beta_1)U_{1,3}(\theta_2, \beta_2)U_{2,3}(\theta_3, \beta_3) \cdots \times \tag{6.22}$$

$$U_{1,n}(\theta_{\frac{(n-1)(n-2)}{2}+1}, \beta_{\frac{(n-1)(n-2)}{2}+1})U_{2,n}(\theta_{\frac{(n-1)(n-2)}{2}+2}, \beta_{\frac{(n-1)(n-2)}{2}+2}) \cdots \times$$

$$U_{n-1,n}(\theta_{\frac{n(n-1)}{2}}, \beta_{\frac{n(n-1)}{2}}).$$

Notice the planar rotation factors on the right-hand side of (6.22) are ordered into subsequences according to the second index k, starting from $k = 2$. The subsequence corresponding to k contains factors $U_{1,k}, U_{2,k}, ..., U_{k-1,k}$, in that order.

Proof. Consider a matrix $X \in SU(n)$ and a planar rotation $U_{n-1,n}(\bar{\theta}, \bar{\beta})$. Multiplication of X on the right by $U_{n-1,n}(\bar{\theta}, \bar{\beta})$ only affects the $(n-1)$-th and n-th columns of X. In particular, we select $\bar{\theta}$ and $\bar{\beta}$ so that element $(n, n-1)$ of $XU_{n-1,n}(\bar{\theta}, \bar{\beta})$ is zero. Now consider multiplication of $XU_{n-1,n}$ on the right by a planar rotation $U_{n-2,n}(\tilde{\theta}, \tilde{\beta})$. This only affects the $(n-2)$-th and n-th columns of $XU_{n-1,n}$. In particular, it does not affect the zero in the position $n, n-1$ which was previously introduced. We choose $\tilde{\theta}$ and $\tilde{\beta}$ so that the resulting matrix $XU_{n-1,n}U_{n-2,n}(\tilde{\theta}, \tilde{\beta})$ has a zero in the position $(n, n-2)$. Continuing this way, we multiply on the right by matrices $U_{n-3,n}$, $U_{n-4,n}$,..., $U_{1,n}$ and introduce zeros in the positions $(n, n-3)$, $(n, n-4)$,...,$(n, 1)$, in that order. The matrix $XU_{n-1,n}U_{n-2,n} \cdots U_{1,n}$ has the n-th row equal to zero except for the (n, n)-th element which must have magnitude equal to one, since the matrix is unitary. This also shows that the last column is zero except for the (n, n)-th element. Therefore, we have

$$XU_{n-1,n}U_{n-2,n} \cdots U_{1,n} = \begin{pmatrix} \tilde{X}_{(n-1) \times (n-1)} & 0 \\ 0 & e^{i\eta} \end{pmatrix},$$

for some $\eta \in \mathbf{R}$ and $\tilde{X}_{(n-1) \times (n-1)}$ in $U(n-1)$. Now, further multiplication by factors $U_{j,k}$, with $k < n$, only affects $\tilde{X}$ and therefore, as above, we can choose $U_{n-2,n-1}$, $U_{n-3,n-1}$, ..., $U_{1,n-1}$, to make the last row and column of $\tilde{X}_{(n-1) \times (n-1)}$ equal to zero except for the diagonal element. Continuing this way, after multiplication by $U_{k-1,k}$, $U_{k-2,k}$,...,$U_{1,k}$, for all the k's, $k = n, n-1, n-2, ..., 2$, we obtain a diagonal unitary matrix with determinant equal to one. Therefore, we can write

$$XU_{n-1,n}U_{n-2,n} \cdots U_{1,n}U_{n-2,n-1}U_{n-3,n-1} \cdots U_{1,n-1} \cdots U_{2,3}U_{1,3}U_{1,2}$$

$$= D(\alpha_1, \alpha_2, ..., \alpha_{n-1}).$$

Multiplying this expression by $U_{1,2}^{-1}U_{1,3}^{-1}U_{2,3}^{-1} \cdots U_{n-1,n}^{-1}$, and using (6.21), we obtain (6.22) with appropriate parameters. $\square$

We remark that we could have proceeded by multiplying X by the planar rotations on the left and placing zeros in columns rather than rows. This way, we would have obtained a formula similar to (6.22) but with the planar rotations $U_{j,k}$ on the left of D. The procedure is reminiscent of row (or column) reduction in linear algebra by Gaussian elimination (see, e.g., [176]).

6.3 Cartan Decompositions

In this section we shall discuss Cartan decompositions of Lie groups. We shall start giving some definitions and facts for general Lie groups associated to semisimple Lie algebras and then focus on $SU(n)$. Cartan decompositions appear to be ubiquitous in quantum control and dynamics and in quantum information theory, and we shall see several examples of this in the remainder of this book.

6.3.1 Cartan decomposition of semisimple Lie algebras

Consider a semisimple (cf. Definition 3.1.6) Lie algebra $\mathcal{L}$ with an inner product given by the Killing form in Definition 3.2.2, which in the case of $su(n)$ can be taken equal to the trace inner product defined in (6.7). A decomposition of $\mathcal{L}$ of the form

$$\mathcal{L} = \mathcal{K} \oplus \mathcal{P} \tag{6.23}$$

with $\mathcal{P} := \mathcal{K}^\perp$ is called a **Cartan decomposition** of $\mathcal{L}$ if the following commutation relations are verified:

$$[\mathcal{K}, \mathcal{K}] \subseteq \mathcal{K}, \tag{6.24}$$

i.e., $\mathcal{K}$ is a subalgebra of $\mathcal{L}$,

$$[\mathcal{K}, \mathcal{P}] \subseteq \mathcal{P}, \tag{6.25}$$

and

$$[\mathcal{P}, \mathcal{P}] \subseteq \mathcal{K}. \tag{6.26}$$

Associated with a Cartan decomposition (6.23) is a **Cartan involution**, that is, a Lie algebra isomorphism $\theta : \mathcal{L} \to \mathcal{L}$ which is equal to the identity on $\mathcal{K}$ and multiplies by -1 the elements of $\mathcal{P}$, i.e.,

$$\theta(K) = K, \qquad \forall K \in \mathcal{K}, \tag{6.27}$$

$$\theta(P) = -P, \qquad \forall P \in \mathcal{P}. \tag{6.28}$$

More specifically, given a Cartan decomposition, relations (6.27) and (6.28) determine a Cartan involution θ and conversely given a Cartan involution θ, the $+1$ and -1 eigenspaces of θ determine a Cartan decomposition (cf. Exercise 6.8). Cartan involutions for the Lie algebra $su(n)$ play an important role in the analysis of quantum dynamics as they have a direct correspondence with quantum symmetries as we shall see below.

6.3.2 The decomposition theorem for Lie groups

The importance of a Cartan decomposition for control purposes lies in the following theorem whose proof can be found in standard Lie algebra and Lie group textbooks such as [130], [131].

Theorem 6.3.1 *Given a Cartan decomposition of a semisimple Lie algebra $\mathcal{L}$ as above, every element $X \in e^{\mathcal{L}}$ can be written as*

$$X = PK, \tag{6.29}$$

where P is the exponential of an element of $\mathcal{P}$ and K is an element of the Lie group corresponding to $\mathcal{K}$, i.e., $e^{\mathcal{K}}$.

The coset space (cf. subsection 3.3.1.2), associated with the Cartan decomposition, is $e^{\mathcal{L}}/e^{\mathcal{K}}$. It is called the *globally Riemannian symmetric space*. Cartan classified all the symmetric spaces of the *classical Lie groups*, i.e., $SU(n)$, $Sp(n)$ and $SO(n)$ (see [56], [57], [130]). We shall review the results for $SU(n)$ in the following subsection.

Example 6.3.2 Consider the decomposition of $su(n)$

$$su(n) = so(n) \oplus \mathcal{I}, \tag{6.30}$$

where $\mathcal{I}$ is the vector space in $su(n)$ of purely imaginary matrices and clearly $\mathcal{I} = so(n)^{\perp}$. It is obvious that the following commutation relations hold

$$[so(n), so(n)] \subseteq so(n),$$

$$[so(n), \mathcal{I}] \subseteq \mathcal{I},$$

$$[\mathcal{I}, \mathcal{I}] \subseteq so(n),$$

and therefore the decomposition (6.30) is a Cartan decomposition of $su(n)$ of the form (6.23), where the role of $\mathcal{K}$ is played by $so(n)$ and the role of $\mathcal{P}$ is played by $\mathcal{I}$. An application of Theorem 6.3.1 says that any matrix $X \in SU(n)$, i.e., any $n \times n$ unitary matrix with determinant equal to one, can be written as $X = PK$, where P is the exponential of a skew-Hermitian, purely imaginary, matrix and K is a matrix in $SO(n)$, namely an $n \times n$ orthogonal matrix with determinant equal to one.

6.3.3 Refinement of the decomposition; Cartan subalgebras

Consider now a subalgebra $\mathcal{A}$ of $\mathcal{L}$ which is a subspace of $\mathcal{P}$. Since $[\mathcal{A}, \mathcal{A}] \subseteq \mathcal{A} \subseteq \mathcal{P}$ and $[\mathcal{A}, \mathcal{A}] \subseteq [\mathcal{P}, \mathcal{P}] \subseteq \mathcal{K}$, we have $[\mathcal{A}, \mathcal{A}] = \{0\}$, i.e., $\mathcal{A}$ has to be necessarily Abelian. A maximal Abelian subalgebra in $\mathcal{P}$ (i.e., an Abelian subalgebra such that any element of $X \in \mathcal{P}$, $X \notin \mathcal{A}$ is such that $[X, \mathcal{A}] \neq \{0\}$) is called a **Cartan subalgebra** of $\mathcal{L}$, for the given decomposition.[4] Its dimension is called the **rank** of the symmetric space $e^{\mathcal{L}}/e^{\mathcal{K}}$.

---

[4]This concept should not be confused with the Cartan subalgebra introduced for a general semisimple Lie algebra in Chapter 4 in Definition 4.1.2 which is a concept for the whole Lie algebra $\mathcal{L}$ not referring to any decomposition (cf. Exercise 6.4).

Example 6.3.3 In Example (6.3.2), for the case of $su(n)$, a Cartan subalgebra is spanned by the set of diagonal matrices and therefore the rank of the associated symmetric space is $n - 1$. There is no larger set of commuting linearly independent matrices, since every set of mutually commuting skew-Hermitian matrices are simultaneously diagonalizable (cf., e.g., [135]).

The following theorem [130] will allow us to refine the Cartan decomposition for the Lie group $e^{\mathcal{L}}$.

Theorem 6.3.4 *Let $\mathcal{A}$ be a Cartan subalgebra relative to the decomposition $\mathcal{L} = \mathcal{K} \oplus \mathcal{P}$. Then*

$$\bigcup_{K \in e^{\mathcal{K}}} KAK^{-1} = \mathcal{P}.$$

We also have [130].

Theorem 6.3.5 *Given two Cartan subalgebras in $\mathcal{P}$, $\mathcal{A}$ and $\mathcal{A}'$, there exists a $K \in e^{\mathcal{K}}$ such that*

$$KAK^{-1} = \mathcal{A}'.$$

Theorems 6.3.4 and 6.3.5 say that every element of $\mathcal{P}$ belong to a Cartan subalgebra and that all the Cartan subalgebras are conjugate via an element K of $e^{\mathcal{K}}$.

Example 6.3.6 For the Cartan decomposition of $su(n)$ treated in Example 6.3.2, Theorem 6.3.4 says that every purely imaginary matrix $\tilde{P}$ can be written as

$$\tilde{P} = K\tilde{A}K^{T},$$

where $\tilde{A}$ is a skew-Hermitian purely diagonal matrix and K is a special orthogonal matrix ($\in SO(n)$). In other words, every skew-Hermitian purely imaginary matrix can be diagonalized using an orthogonal matrix.

Theorem 6.3.7 *Consider a semisimple Lie algebra $\mathcal{L}$ with a Cartan decomposition (6.23). Then, every element $X \in e^{\mathcal{L}}$ can be written as[5]*

$$X = K_1 A K_2, \tag{6.31}$$

where K_1 and K_2 are elements of $e^{\mathcal{K}}$ and A is an element of $e^{\mathcal{A}}$.

---

[5]We shall refer to the decomposition (6.31) as the Cartan KAK decomposition while (6.29) is called the Cartan PK decomposition.

Proof. If $X \in e^{\mathcal{L}}$, we know from Theorem 6.3.1 that (6.29) holds with $P = e^{\tilde{P}}$, $\tilde{P} \in \mathcal{P}$ and $K \in e^{\mathcal{K}}$. Using Theorem 6.3.4 we can write P as $P = \tilde{K}e^{\tilde{A}}\tilde{K}^{-1}$, with $\tilde{K} \in e^{\mathcal{K}}$ and $\tilde{A} \in \mathcal{A}$ which in (6.29) gives

$$X = \tilde{K}e^{\tilde{A}}\tilde{K}^{-1}K,$$

which is (6.31) with $K_1 = \tilde{K}$, $K_2 = \tilde{K}^{-1}K$ and $A = e^{\tilde{A}}$. □

Example 6.3.8 For Example 6.3.2, Theorem 6.3.7 says that every special unitary matrix $X \in SU(n)$ can be written as in (6.31) where K_1 and K_2 are orthogonal matrices and A is a diagonal matrix of the type

$$A := \text{diag}\big(e^{i\alpha_1}, e^{i\alpha_2}, ..., e^{i\alpha_{n-1}}, e^{-i\sum_{k=1}^{n-1}\alpha_k}\big).$$

The theorem is merely an existence theorem and does not address the problem of finding the factors in (6.31). Several algorithms of numerical linear algebra exist to calculate the factors and in low-dimensional cases the factors can be calculated by elementary matrix manipulations. For example, the Euler decomposition and the methods to calculate it illustrated in the previous section are a special case of a Cartan decomposition, as discussed in the following example.

Example 6.3.9 In the special case of $su(2)$, a Cartan decomposition is

$$su(2) = so(2) \oplus so(2)^{\perp},$$

with

$$so(2) := \text{span}\{\bar{\sigma}_y\}, \qquad so(2)^{\perp} := \text{span}\{\bar{\sigma}_x, \bar{\sigma}_z\}$$

(cf. (6.5)). If we choose the Cartan subalgebra $\mathcal{A} := \text{span}\{\bar{\sigma}_z\}$, then the decomposition (6.31) reads as

$$X = e^{\bar{\sigma}_y t_3} e^{\bar{\sigma}_z t_2} e^{\bar{\sigma}_y t_1}$$

since all the Lie groups involved are one-dimensional. This is an Euler decomposition since $\bar{\sigma}_y$ and $\bar{\sigma}_z$ are orthogonal to each other.

6.3.4 Cartan decompositions of $su(n)$

A consequence of Cartan classification of the symmetric spaces of $SU(n)$ is the classification of all the possible Cartan decompositions for the Lie algebra $su(n)$ up to conjugation. In fact, up to conjugation, there are only three types of decompositions, denoted by **AI, AII, AIII**. The corresponding subalgebras $\mathcal{K}$ are (conjugate to) the following:

AI

$$\mathcal{K} = so(n),$$

AII

$$K = sp\left(\frac{n}{2}\right),$$

if n is even,

AIII

$$K = \text{span}\left\{\begin{pmatrix} A & 0 \\ 0 & B \end{pmatrix}, \mid A \in u(p),\ B \in u(q),\ p+q=n,\ \text{Tr}(A)+\text{Tr}(B)=0\right\}. \tag{6.32}$$

The subspace $\mathcal{P}$ in all the cases can be found as the orthogonal complement of K in $su(n)$. The Cartan subalgebras $\mathcal{A}$ are as follows. In the case **AI**, $\mathcal{A}$ is (conjugate via an element of $SO(n)$ to) the subalgebra of diagonal matrices. Therefore, this decomposition has rank $n-1$. In the case **AII**, $\mathcal{A}$ is (conjugate via an element of $Sp(\frac{n}{2})$ to) the subalgebra of repeat diagonal matrices, i.e., the matrices of the form

$$A = \begin{pmatrix} D & 0 \\ 0 & D \end{pmatrix}, \tag{6.33}$$

with $D \in su(\frac{n}{2})$ diagonal. Therefore, the rank of this decomposition is $\frac{n}{2}-1$. In the case **AIII**, $\mathcal{A}$ is (conjugate via an element of the Lie group of 2-block diagonal special unitary matrices (cf. (6.32)) to) the subalgebra of matrices of the form

$$A = \begin{pmatrix} 0 & B \\ -B^T & 0 \end{pmatrix}. \tag{6.34}$$

Here B is a real $p \times q$ matrix which is zero everywhere except for the first p columns if $p \leq q$, or the first q rows if $p > q$, which are occupied by a $p \times p$, or respectively, a $q \times q$, diagonal matrix. Therefore, the rank of this decomposition is $\min\{p,q\}$. In the following, for this decomposition we shall assume, without loss of generality, $p \leq q$.

The Cartan involution in the case **AI**, which we denote by θ_I, has as the $+1$ (-1) eigenspace, up to a unitary similarity transformation T, $so(n)$ $(so(n)^\perp)$. Therefore, in the standard case, where T is taken to be the identity, θ_I is given by complex conjugation. In the general case, it has the form

$$\theta_I(A) := TT^T \bar{A}(TT^T)^\dagger, \tag{6.35}$$

for a given unitary matrix T, and $\bar{A}$ denotes the complex conjugate of the matrix A. Analogously, the Cartan involution in the case **AII**, θ_{II}, is given by

$$\theta_{II}(A) := TJT^T \bar{A}(TJT^T)^\dagger, \tag{6.36}$$

for some unitary matrix T, where J is defined in equation (3.23). In the special case where T is the identity, it is easily seen that the matrices spanning the $+1$ eigenspace of θ_{II} are the skew-Hermitian matrices satisfying equation (3.29), i.e., the matrices in the symplectic Lie algebra $sp(\frac{n}{2})$. In the case **AIII**, the involution θ_{III} has the form

$$\theta_{III}(A) = TI_{p,q}T^{\dagger}A(TI_{p,q}T^{\dagger})^{\dagger}, \qquad (6.37)$$

for some unitary matrix T, where $I_{p,q}$ is the matrix

$$I_{p,q} = \begin{pmatrix} \mathbf{1}_{p\times p} & 0 \\ 0 & -\mathbf{1}_{q\times q} \end{pmatrix},$$

with $\mathbf{1}_{p\times p}$ and $\mathbf{1}_{q\times q}$ the $p \times p$ and $q \times q$ identity matrices, respectively, and $p+q = n$. In the special case where T is the identity matrix the matrices in the $+1$ eigenspace of θ_{III} are the matrices in the subalgebra of the skew-Hermitian matrices defined in (6.32).

6.3.5 Cartan involutions of $su(n)$ and quantum symmetries

6.3.5.1 Discrete quantum symmetries

Given a quantum system with underlying Hilbert space $\mathcal{H}$, we define a **quantum mechanical discrete symmetry** as a map Θ, $\Theta : \mathcal{H} \to \mathcal{H}$, of the form

$$\Theta := e^{i\phi}U,$$

where ϕ is a constant, physically irrelevant, real parameter, and U is either a *unitary* operator or an *anti-unitary* one. An anti-unitary operator U, $|\alpha\rangle \to |\tilde{\alpha}\rangle := U|\alpha\rangle$ is defined as satisfying

$$\langle \tilde{\beta}|\tilde{\alpha}\rangle = \langle \beta|\alpha\rangle^{*}, \qquad (6.38)$$

$$U\left(c_1|\alpha\rangle + c_2|\beta\rangle\right) = c_1^{*}U|\alpha\rangle + c_2^{*}U|\beta\rangle. \qquad (6.39)$$

Once a basis of the Hilbert space $\mathcal{H}$ is chosen, an anti-unitary operator U can always be written as

$$U|\alpha\rangle = X|\bar{\alpha}\rangle, \qquad (6.40)$$

where $|\bar{\alpha}\rangle$ is the operation which conjugates all the components of the vector $|\alpha\rangle$ and X is unitary.

A symmetry Θ, whether unitary or anti-unitary, induces a linear transformation on the space of Hermitian operators (observables) A, given by

$$A \to \Theta A\Theta^{-1} := \tilde{\theta}(A). \qquad (6.41)$$

It is in fact easily verified that $\tilde{\theta}(A)$ is a linear and Hermitian operator. Moreover the eigenvalues of $\tilde{\theta}(A)$ are the same as those of A and a set of orthonormal eigenvectors is given by $\{\Theta|\alpha_j\rangle\}$, where $\{|\alpha_j\rangle\}$ is an orthonormal basis of eigenvectors of A.

Description of the symmetry Θ is usually done by specifying how $\tilde{\theta}$ acts on Hermitian operators rather than how Θ acts on states. This is because Hermitian operators represent physical observables and therefore the action of $\tilde{\theta}$ on observables is suggested by physical considerations. For example, the *parity* or *space inversion symmetry* is defined such that

$$\tilde{\theta}(\hat{x}) = -\hat{x},$$

where $\hat{x}$ denotes the position operator. On the other hand, specification of $\tilde{\theta}$ on an irreducible set of observables uniquely determines Θ up to a phase factor [113]. An irreducible set of observables $\{A_j\}$ is defined such that, if an observable B commutes with all of the $\{A_j\}$'s, then B is a multiple of the identity.

6.3.5.2 The Jordan algebra $iu(n)$

Consider now the vector space of Hermitian matrices of dimension n, which represent observables for an n-dimensional quantum system. This vector space (over the reals) will be denoted by $iu(n)$ as it may be obtained by multiplying by the imaginary unit i all elements of $u(n)$. With the anti-commutation operation

$$\{A, B\} := AB + BA,$$

it has the structure of a *Jordan algebra* (cf. Exercise 6.7). A homomorphism $\phi : iu(n) \to iu(n)$ over this Jordan algebra is a linear map which preserves the multiplication in the algebra, i.e.,

$$\phi(\{A, B\}) = \{\phi(A), \phi(B)\}.$$

In particular, quantum symmetries $\tilde{\theta}$ from Definition (6.41) are Jordan algebra homomorphisms.

6.3.5.3 Correspondence between quantum symmetries and Cartan involutions; Cartan decompositions of the Jordan algebra $iu(n)$

Special types of quantum symmetries $\tilde{\theta}$ are the ones which are equal to the identity when applied twice, i.e., $\tilde{\theta}^2 = 1$. We can call these symmetries *Cartan involutory symmetries* or simply *Cartan symmetries*. A Cartan symmetry for us is a Jordan algebra homomorphism on $iu(n)$ such that $\tilde{\theta}^2 = 1$. Consider the formula

$$\theta(A) := i\tilde{\theta}(iA). \tag{6.42}$$

If $\mathcal{A}_+$ and $\mathcal{A}_-$ are the $+1$ and -1 eigenspaces of $\tilde{\theta}$ in $iu(n)$ then $-i\mathcal{A}_+$ and $-i\mathcal{A}_-$ are the -1 and $+1$ eigenspaces of θ, i.e., θ is defined as the Lie algebra homomorphism which multiplies by -1 on $-i\mathcal{A}_+$ and by $+1$ on $-i\mathcal{A}_-$. Consistently with the notation above we can set $\mathcal{P} := -i\mathcal{A}_+$ and $\mathcal{K} = -i\mathcal{A}_-$ (notice the eigenspaces are inverted). Vice versa given $\mathcal{K}$ and $\mathcal{P}$ the $+1$ and

-1, respectively, eigenspaces of θ, $\mathcal{A}_+ = i\mathcal{P}$ and $\mathcal{A}_- = i\mathcal{K}$ are the $+1$ and -1 eigenspaces of $\tilde{\theta}$, i.e., $\tilde{\theta}$ is defined as the Jordan algebra homomorphism which multiplies by $+1$ on $i\mathcal{P}$ and by -1 on $i\mathcal{K}$. This definition is consistent in the cases **AI** and **AII** above because in these cases we have (Exercise 6.9)

$$\{i\mathcal{P}, i\mathcal{P}\} \subseteq i\mathcal{P}, \qquad \{i\mathcal{K}, i\mathcal{P}\} \subseteq i\mathcal{K}, \qquad \{i\mathcal{K}, i\mathcal{K}\} \subseteq i\mathcal{P}, \tag{6.43}$$

that is, the isomorphism which multiplies by $+1$ elements in $i\mathcal{P}$ and by -1 elements in $i\mathcal{K}$ is in fact a Jordan algebra homomorphism. In the case **AIII** we use the correspondence

$$\theta(A) := -i\tilde{\theta}(iA) \tag{6.44}$$

Defined again $\mathcal{A}_+$ and $\mathcal{A}_-$ as the $+1$ and -1 eigenspaces of $\tilde{\theta}$ in $iu(n)$, then $-i\mathcal{A}_+$ and $-i\mathcal{A}_-$ are the $+1$ and -1 eigenspaces of θ, i.e., θ is defined as the Lie algebra homomorphism which multiplies by $+1$ on $-i\mathcal{A}_+$ and by -1 on $-i\mathcal{A}_-$. In this case, the eigenspaces are not reversed. Consistently with the notation above we can set $\mathcal{P} := -i\mathcal{A}_-$ and $\mathcal{K} = -i\mathcal{A}_+$. Vice versa given $\mathcal{K}$ and $\mathcal{P}$ the $+1$ and -1, respectively, eigenspaces of θ, $\mathcal{A}_- = i\mathcal{P}$ and $\mathcal{A}_+ = i\mathcal{K}$ are the -1 and $+1$ eigenspaces of $\tilde{\theta}$, i.e., $\tilde{\theta}$ is defined as the Jordan algebra homomorphism which multiplies by -1 on $i\mathcal{P}$ and by $+1$ on $i\mathcal{K}$. This definition is consistent in the cases of Cartan decomposition of the type **AIII** since we have (Exercise 6.9)

$$\{i\mathcal{K}, i\mathcal{K}\} \subseteq i\mathcal{K}, \qquad \{i\mathcal{K}, i\mathcal{P}\} \subseteq i\mathcal{P}, \qquad \{i\mathcal{P}, i\mathcal{P}\} \subseteq i\mathcal{K}. \tag{6.45}$$

In conclusion, to every Cartan decomposition (involution) of the Lie algebra $u(n)$ there corresponds, via formula (6.42) or (6.44), a Cartan decomposition (symmetry) of the Jordan algebra $iu(n)$.

Tables 6.1 and 6.2 summarize the Cartan involutions of $u(n)$ and the corresponding Cartan symmetries of the various types. In order to be consistent with the involutions defined in (6.35), (6.36) and (6.37), we have included span$\{i\mathbf{1}\}$ in the -1 eigenspace of the involution in the cases **AI** and **AII** and in the $+1$ eigenspace for the case **AIII**. In Tables 6.1 and 6.2, the matrix T represents a general unitary matrix, and the orthogonal complement is taken in $u(n)$ or $iu(n)$.

6.3.6 Computation of the factors in the Cartan decompositions of $SU(n)$

There exist several linear algebra algorithms to calculate the factors in the Cartan KAK decomposition (6.31). We shall summarize here some of the main ideas, omitting the details concerning numerical implementation. References can be found in [50], [51]. These algorithms treat the decompositions of the various types, **AI**, **AII** and **AIII**, in the standard form. This gives for example $su(n) = so(n) \oplus so(n)^\perp$ for the type **AI**. Therefore, given a decomposition of $su(n)$ the first task is to find a similarity transformation T

TABLE 6.1: Cartan involutions and symmetries for the various types of symmetric spaces: Up to a conjugacy T, which is a general unitary matrix, there are three types of Cartan decompositions of of the Lie algebra $u(n)$ and of the Jordan algebra $iu(n)$. In the table θ is the Cartan involution, $\tilde\theta$ the corresponding Cartan symmetry acting on $iu(n)$ (i.e., acting on observables) given by formula (6.42) for the case **AI** and **AII** and Θ is the corresponding quantum symmetry on states. Notice Θ_I and Θ_{II} are anti-unitary while Θ_{III} is unitary.

Type	θ	$\tilde\theta$	Θ
AI	$\theta_I(A) := TT^T \bar{A}(TT^T)^\dagger$	$\tilde\theta_I(B) = TT^T \bar{B}(TT^T)^\dagger$	$\Theta_I(\lvert\psi\rangle) = TT^T\lvert\bar\psi\rangle$
AII	$\theta_{II}(A) := TJT^T \bar{A}(TJT^T)^\dagger$	$\tilde\theta_{II}(B) = TJT^T \bar{B}(TJT^T)^\dagger$	$\Theta_{II}(\lvert\psi\rangle) = TJT^T\lvert\bar\psi\rangle$
AIII	$\theta_{III}(A) = TI_{p,q}T^\dagger A(TI_{p,q}T^\dagger)^\dagger$	$\tilde\theta_{III}(B) = TI_{p,q}T^\dagger B(TI_{p,q}T^\dagger)^\dagger$	$\Theta_{III}(\lvert\psi\rangle) = TI_{p,q}T^\dagger\lvert\psi\rangle$

TABLE 6.2: Eigenspaces of the Cartan involution θ and Cartan symmetry $\tilde{\theta}$ for various types of Cartan decompositions.

Type	$\mathcal{K}$, +1 eigenspace of θ	+1 eigenspace of $\tilde{\theta}$
AI	$Tso(n)T^\dagger$	$Tiso(n)^\perp T^\dagger$
AII	$Tsp(\frac{n}{2})T^\dagger$	$Tisp(\frac{n}{2})^\perp T^\dagger$
AIII	$\{K \in u(n) : K = \begin{pmatrix} A & 0 \\ 0 & B \end{pmatrix}, A \in u(p), B \in u(q), p+q=n\}$	$\{K \in iu(n) : K = \begin{pmatrix} A & 0 \\ 0 & B \end{pmatrix}, A \in iu(p), B \in iu(q), p+q=n\}$

(the matrix T in Tables 6.1 and 6.2) to put the two subspaces $\mathcal{K}$ and $\mathcal{P}$ in (6.23) in the standard form. Then, once the factors in the decomposition are found, the same similarity transformation T is used to calculate the factors in the original representation. The value of T is more easily obtained when the involution θ (or equivalently, the quantum symmetry $(\tilde{\theta}, \Theta)$) associated with the decomposition is available. This is illustrated in the following example for two coupled two-level systems. This example is a special case of a decomposition (the *Odd-Even Decomposition* (OED)) which will be further discussed in Chapter 9.

Example 6.3.10 Consider $su(4)$, i.e., the dynamical Lie algebra associated with a controllable system of two coupled two-level systems and consider a Cartan decomposition of the form

$$su(4) = \mathcal{K} \oplus \mathcal{P},$$

with

$$\mathcal{K} := \text{span}\{\mathbf{1} \otimes \sigma, \sigma \otimes \mathbf{1}\}, \qquad \sigma = \bar{\sigma}_{x,y,z},$$

with $\bar{\sigma}_{x,y,z}$ defined in (6.5) and $\mathcal{P} = \mathcal{K}^{\perp}$. A simple dimension count shows that this decomposition must be of type **AI**. In particular, this means that $\mathcal{K}$ is conjugate to $so(4)$. However, the matrices of $\mathcal{K}$ are not purely real, i.e., $\mathcal{K} \neq so(n)$. To find the unitary matrix T such that $\mathcal{K} = Tso(4)T^{\dagger}$, we first find the involution corresponding to the decomposition. Consider the Cartan symmetry $\tilde{\theta}_{II}$, in Table 6.1 with T equal to the identity, on $iu(2)$. It is easy to verify that the $+1$ eigenspace is $\text{span}\{\mathbf{1}\}$ and the -1 eigenspace is $\text{span}\{i\bar{\sigma}_x, i\bar{\sigma}_y, i\bar{\sigma}_z\}$. Consider now $\tilde{\theta} := \tilde{\theta}_{II} \otimes \tilde{\theta}_{II}$ on $iu(4)$. This is a Cartan quantum symmetry on $iu(4)$ which acts as (see Remark 6.3.11 below)

$$\tilde{\theta}(A) := \tilde{\theta}_{II} \otimes \tilde{\theta}_{II}(A) = J \otimes J\bar{A}J^{-1} \otimes J^{-1}. \tag{6.46}$$

Therefore, the matrix T which gives $Tso(4)T^{\dagger} = \mathcal{K}$ has to be unitary and, from Table 6.1, such that $TT^{T} = J \otimes J$. This matrix can be chosen up to an orthogonal factor. One possible choice is

$$T := \frac{1}{\sqrt{2}} \begin{pmatrix} 0 & i & 1 & 0 \\ i & 0 & 0 & -1 \\ i & 0 & 0 & 1 \\ 0 & -i & 1 & 0 \end{pmatrix}. \tag{6.47}$$

The columns of the matrix (6.47) are often referred to as the *magic basis*.

Remark 6.3.11 The operator $\tilde{\theta}$ is the tensor product of two anti-linear operators $\tilde{\theta}_{II}$ each acting on $iu(2)$. The vector space $iu(4)$ is the tensor space of two copies of $iu(2)$ defined according to the general definition in subsection 1.1.3. A linear operator which is the tensor product of two operators is defined according to the rules in 1.1.3.3. In particular (cf. (1.14)), if θ_1 and θ_2 are linear on $iu(2)$, then for any A in $iu(4)$, $A := \sum_k \alpha_k B_k \otimes C_k$, $B_k, C_k \in iu(2)$ we have

$$\theta_1 \otimes \theta_2(A) := \sum_k \alpha_k \theta_1(B_k) \otimes \theta_2(C_k).$$

If θ_1 and θ_2 are *anti*-linear then the definition is

$$\theta_1 \otimes \theta_2(A) := \sum_k \alpha_k^* \theta_1(B_k) \otimes \theta_2(C_k).$$

Applying this with $\theta_1 = \theta_2 = \tilde{\theta}_{II}$, with $\tilde{\theta}_{II}(B) := J\bar{B}J^{-1}$, one obtains formula (6.46). Let Θ_1 (Θ_2) be the anti-linear operator on the vectors space $\mathcal{H}_1$ ($\mathcal{H}_2$) corresponding to θ_1 (θ_2). Then $\Theta_1 \otimes \Theta_2$ is defined on a basis $|e_j\rangle \otimes |f_k\rangle$ as $\Theta_1 e_j\rangle \otimes \Theta_2 |f_k\rangle$ and extended by anti-linearity on the full space $\mathcal{H}_1 \otimes \mathcal{H}_2$.

In the rest of this section, we shall assume that we are dealing with a decomposition in the standard form. The main idea of the algorithms for the computation of the factors in Cartan decomposition is to reduce the computation of the factors to a *structured eigenvalue problem* for which there are several algorithms available [52], [100]. We show here how to obtain such a reduction and refer to the literature on structured eigenvalue problems for this particular aspect of the algorithms.

Consider first the case of the KAK decomposition of the type **AI** which says that every unitary matrix U can be written as

$$U = O_1 D O_2, \tag{6.48}$$

where O_1 and O_2 are orthogonal matrices (with determinant equal to one) and D is diagonal. From (6.48) we obtain

$$U^T = O_2^T D O_1^T,$$

which combined with (6.48) gives

$$UU^T = O_1 D^2 O_1^T,$$

or equivalently

$$UU^T O_1 = O_1 D^2.$$

Therefore, the problem amounts to finding the eigenvalues and a set of orthonormal real eigenvectors of the symmetric matrix UU^T. Once the matrices O_1 and D^2 have been found such that $UU^T O_1 = O_1 D^2$, then one can use (6.48) to find the factor O_2.

Example 6.3.12 Consider the matrix $U \in SU(3)$,

$$U = \begin{pmatrix} \frac{1}{\sqrt{2}} & 0 & \frac{-i}{\sqrt{2}} \\ \frac{1}{\sqrt{2}} & 0 & \frac{i}{\sqrt{2}} \\ 0 & i & 0 \end{pmatrix}.$$

We calculate

$$UU^T = \begin{pmatrix} 0 & 1 & 0 \\ 1 & 0 & 0 \\ 0 & 0 & -1 \end{pmatrix}.$$

This matrix has eigenvector $\vec{v}_1 = [\frac{1}{\sqrt{2}}, \frac{1}{\sqrt{2}}, 0]^T$ corresponding to eigenvalue 1, and eigenvectors $\vec{v}_2 = [-\frac{1}{\sqrt{2}}, \frac{1}{\sqrt{2}}, 0]^T$ and $\vec{v}_3 = [0, 0, 1]^T$ both corresponding to the eigenvalue -1. Moreover $\vec{v}_{1,2,3}$ form an orthonormal basis. Therefore, we have in (6.48) $O_1 = [\vec{v}_1, \vec{v}_2, \vec{v}_3]$. The first diagonal entry of D is a square root of 1 while the other two diagonal entries are square roots of -1. Choosing 1, i and $-i$, we find that (6.48) is verified with

$$O_2 = \begin{pmatrix} 1 & 0 & 0 \\ 0 & 0 & 1 \\ 0 & -1 & 0 \end{pmatrix}.$$

In the case **AII**, the KAK decomposition reads as

$$U = K_1 A K_2, \tag{6.49}$$

where K_1 and K_2 are in $Sp(\frac{n}{2})$ and A is a repeat two-block (see (6.33)) diagonal unitary matrix. The matrices K_1 and K_2 satisfy equation (3.22), i.e., we have

$$K_{1,2} J K_{1,2}^T = J.$$

From (6.49), we have

$$U^T J U = K_2^T A K_1^T J K_1 A K_2. \tag{6.50}$$

Using the fact that equation (3.22) also holds true for $K_{1,2}^\dagger$, we get

$$K_{1,2}^T J K_{1,2} = J, \tag{6.51}$$

which placed into (6.50) gives

$$U^T J U = K_2^T J J^T A J A K_2.$$

As A is a repeat block diagonal matrix, we have $J^T A J = A$, which gives

$$J^T U^T J U = J^T K_2^T J A^2 K_2.$$

From (6.51), we obtain $J^T K_2^T J = K_2^\dagger$, which finally gives

$$J^T U^T J U = K_2^\dagger A^2 K_2.$$

Therefore, the problem consists of diagonalizing, via a symplectic matrix K_2, the matrix $J^T U^T J U$. It is again a structured eigenvalue problem.

Finally, for the KAK decomposition of type **AIII**, we can write the decomposition (6.31) in the block form

$$\begin{pmatrix} U_{11} & U_{12} \\ U_{21} & U_{22} \end{pmatrix} = \begin{pmatrix} K_{11} & 0 \\ 0 & K_{12} \end{pmatrix} \begin{pmatrix} C_1 & S \\ -S^T & C_2 \end{pmatrix} \begin{pmatrix} K_{21} & 0 \\ 0 & K_{22} \end{pmatrix}. \qquad (6.52)$$

We refer here to the decomposition and the choice of the Cartan subalgebra in subsection 6.3.4, with $p \leq q$. The matrices K_{jk}, $j, k = 1, 2$, are unitary. The matrix $\begin{pmatrix} C_1 & S \\ -S^T & C_2 \end{pmatrix}$ is the exponential of a matrix of the type A in (6.34). The submatrices C_1 and C_2 are diagonal of dimensions $p \times p$ and $q \times q$, respectively. Assume $p \leq q$. Then $C_2 = \begin{pmatrix} C_1^a & 0 \\ 0 & 1 \end{pmatrix}$, where we denote by C_1^a the anti-transposed of the square matrix C_1, i.e., the matrix obtained by reflecting along the secondary diagonal. The elements in C_1 and C_2 are cosines of some angles. Under the assumption that $p \leq q$, only the first p columns of S are nonzero and are occupied by a $p \times p$ diagonal matrix, with $C_1^2 + SS^T = 1_{p \times p}$. From equation (6.52), the first block gives

$$U_{11}U_{11}^\dagger = K_{11}C_1^2 K_{11}^\dagger.$$

This is again an eigenvalue problem and C_1^2 and K_{11} are chosen to satisfy this equation. This determines C_1 up to signs and therefore C_2 as well as S, up to signs, as well as K_{11}. Once we have C_1 and K_{11}, the equation of block $(1,1)$ allows us to find K_{21}. Then, the equation of block $(2,1)$ allows us to find K_{12} and the equation for the block $(1,2)$ allows us to find K_{22}. The last equation for the block $(2,2)$ should be used to resolve ambiguity due to degeneracy of eigenvalues and or choices of the signs in the previous steps.

6.4 Examples of Application of Decompositions to Control

In this section we present two examples of quantum systems which can be controlled using the decompositions described in the previous sections. Both these models concern a pair of two coupled spin-$\frac{1}{2}$ particles in a driving electromagnetic field. In the first case, we assume a weak interaction of the Ising type (2.65) and we apply Cartan decomposition. In the second case we consider a Heisenberg interaction (2.64), not necessarily weak, and we apply a combination of the previously described techniques.

6.4.1   Control of two coupled spin-$\frac{1}{2}$ particles with Ising interaction

A possible (simplified) model for two coupled spin-$\frac{1}{2}$'s with Ising interaction (2.65) (cf. subsection 2.4.2) is given by the Schrödinger operator equation

$$\dot{X} = -i\left(J_{12}\sigma_z \otimes \sigma_z + \gamma_1\left(\sum_{l=x,y,z} \sigma_l \otimes \mathbf{1}u_l(t)\right) + \gamma_2\left(\sum_{l=x,y,z} \mathbf{1} \otimes \sigma_l u_l(t)\right)\right)X.$$
(6.53)

The Pauli matrices $\sigma_{x,y,z}$ were defined in (1.20), J_{12} is the coupling constant which is assumed small (weak coupling) and γ_1 and γ_2 are the gyromagnetic ratios of spin 1 and 2, respectively. Equation 6.53 can be further simplified. If we scale the time by a factor J_{12}, we can eliminate J_{12} from the equation. Also, by scaling the controls by a factor γ_1, we can eliminate γ_1 from the equations and replace γ_2 with the ratio $r := \frac{\gamma_2}{\gamma_1}$. Another simplification is obtained in the case, which is common in experiments, where the z component of the magnetic field u_z is constant (cf. section 10.1). In this case one defines $\tilde{X} = e^{i(\sigma_z \otimes \mathbf{1} + r\mathbf{1} \otimes \sigma_z)u_z t}X$, and after redefining the controls u_x and u_y the equation for $\tilde{X}$ is given by

$$\dot{\tilde{X}} = -i\left(\sigma_z \otimes \sigma_z + \left(\sum_{l=x,y} \sigma_l \otimes \mathbf{1}u_l(t)\right) + r\left(\sum_{l=x,y} \mathbf{1} \otimes \sigma_l u_l(t)\right)\right)\tilde{X}.$$
(6.54)

We shall make reference to equation (6.54), assuming $|r| \neq 1$ by keeping in mind that, if we solve the control problem to drive $\tilde{X}$ from the identity to a given value $\tilde{X}_f$ in time T, the actual value for $X(T)$ is $X(T) = e^{-i(\sigma_z \otimes \mathbf{1} + r\mathbf{1} \otimes \sigma_z)u_z T}\tilde{X}_f$.

The Lie algebra generated by the matrices which multiply the controls in (6.54), if $r \neq \pm 1$, is given by

$$\mathcal{K} := \text{span}\{i\mathbf{1} \otimes \sigma_{x,y,z}, i\sigma_{x,y,z} \otimes \mathbf{1}\}.$$

This would be the dynamical Lie algebra of the system if there were no interaction. The corresponding Lie group is the group[6]

$$e^{\mathcal{K}} = \{X_1 \otimes X_2 | X_1, X_2 \in SU(2)\}.$$

This means that if we did not have interaction we would be able to induce arbitrary transformations on the two spins separately, but no dynamics would 'entangle' the two spins. Since we have $r \neq \pm 1$, the two spins would react differently to the common magnetic field and this is sufficient to induce arbitrary

---

[6]This can be easily seen by applying the formula for exponentials of Kronecker products (1.22).

transformations on them separately. Notice that if $r = 1$, then the two spins would interact exactly the same way with the external field. The corresponding Lie algebra of the matrices multiplying the controls is three-dimensional, and it is given by

$$\tilde{\mathcal{K}} := \text{span}\{i\mathbf{1} \otimes \sigma_{x,y,z} + i\sigma_{x,y,z} \otimes \mathbf{1}\},$$

and the corresponding Lie group $e^{\tilde{\mathcal{K}}} = \{X \otimes X | X \in SU(2)\}$. In this case, the two spins are perfectly equivalent, and the transformation induced on one spin is the same as the one induced on the other.

Going back to the case $r \neq 1$, it is easy to verify that

$$\mathcal{P} := \mathcal{K}^{\perp} = \text{span}\{i\sigma_l \otimes \sigma_j | l, j = x, y, z\},$$

where the orthogonal complement is taken in $su(4)$. The interaction matrix is in $\mathcal{P}$, and system (6.54) is controllable since the dynamical Lie algebra turns out to be equal to $su(4)$. Moreover, we have that the conditions (6.24)-(6.26) are verified and therefore the decomposition

$$su(4) = \mathcal{K} \oplus \mathcal{P},$$

is a Cartan decomposition of $su(4)$. It is in fact the decomposition of type **AI** which we have considered in Example 6.3.10. The subalgebra in $\mathcal{P}$,

$$\mathcal{A} := \text{span}\{i\sigma_x \otimes \sigma_x, i\sigma_y \otimes \sigma_y, i\sigma_z \otimes \sigma_z\},$$

is a Cartan subalgebra (it is Abelian and it has dimension 3). Therefore, every element X of $SU(4)$ can be written as

$$X = X_1 \otimes X_2 e^{-i\sigma_x \otimes \sigma_x t_1} e^{-i\sigma_y \otimes \sigma_y t_2} e^{-i\sigma_z \otimes \sigma_z t_3} Y_1 \otimes Y_2, \qquad (6.55)$$

for some real parameters t_1, t_2, t_3, and $X_1, X_2, Y_1, Y_2 \in SU(2)$. The decomposition (6.55) displays the two main ingredients in the evolution of the two spins. The factors $X_1 \otimes X_2$ and $Y_1 \otimes Y_2$ describe the evolutions of the spins by themselves while the three factors $e^{-i\sigma_x \otimes \sigma_x t_1}, e^{-i\sigma_y \otimes \sigma_y t_2}, e^{-i\sigma_z \otimes \sigma_z t_3}$ are responsible for the interaction between the two spins. Every unitary evolution can be decomposed in a way that makes transparent the contributions of the 'local' transformation on the single spins and the 'entangling' transformation on the pair of two spins. This will be further explored and generalized in Chapter 9.

The decomposition (6.55) can be further refined by recalling that the Pauli matrices are unitarily equivalent to each other. Therefore, there exists a matrix $U_x \in SU(2)$ such that $U_x \sigma_z U_x^{\dagger} = \sigma_x$ and a matrix $U_y \in SU(2)$ such that $U_y \sigma_z U_y^{\dagger} = \sigma_y$. Therefore, we have

$$U_x \otimes U_x e^{-i\sigma_z \otimes \sigma_z t_1} U_x^{\dagger} \otimes U_x^{\dagger} = e^{-i\sigma_x \otimes \sigma_x t_1}, \ U_y \otimes U_y e^{-i\sigma_z \otimes \sigma_z t_2} U_y^{\dagger} \otimes U_y^{\dagger} = e^{-i\sigma_y \otimes \sigma_y t_2}.$$

Plugging this into (6.55), we obtain a decomposition of X of the type

$$X = X_1 \otimes X_2 e^{-i\sigma_z \otimes \sigma_z t_1} X_3 \otimes X_4 e^{-i\sigma_z \otimes \sigma_z t_2} X_5 \otimes X_6 e^{-i\sigma_z \otimes \sigma_z t_3} X_7 \otimes X_8, \quad (6.56)$$

with $X_j \in SU(2)$, $j = 1, ..., 8$. Now, assume that a decomposition (6.56) is known for the target state in our control problem. Methods such as the ones described in subsection 6.3.6 can be used for its computation. Then the factors of the type $X_j \otimes X_{j+1}$, $j = 1, 3, 5, 7$, can be obtained with the system (6.54) by very fast, high amplitude controls which would make the effect of the interaction term negligible in the whole evolution. The design of these controls is not specified here but notice that the problem without the interaction term is much easier as the dynamical Lie algebra $\mathcal{K}$ is only six-dimensional. In order to obtain factors of the type $e^{-i\sigma_z \otimes \sigma_z t}$, we only have to set the controls equal to zero for time t (notice that t_j, $j = 1, 2, 3$ in (6.56) can always be taken positive as the orbit $\{e^{-i\sigma_z \otimes \sigma_z t} | t \in \mathbf{R}\}$ is periodic). Alternating fast high amplitude controls with controls identically equal to zero, we drive the state from the identity to the desired target. This control has important properties as for time optimality as we shall see in the next chapter.

6.4.2   Control of two coupled spin-$\frac{1}{2}$ particles with Heisenberg interaction

We consider now the model of two interacting spin-$\frac{1}{2}$ particles with Heisenberg interaction (2.64). This model will give us the opportunity to apply much of the machinery introduced in the previous sections, including some topics from Chapter 3 and the Cartan decomposition of $so(3)$. In many cases, the Heisenberg interaction cannot be considered weak compared to the control strength and it has to be constructively used in the control algorithm. The Schrödinger operator equation can be written as (6.53) where the Ising term $J_{12}\sigma_z \otimes \sigma_z$ is replaced by the isotropic Heisenberg term, $J_{12}(\sigma_x \otimes \sigma_x + \sigma_y \otimes \sigma_y + \sigma_z \otimes \sigma_z)$. We perform a change of coordinates $X \to TXT^\dagger$, where T is the matrix (6.47). This matrix T diagonalizes the isotropic Heisenberg term. We re-scale the time by a factor J_{12} and the controls by a factor γ_1 so that only the ratio $r := \frac{\gamma_2}{\gamma_1}$ appears explicitly in the equation. The Schrödinger operator equation can then be written as

$$\dot{X} = (A + B_x u_x + B_y u_y + B_z u_z)X, \quad (6.57)$$

where

$$A = \text{diag}(3i, -i, -i, -i), \quad (6.58)$$

and assuming for simplicity $r = 2$,

$$B_x = \begin{pmatrix} 0 & 0 & 0 & 1 \\ 0 & 0 & -3 & 0 \\ 0 & 3 & 0 & 0 \\ -1 & 0 & 0 & 0 \end{pmatrix},$$

$$B_y = \begin{pmatrix} 0 & 0 & 1 & 0 \\ 0 & 0 & 0 & 3 \\ -1 & 0 & 0 & 0 \\ 0 & -3 & 0 & 0 \end{pmatrix},$$

$$B_z = \begin{pmatrix} 0 & 1 & 0 & 0 \\ -1 & 0 & 0 & 0 \\ 0 & 0 & 0 & -3 \\ 0 & 0 & 3 & 0 \end{pmatrix}.$$

We, first of all, notice that this system is operator controllable, as the dynamical Lie algebra is equal to $su(4)$. The general strategy for control is to set alternatively two of the components of the control equal to zero and use the other one to control on a subgroup of the group $SU(4)$. In this way, the problem of control is decomposed into subproblems on lower-dimensional manifolds. We start by setting $u_x \equiv u_y \equiv 0$ (the other cases are analogous), i.e., we study the equation

$$\dot{X} = AX + B_z X u_z. \tag{6.59}$$

The dynamical Lie algebra associated with this equation, which we denote by $\mathcal{L}_z$, can be calculated with the algorithm in subsection 3.2.2 of Chapter 3. It is five-dimensional and it is spanned by the matrices

$$S_{x,y,z} := \begin{pmatrix} i\sigma_{x,y,z} & 0 \\ 0 & 0 \end{pmatrix}, \quad R_1 := \begin{pmatrix} i1_{2\times2} & 0 \\ 0 & -i1_{2\times2} \end{pmatrix}, \quad R_2 := \begin{pmatrix} 0 & 0 \\ 0 & i\sigma_y \end{pmatrix}. \tag{6.60}$$

As expected from Theorem 3.1.8, the Lie algebra $\mathcal{L}_z$ is reductive. The semisimple (simple in this case) part $\mathcal{S}$ is spanned by $S_{x,y,z}$ and the Abelian center $\mathcal{Z}$ is spanned by R_1 and R_2. We have

$$\mathcal{L}_z = \mathcal{S} \bar{\oplus} \mathcal{Z}. \tag{6.61}$$

The semisimple part, $\mathcal{S}$, is isomorphic to $su(2)$. The Lie group $e^{\mathcal{S}}$ consists of products of matrices of the form $\begin{pmatrix} X & 0 \\ 0 & 1_{2\times2} \end{pmatrix}$ with $X \in SU(2)$, while the Lie group $e^{\mathcal{Z}}$ consists of product of matrices of the form

$$\begin{pmatrix} e^{it_1}1_{2\times2} & 0 \\ 0 & e^{-it_1}1_{2\times2} \end{pmatrix},$$

and

$$\begin{pmatrix} 1_{2\times2} & 0 \\ 0 & Y \end{pmatrix},$$

with

$$Y = \begin{pmatrix} \cos(t_2) & -\sin(t_2) \\ \sin(t_2) & \cos(t_2) \end{pmatrix},$$

for t_1 and t_2 in $\mathbf{R}$. We rewrite (6.59) as

$$\dot{X} = (A_S + B_{zS}u_z)X + (A_R + B_{zR}u_z)X, \qquad X(0) = 1_{4\times 4}$$

with $A = A_S + A_R$ and $B_z = B_{zS} + B_{zR}$, with $\{A_S, B_{zS}\} \in \mathcal{S}$, $\{A_R, B_{zR}\} \in \mathcal{Z}$. Consider now the two equations

$$\dot{U} = (A_S + B_{zS}u_z)U, \qquad U(0) = 1_{4\times 4}, \tag{6.62}$$

$$\dot{V} = (A_R + B_{zR}u_z)V, \qquad V(0) = 1_{4\times 4}.$$

Because of (6.61), for the solution X of (6.59), we have $X = UV = VU$. The problem of control for system (6.62) is essentially a problem for a system on $SU(2)$ of the type discussed in subsection 6.1.4. The control u_z can be chosen to drive U from the identity to any value $\begin{pmatrix} U_f & 0 \\ 0 & 1_{2\times 2} \end{pmatrix}$ with $U_f \in SU(2)$. The corresponding solution of (6.59) is given by

$$X_f := \begin{pmatrix} e^{it_1}1_{2\times 2} & 0 \\ 0 & e^{-it_1}1_{2\times 2} \end{pmatrix} \begin{pmatrix} 1_{2\times 2} & 0 \\ 0 & e^{i\sigma_y t_2} \end{pmatrix} \begin{pmatrix} U_f & 0 \\ 0 & 1_{2\times 2} \end{pmatrix}, \tag{6.63}$$

for some $t_1, t_2 \in \mathbf{R}$.

Now we show how the above considerations on the Lie algebraic structure of the equation (6.59) can be used to achieve control objectives for system (6.57). Assume we want to transfer a (known) state $\vec{\psi}_0$ to an eigenvector of A, say $\vec{\psi}_1 := [e^{i\phi}, 0, 0, 0]^T$, $\phi \in \mathbf{R}$. It is natural to require that the final state is an eigenvector of A since we want the state to remain in the desired value after we switch the control to zero. Since we can achieve any arbitrary $U_f \in SU(2)$ in (6.63), we can control at will the first two components of the state vector. We choose U_f so as to introduce a zero in the position 2. The remaining two factors in (6.63) will introduce a common phase factor in the first two components and will modify the remaining components 3 and 4 without changing their total length. Now a similar analysis with $u_y \neq 0$ and $u_x \equiv u_z \equiv 0$ shows that, with u_y, we can modify the components 1 and 3 at will, while the components 2 and 4 also change but their total length remains constant. Analogously, using u_x, we can transform at will the components 1 and 4, while the components 2 and 3 evolve. We can alternate controls u_x, u_y and u_z and, at each step, transfer magnitude from one of the components 2, 3, 4 into the first component. If at each step the component is chosen as the one with maximum magnitude, it can be shown that the state will converge to the desired state $\vec{\psi}_1 := [e^{i\phi}, 0, 0, 0]^T$.

The problem for the control of the evolution operator is more complicated and more general. We give here some of the ideas and refer to [8] for a complete treatment. By solving a two-level problem on $SU(2)$, with u_z possibly different from zero and $u_x \equiv u_y \equiv 0$, we can control to a value X_f in (6.63) for an arbitrary $U_f \in SU(2)$. By choosing final time and control appropriately, it is possible to make the extra factor $\begin{pmatrix} e^{it_1}\mathbf{1}_{2\times 2} & 0 \\ 0 & e^{-it_1}\mathbf{1}_{2\times 2} \end{pmatrix} \begin{pmatrix} \mathbf{1}_{2\times 2} & 0 \\ 0 & e^{i\sigma_y t_2} \end{pmatrix}$ in (6.63) equal to the identity, so that we obtain matrices which are equal to the identity except for the elements at the intersection of the first and second rows and columns which are occupied by an element of $SU(2)$. In the same way, using u_y and u_x we can obtain matrices with rows and columns 1 and 3 and 1 and 4, respectively, equal to arbitrary matrices in $SU(2)$. Now, we use a decomposition into planar rotations (6.22) for the target operator $X_f \in SU(4)$. This decomposes the problem into subproblems for a diagonal matrix $D \in SU(4)$ and matrices of the type $U_{1,2}$, $U_{1,3}$, $U_{2,3}$, $U_{1,4}$, $U_{2,4}$, $U_{3,4}$. Matrices of the type $U_{1,2}$, $U_{1,3}$ and $U_{1,4}$ can be obtained as above, while matrices of the form $U_{2,3}$ can be obtained as $U_{1,2}U_{1,3}U_{1,2}^\dagger$, for appropriate elements $U_{1,2}$ and $U_{1,3}$. Analogously, one can obtain matrices of the form $U_{2,4}$ and $U_{3,4}$. Diagonal matrices can be obtained as exponentials of A in (6.58) and exponentials of matrices that are obtained by permuting the diagonal elements of A using similarity transformations of the type $U_{1,2}$, $U_{1,3}$ and $U_{1,4}$.

6.5   Notes and References

Techniques of Lie group decompositions for the control of quantum systems have been used in many papers. In particular, for the two-level problem more sophisticated results exist which incorporate in the design arbitrary bounds on the control [227] and/or minimize number of required switches [78]. The fact that Cartan decomposition could be used for the control of two interacting spins was recognized by several authors [73], [161]. In particular the paper [161] presents a geometric treatment which allows one to conclude that the controls based on Cartan's decompositions are the ones that infimize the time to reach a particular target. This will be further discussed in the next chapter, which is devoted to the optimal control of quantum systems. Applications of Cartan decomposition to the three-spin case and general n-spin case are presented in [162], [163], while the paper [228] contains a treatment of the two-spin problem with Ising interaction which avoids very large fast control pulses and, in fact, allows for arbitrarily bounded controls. The problem with Heisenberg interaction of subsection 6.4.2 was considered in [280] and [8]. For further applications and examples of Lie group decompositions applied to quantum control we refer to [76], [249], [250] and references therein. Our presentation of the Cartan decompositions of $su(n)$ mainly follows [81], in particular for the relation between decompositions and quantum symmetries.

One important problem in quantum information theory is to design universal quantum circuits which perform an arbitrary unitary evolution on a register of n-qubits, as a cascade of given elementary gates, typically, one qubit and two qubits gates. This is a question of decomposition of the Lie group $U(2^n)$. One problem that was considered by researchers in the past decades was how to devise design methods which would minimize the number of gates used as a function of n, i.e., the complexity of the circuit. The techniques used are in many cases again variations of Cartan decomposition. One survey paper on the subject is [200]

6.6 Exercises

Exercise 6.1 Consider the problem of driving the state $\psi_0 = \begin{pmatrix} 1 \\ 0 \end{pmatrix}$ to the state $\psi_f = \begin{pmatrix} \frac{1}{2} \\ \frac{\sqrt{3}}{2} \end{pmatrix}$, for the system

$$\dot{X} = AX + BXu,$$

with $A = \begin{pmatrix} i & 0 \\ 0 & -i \end{pmatrix}$ and $B = \begin{pmatrix} 0 & 1 \\ -1 & 0 \end{pmatrix}$. Assume there is no a priori bound on the control. Use the Euler decomposition technique of subsection 6.1.4 to find a control law for this problem.

Exercise 6.2 Find a decomposition into polar rotations for the special unitary matrix

$$X_f = \begin{pmatrix} \frac{i}{\sqrt{2}} & -\frac{i}{\sqrt{2}} & 0 \\ \frac{\sqrt{3}i}{2\sqrt{2}} & \frac{\sqrt{3}i}{2\sqrt{2}} & \frac{1}{2} \\ \frac{-i}{2\sqrt{2}} & \frac{-i}{2\sqrt{2}} & -\frac{\sqrt{3}}{2} \end{pmatrix}.$$

Exercise 6.3 Let $\mathcal{K}$ be a subalgebra of $\mathcal{L} \subseteq su(n)$. Recall the inner product on $\mathcal{L}$ is the one inherited from $su(n)$ namely $\langle A, B \rangle = \mathrm{Tr}(AB^\dagger)$. Prove that

$$[\mathcal{K}, \mathcal{K}^\perp] \subseteq \mathcal{K}^\perp.$$

Compare with (6.25).

Exercise 6.4 Consider a Cartan subalgebra $\mathcal{A}$ as a maximal Abelian subalgebra in $\mathcal{P}$, for a given Cartan decomposition of $su(n)$. Give an example where this coincides with the concept of Cartan subalgebra of $su(n)$ introduced in Chapter 4 and an example where it does not coincide.

Exercise 6.5 Find the factors in the **AI**, **AII** and **AIII**, with $p = 2$, $q = 2$, Cartan decomposition of the unitary matrix T in (6.47).

Exercise 6.6 Use the methods of Chapters 3 and 4 to deduce the decomposition of $\mathcal{L}_z$ given in (6.61).

Exercise 6.7 Check that the vector space $iu(n)$ with the product $A \diamond B := \{A, B\} := AB + BA$ has the structure of a nonassociative algebra. This means that it is a vector space with a multiplication $A \diamond B$ which is nonassociative, i.e., there are elements A, B, C such that

$$A \diamond (B \diamond C) \neq (A \diamond B) \diamond C,$$

and that the multiplication $A \diamond B$ satisfies the *Jordan identity*

$$(A \diamond A) \diamond (A \diamond B) = A \diamond ((A \diamond A) \diamond B).$$

Exercise 6.8 Prove that there exists a one to one correspondence between Cartan involutions and Cartan decompositions as stated at the end of subsection 6.3.1.

Exercise 6.9 Verify formulas (6.43) in the cases **AI AII** and formulas (6.45) in the cases **AIII** of Cartan decompositions.

Chapter 7

Optimal Control of Quantum Systems

Optimal control theory provides a powerful set of tools and concepts that can be applied to quantum control systems. In general terms, optimal control theory (see, e.g., [110], [177], [181], [214]) is concerned with the control of a system

$$\dot{x} = f(x, u), \tag{7.1}$$

and the simultaneous minimization of a functional of the state x and the control u. Problems of control of quantum systems can naturally be formulated in this setting.

Section 7.1 presents the general formulation of optimal control problems and section 7.2 gives the basic necessary conditions of optimality, the *Pontryagin maximum principle* (PMP). An outline of the derivation of the PMP in the case of a fixed final time and free final state is presented in Appendix F. The derivation in this case is particularly simple and instructive. Section 7.3 presents an example of application of optimal control techniques to a quantum control problem that can be solved analytically. In section 7.4, we turn our attention to the optimal control problem of minimizing the *time* for a prescribed state transfer. This problem is very much motivated for quantum systems. In fact, one of the major problems in this context is that quantum systems are very sensitive to the presence of the environment which often destroys the main features of quantum dynamics. This is called *de-coherence*. One method to prevent de-coherence is to obtain the desired state transfer in the least possible time so that the interaction with the environment becomes negligible. The treatment of the time optimal control problem uses different tools depending on whether or not there is a prescribed bound on the control magnitude. For unbounded controls, the time optimal control does not exist. However, an infimum can be obtained in several cases (see subsection 7.4.2) using a geometric argument based on the theory of Riemannian symmetric spaces, some of which was discussed in the previous chapter.

Very rarely, and typically only for low-dimensional cases, or under special symmetries, the solution of an optimal control problem can be obtained explicitly in the form of a given function of time. Much more often, in particular for higher-dimensional control problems in molecular dynamics, numerical and

DOI: 10.1201/9781003051268-7

iterative algorithms are used in order to find the optimal control. We shall discuss some of these methods in section 7.5.

In more recently introduced terminology, the map from the class of control functions used to the cost to be minimized is referred to as the *optimal control landscape*. Remarkable regularity properties of such a map have been recently discovered. These properties explain the fact that, in laboratory experiments, finding optimal control functions for quantum systems by numerical algorithms has proven particularly easy. We summarize some of these facts in section 7.6.

7.1 Formulation of the Optimal Control Problem

A general optimal control problem can be formulated as follows:

Optimal Control Problem:

Given a set $\mathcal{X}$ of (state) functions $x : \mathbf{R} \to \mathbf{R}^n$, and a set $\mathcal{U}$ of (control) functions $u : \mathbf{R} \to \mathbf{R}^m$, find the functions $x \in \mathcal{X}$ and $u \in \mathcal{U}$ which minimize a cost functional $J : \mathcal{X} \times \mathcal{U} \to \mathbf{R}$ and satisfy the *dynamical constraint* (7.1), almost everywhere.

Virtually every control problem can be formulated as a special case of the above general optimal control problem. Here is a simple example.

Example 7.1.1 Assume that we want to find a piecewise continuous control u to drive the state x of (7.1) from a given value x_0 to a desired state x_f in time T. Then $\mathcal{U}$ is the set of piecewise continuous functions on $[0, T]$, $\mathcal{X}$ can be taken as the set of continuous functions on $[0, T]$ satisfying $x(0) = x_0$, and (7.1), almost everywhere (a.e.), and the cost functional we seek to minimize is

$$J(u) := ||x(T) - x_f||^2, \tag{7.2}$$

i.e., a measure of the distance of the final state from the desired one. We can also incorporate in the cost a term which takes into consideration the energy used during the given interval of time. For example, a cost of the form

$$J(u) = \lambda ||x(T) - x_f||^2 + \int_0^T ||u(t)||^2 dt, \tag{7.3}$$

with $\lambda > 0$, will incorporate a penalty on the final state and an energy like term for the control. We can increase or decrease λ according to whether or not we are willing to use more energy to obtain a better final state or not.

Problems of optimal control are rarely expressed in the general form above discussed. They are expressed in one of the three equivalent forms described in the following subsection.

7.1.1 Optimal control problems of Mayer, Lagrange and Bolza

In the following, it is assumed that the initial value for the state x is given. Therefore, the choice of the control u and the requirement that (7.1) is satisfied determine x uniquely, assuming that the space of control functions $\mathcal{U}$ consists of functions regular enough to guarantee the existence and uniqueness of the solution of (7.1), a.e.

Given the system

$$\dot{x} = f(x, u), \qquad x \in \mathbf{R}^n, u \in \mathbf{R}^m, \tag{7.4}$$

the **problem of Mayer** is to determine a control function u, in an appropriate set of functions on $[0, T]$, to minimize the cost functional in the form

$$J(u) := \phi(x(T), T), \qquad \cdot \tag{7.5}$$

where ϕ is a smooth function : $\mathbf{R}^n \times \mathbf{R} \to \mathbf{R}$. Examples of Mayer problems are *minimum time problems* where the problem is to drive the state to a given value in minimum time. In this case the set of admissible controls consists of the ones steering to the desired target state and the cost has the form (7.5), with $J = T$. Another example is a problem with the cost of the form of $J(u)$ in (7.2). Mayer problems arise when there is a particular emphasis on the final state and/or time.

In the **problem of Lagrange**, the cost functional takes the form

$$J(u) := \int_0^T L(x(t), u(t), t)dt, \tag{7.6}$$

where L is a (smooth) function : $\mathbf{R}^n \times \mathbf{R}^m \times \mathbf{R} \to \mathbf{R}$. A problem of Lagrange describes a situation where the cost accumulates with time. This is the case, for example, when one wants to minimize the energy used during the control action and/or the average distance of the trajectory from a given point.

A **problem of Bolza** is a combination of problems of Mayer and Lagrange as the cost takes the form

$$J(u) := \phi(x(T), T) + \int_0^T L(x(t), u(t), t)dt, \tag{7.7}$$

with ϕ and L (smooth) functions : $\mathbf{R}^n \times \mathbf{R} \to \mathbf{R}$, $\mathbf{R}^n \times \mathbf{R}^m \times \mathbf{R} \to \mathbf{R}$, respectively. Bolza problems arise when there is a cumulative cost which increases during the control action but special emphasis is placed on the situation at the final time. An example is the cost of the form (7.3) in Example 7.1.1.

Mayer, Lagrange and Bolza problems are all equivalent in that each of them can be converted to any other one. It is obvious that Lagrange and Mayer problems are special cases of Bolza problems. A Bolza problem for the system (7.4) with cost (7.7) can be transformed into a Mayer problem by introducing an extra component for the state vector y which satisfies the equation

$$\dot{y} = L(x, u, t), \qquad y(0) = 0.$$

Using this extra variable the cost takes the Mayer form

$$J(u) = \phi(x(T), T) + y(T).$$

A Mayer problem with cost (7.5) can be converted into a Lagrange problem by rewriting the cost as

$$J(u) = \phi(x(T), T) = \phi(x(0), 0) + \int_0^T \frac{d}{dt}\phi(x(t), t)\, dt$$

$$= \phi(x(0), 0) + \int_0^T \left(\phi_x(x(t), t) f(x(t), u(t)) + \frac{\partial}{\partial t}\phi(x(t), t) \right) dt.$$

Since $x(0)$ is fixed, the problem is to minimize the cost

$$\bar{J}(u) = \int_0^T L(x(t), u(t), t)\, dt,$$

where $L(x(t), u(t), t)$ is given by

$$L(x(t), u(t), t) = \phi_x(x(t), t) f(x(t), u(t)) + \frac{\partial}{\partial t}\phi(x(t), t),$$

which is a problem of Lagrange.

Remark 7.1.2 The optimal control problem and its solution are closely related to the concept of reachable set $\mathcal{R}(T)$ introduced in Chapter 3 (cf. formula (3.2) and the discussion there). In fact, much of the geometric treatment concerning necessary conditions of optimality is based on this concept [3]. Consider for instance a problem of Mayer with fixed terminal time T and cost function $\phi(\cdot, T)$ as in (7.5). Then the minimum coincides with the minimum of the function $\phi(\cdot, T)$ over the reachable set $\mathcal{R}(T)$ at time T. We can therefore, in principle, separate the two aspects of the optimal control problem. The reachable set $\mathcal{R}(T)$ describes the possible *dynamics* we can have over the interval $[0, T]$ and, on this set, we then perform a *static optimization* for the function $\psi(\cdot, T)$.

7.1.2 Optimal control problems for quantum systems

For quantum control systems, the state (x) may be the density matrix, the pure state vector, or one may consider the dynamics of the evolution operator. We shall focus on the case where equation (7.1) describes the quantum state. Some care needs to be taken because the state (whether one considers a pure state or a density matrix) is complex.

In particular, consider the dynamics (7.1) described by the (controlled) Schrödinger equation

$$\frac{d}{dt}\vec{\psi} = -iH(u)\vec{\psi}, \tag{7.8}$$

and a general cost of the Bolza type which can be written as

$$J(u) = \phi(\vec{\psi}(T), T) + \int_0^T L(\vec{\psi}(t), u(t), t) \, dt. \tag{7.9}$$

In this case, both the state $\vec{\psi}$ and the matrix $-iH(u)$ are complex quantities. To express the problem in terms of real quantities only, we can write

$$\vec{\psi} = \vec{\psi}_R + i\vec{\psi}_I$$

and

$$-iH(u) = R(u) + iI(u),$$

with $\vec{\psi}_R$ and $\vec{\psi}_I$ real vectors of dimension n and $R(u)$ and $I(u)$ real $n \times n$ matrix functions of u, $R(u)$ skew-symmetric and $I(u)$ symmetric for every value of u. Placing this into (7.8) and separating the imaginary and real part, we obtain the two real differential equations

$$\frac{d}{dt} \vec{\psi}_R = R(u)\vec{\psi}_R - I(u)\vec{\psi}_I,$$

$$\frac{d}{dt} \vec{\psi}_I = I(u)\vec{\psi}_R + R(u)\vec{\psi}_I.$$

By defining

$$x := [\vec{\psi}_R^T, \vec{\psi}_I^T]^T, \tag{7.10}$$

and

$$\tilde{H}(u) := \begin{pmatrix} R(u) & -I(u) \\ I(u) & R(u) \end{pmatrix}, \tag{7.11}$$

we can write the differential equation describing the dynamics involving only real quantities as

$$\dot{x} = \tilde{H}(u)x. \tag{7.12}$$

$\tilde{H}(u)$ is skew-symmetric and symplectic for every u, i.e., it belongs to $so(2n) \cap sp(n)$ (cf. Exercise 7.4). Also, the cost (7.9) can be rewritten as

$$J(u) = \tilde{\phi}(x(T), T) + \int_0^T \tilde{L}(x(t), u(t), t) \, dt, \tag{7.13}$$

for appropriate functions $\tilde{\phi}$ and $\tilde{L}$.

We may also consider the dynamics for the density matrix ρ given by the Liouville's equation (1.51), in which case the cost (7.13) can be written with ρ replacing x, possibly splitting ρ into its its real and imaginary part, as before, if we insist on dealing with real variables only.

A common choice for the cost functional (7.9) in molecular control is the *laser electric field fluence*

$$J(u) = k \int_0^T u^2(t) \, dt, \qquad k > 0,$$

which measures the energy of the electric field in the interval $[0, T]$ (cf. (2.3), (2.28)). Another possibility is to choose a cost of the type[1]

$$J(u) := k \int_0^T \left(\frac{du}{dt} \right)^2 dt, \qquad k > 0,$$

which filters the high frequency components of the control field. When emphasis is placed on the final state also, one can minimize a cost functional in the form (7.3), or, in the form,

$$J = \frac{1}{2} \langle O \rangle_\psi + \frac{k}{2} \int_0^T u^2(t) dt := \frac{1}{2} \vec{\psi}^\dagger O \vec{\psi} + \frac{k}{2} \int_0^T u^2(t) dt, \qquad k > 0, \quad (7.14)$$

where O is a Hermitian, negative definite, matrix. For example, we may choose $O = -\vec{\psi}_f \vec{\psi}_f^\dagger$ if $\vec{\psi}_f$ is the desired state. If the state is described in terms of the density matrix ρ, the cost corresponding to (7.14) can be written more elegantly in a form linear in ρ, rather than quadratic, i.e.,

$$J = \frac{1}{2} \langle O \rangle_\rho + \frac{k}{2} \int_0^T u^2(t) dt = \frac{1}{2} Tr(O\rho) + \int_0^T u^2(t) dt. \qquad (7.15)$$

7.2   The Necessary Conditions of Optimality

7.2.1   General necessary conditions of optimality

In every derivation of **necessary conditions** of optimality, one assumes that the control u is optimal and replaces u with a slightly different control u^ϵ, where ϵ is a small parameter. The control u^ϵ is still admissible, and it is called a *variation* of u. Then one imposes that, for the cost $J = J(u)$,

$$J(u^\epsilon) - J(u) \geq 0, \qquad (7.16)$$

which gives the desired necessary conditions on the control u. According to what type of variation one considers, different optimality conditions are obtained.

The basic necessary conditions of optimality in optimal control, known as the Pontryagin maximum principle (PMP), are obtained using a so-called *strong variation* (cf. Figure 7.1). A *strong variation* u^ϵ of u is defined as

---

[1]In this case the Lagrange functional does not strictly have the form (7.6). One either allows the cost to be a functional of the derivative of the control or may include an extra state variable in the equations of the dynamics, $\dot{u} = v$ and uses v as the new control.

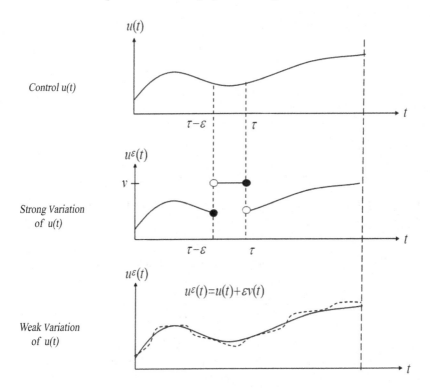

FIGURE 7.1: Strong and weak variation of a control function $u = u(t)$.

follows: Fix $\tau \in (0, T]$, then u^ϵ is a function equal to u for $t \in [0, \tau - \epsilon]$ and for $t \in (\tau, T]$, and constant and equal to v for $t \subseteq (\tau - \epsilon, \tau]$. The value v is any value in the set of admissible values for the control, $\mathcal{U} \subseteq \mathbf{R}^m$. We state below a version of the PMP for a Mayer problem for fixed final time and free final state. An outline of the derivation of the PMP in this case is presented in Appendix F. This derivation uses a strong variation.

Theorem 7.2.1 *(Pontryagin maximum principle for Mayer problems. Fixed final time and free final state) Assume u is the optimal control and x the corresponding trajectory solution of (7.1). Then, there exists a nonzero vector λ solution of the adjoint equations*

$$\dot{\lambda}^T = -\lambda^T f_x(x(t), u(t)) \tag{7.17}$$

with terminal condition[2]

$$\lambda^T(T) = -\phi_x(x(T)), \tag{7.18}$$

---

[2] We drop the argument T in the function ϕ, since the final time T is fixed.

such that, for almost every $\tau \in (0, T]$, we have

$$\lambda^T(\tau) f(x(\tau), u(\tau)) \geq \lambda^T(\tau) f(x(\tau), v), \tag{7.19}$$

for every v in the set of the admissible values for the control $\mathcal{U}$.
  Furthermore, for every $\tau \in [0, T]$,

$$\lambda^T(\tau) f(x(\tau), u(\tau)) = c, \tag{7.20}$$

for a constant c.

The vector λ is called the **costate**.

Remark 7.2.2 The condition $\lambda \neq 0$ is due to the fact that it is implicitly assumed that, at time T, the cost $\phi(x(T))$ has nontrivial dependence on the final condition as it varies in the reachable set (cf. (7.18)).

Remark 7.2.3 If there is no bound on the possible value of the control u, equation (7.19) can be replaced by

$$\frac{\partial(\lambda^T f(x, u))}{\partial u} = \lambda^T f_u(x, u) = 0. \tag{7.21}$$

It is instructive to derive this condition without Theorem 7.2.1. In this case the necessary condition of optimality can be obtained by using a *weak variation* of the control. A weak variation is defined (cf. Figure 7.1) as

$$u(t) \to u^\epsilon := u(t) + \epsilon v(t),$$

where $v(t)$ is a given function. If we denote by x^ϵ the trajectory corresponding to u^ϵ for ϵ small, we have for a Mayer problem

$$\Delta J := J(x^\epsilon(T)) - J(x(T)) \approx \phi_x(x(T))(x^\epsilon(T) - x(T)). \tag{7.22}$$

Defining $z(t) := \frac{d}{d\epsilon} x^\epsilon(t)|_{\epsilon=0}$, we have

$$x^\epsilon(T) - x(T) \approx z(T)\epsilon.$$

Plugging this into (7.22), we obtain

$$\Delta J \approx \phi_x(x(T)) z(T)\epsilon. \tag{7.23}$$

Consider the differential equation for x^ϵ,

$$\dot{x}^\epsilon = f(x^\epsilon, u + \epsilon v).$$

A differential equation for $z(t)$ is obtained by differentiating this equation with respect to ϵ, at $\epsilon = 0$ and switching the order of differentiation between time t and ϵ. We obtain

$$\dot{z} = f_x(x, u)z + f_u(x, u)v, \tag{7.24}$$

with initial condition $z(0) = 0$. Now consider the adjoint equations (same as (7.17))

$$\dot{\lambda}^T = -\lambda^T f_x(x(t), u(t)), \qquad (7.25)$$

with final condition (same as (7.18)) $\lambda^T(T) = -\phi_x(x(T))$. We can write formula (7.23) as

$$\Delta J \approx -\lambda^T(T)z(T)\epsilon. \qquad (7.26)$$

Since $z(0) = 0$, we can write $\lambda(T)z(T)$ as $\lambda^T(T)z(T) = \int_0^T \frac{d}{dt}\lambda^T(t)z(t)dt$ which, using (7.24) and (7.25), gives

$$\lambda^T(T)z(T) = \int_0^T \lambda^T(t)f_u(x(t), u(t))v(t)dt.$$

This, placed in (7.26), gives

$$\Delta J \approx -\epsilon \int_0^T \lambda^T(t)f_u(x(t), u(t))v(t)dt.$$

If u is a local minimum for J, then it should not be possible by a choice of the perturbation v to obtain a smaller value. Since there is no bound on v, this requires $\lambda^T(t)f_u(x(t), u(t)) \equiv 0$, i.e., (7.21).

The Pontryagin maximum principle of Theorem 7.2.1 is most often used in the following way. One defines a function

$$h = h(\lambda, x, u) := \lambda^T f(x, u),$$

which is called the **optimal control Hamiltonian**.[3] With this definition, condition (7.20) has the form

$$h(\lambda(\tau), x(\tau), u(\tau)) = c. \qquad (7.27)$$

Furthermore, for every λ and x, one maximizes the function h over $u \in \mathcal{U}$, i.e., u is chosen so that (cf. (7.19))

$$h(\lambda, x, u) \geq h(\lambda, x, v), \qquad (7.28)$$

for every $v \in \mathcal{U}$. This is a *static* optimization problem for all the values of the parameters λ and x. The solution will depend on these parameters and we can write it as $u := u(x, \lambda)$. Now, by placing this form of u into equations (7.1) and (7.17) we obtain the system of equations

$$\dot{x} = f(x, u(x, \lambda)), \qquad (7.29)$$

---

[3]We use the notation h to avoid confusion with the quantum mechanical Hamiltonian H. The two Hamiltonians are, however, related as both the optimal control problem and Lagrangian mechanics are instances of variational problems (cf. Remark B.1.5).

$$\dot{\lambda}^T = -\lambda^T f_x(x, u(x, \lambda)), \tag{7.30}$$

which has to be solved with the boundary conditions $x(0) = x_0$ and $\lambda^T(T) = -\phi_x(x(T))$. This is a *two points boundary value problem* as the boundary conditions are at time 0 and T. Typically, one leaves the initial condition for λ, $\lambda(0)$, as a free parameter, solves the equations (7.29)-(7.30) and then tries to adjust the value of the parameter $\lambda(0)$ so as to meet condition (7.18). Every control which is obtained with this procedure satisfies the necessary conditions of optimality and it is a candidate to be the optimal control. Every control which satisfies the necessary optimality conditions is called an *extremal*. By comparing the costs obtained with the various extremals one finds the minimum cost and the optimal control. This procedure can typically be carried out analytically only for low-dimensional problems or problems which have special symmetries.

The above procedure assumes that an optimal control exists. The problem of existence of optimal controls is treated, for example, in ([110] Chapter III). Theorems of existence of optimal controls can be given under conditions on the equation in (7.1) and on the set of admissible control functions, which we have assumed here to include at least the set of piecewise continuous functions.[4]

If one has a Bolza problem instead of a Mayer problem, still with free endpoint fixed final time, the conditions of the PMP can be easily adapted.

Recall from subsection 7.1.1, that, by defining the extra variable $y(t) := \int_0^t L(x(s), u(s), s)ds$, the cost $J(u) := \phi(x(T)) + \int_0^T L(x(s), u(s), s)ds$ can be rewritten in the Mayer form

$$J(u) = \bar{\phi}(x(T), y(T)) := \phi(x(T)) + y(T).$$

The costate now has dimension $n + 1$. If we call λ the first n components of the costate and μ the remaining component, the adjoint equations take the form

$$\dot{\lambda}^T = -\lambda^T f_x - \mu L_x, \tag{7.31}$$

$$\dot{\mu} = 0.$$

So μ is a constant, and since $\mu(T) = -\bar{\phi}_y = -1$, we have that condition (7.28) can be rewritten with the Hamiltonian

$$h(\lambda, x, u) := \lambda^T f(x, u) - L(x, u, t). \tag{7.32}$$

Condition (7.20) (7.27) has to be replaced by a more complicated one in general (cf., (5.7) of Theorem II 5.1 of [110]) but still holds if we assume that the Lagrangian L does not depend explicitly on t. For a Bolza problem (7.7) with fixed final time and free final state, if u is optimal, it satisfies (7.28) with h given in (7.32). λ satisfies the first of (7.31) with terminal boundary condition (7.18), and x satisfies (7.1).

---

[4]This allowed us to use a strong variation as an admissible control.

7.2.2 The necessary optimality conditions for quantum control problems

For the quantum optimal control problem (7.12), (7.13), the optimal control Hamiltonian $h(\lambda, x, u)$ takes the form (cf. (7.32))

$$h(\lambda, x, u) = \lambda^T \tilde{H}(u)x - \tilde{L}(x, u). \tag{7.33}$$

The optimal control u has to satisfy (7.28) with this Hamiltonian. The adjoint equations are

$$\dot{\lambda}^T = -\lambda^T \tilde{H}(u) + \tilde{L}_x,$$

which, since $\tilde{H}(u)$ is skew-symmetric, can be written as

$$\dot{\lambda} = \tilde{H}(u)\lambda + \tilde{L}_x^T, \tag{7.34}$$

and the terminal condition is given by $\lambda(T) = -\tilde{\phi}_x^T(x(T))$ (cf. (7.18)). Notice that, if $\tilde{L}_x \equiv 0$, i.e., the integrand in the cost (7.13) does not depend on the state x, state x and costate λ satisfy the *same differential equation* and the equations for λ and x are coupled only through the controls. In fact, these equations can be written in terms of the original *complex* vector, $\vec{\psi}$ and a complex covector $\hat{\lambda} := \hat{\lambda}_R + i\hat{\lambda}_I$ ($\lambda := (\lambda_R^T, \lambda_I^T)^T$, that both satisfy Schrödinger equation (7.8). In terms of the complex variables $\vec{\psi}$ and $\hat{\lambda}$ the optimal control Hamiltonian h in (7.33) reads as (Exercise 7.6)

$$h(\lambda, x, u) = \text{Re}(\hat{\lambda}^\dagger(-iH(u))\vec{\psi}) - L(\vec{\psi}, u). \tag{7.35}$$

In the case where the state is represented by a density matrix ρ, that is, not necessarily a pure state, assume for simplicity L not explicitly dependent on ρ and not explicitly independent of t (so as to make sure that (7.27) holds). Then, using the fact that ad_H is skew symmetric is H is skew-Hermitian (cf. Exercise 4.4) one sees that the costate can be represented by a Hermitian matrix Λ again satisfying Liouville's equation (setting $\hbar = 1$)

$$\dot{\Lambda} = [-iH(u), \Lambda],$$

and the optimal control Hamiltonian h, takes the form

$$h = Tr(K[-iH(u), \rho]).$$

Notice that often the cost takes the Mayer form $J = Tr(O\rho(T))$ assuming we want to minimize the expectation value of of an observable O at the final time T. In this case, the terminal condition (7.18) for the costate Λ takes the simple form

$$\Lambda(T) = -O.$$

7.3   Example: Optimal Control of a Two-Level Quantum System

The differential equation of a spin-$\frac{1}{2}$ particle in an electromagnetic field was discussed in subsection 2.4.2. It can be written, after appropriate scaling of time and magnetic field in (2.60), as

$$\dot{\psi} = (\bar{\sigma}_z u_z + \bar{\sigma}_x u_x(t) + \bar{\sigma}_y u_y(t))\psi. \qquad (7.36)$$

$\bar{\sigma}_{x,y,z}$ are the (multiples of the Pauli) matrices in (6.5) satisfying the commutation relations (6.6). $u_{x,y,z}$ represent the x, y, z components of the magnetic field which is used for control. We assume that the z component is held constant. A scheme is depicted in Figure 7.2.

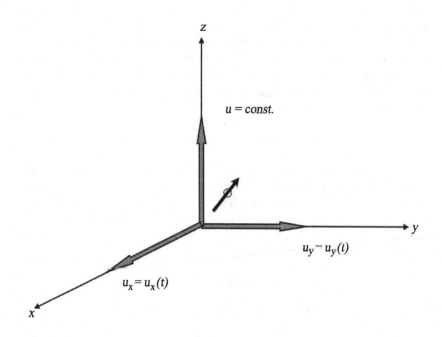

FIGURE 7.2:   Scheme of a spin $\frac{1}{2}$ particle in a magnetic field: The field is varying in the x and y direction and is constant in the z direction.

A possible cost to be minimized is of the form

$$J(u_x, u_y) = -Re(\vec{\psi}^{\,\dagger}(T)\vec{\psi}_f) + \eta \int_0^T u_x^2(t) + u_y^2(t)dt, \qquad (7.37)$$

with the $\eta > 0$ constant and $\vec{\psi}_f$ equal to the desired final state. The cost (7.37) expresses a compromise between the goal of transferring the state to $\vec{\psi}_f$ and the goal of keeping the energy of the field, i.e., the integral in (7.37), small. More importance is given to the second goal if the parameter η is large. Transforming the problem into a real one according to the procedure described in subsection 7.1.2 gives the equation corresponding to (7.36),

$$\dot{x} = (T_z u_z + T_y u_y(t) + T_x u_x(t))x, \tag{7.38}$$

where the matrices $T_{x,y,z}$ are given by

$$T_x := \frac{1}{2}\begin{pmatrix} 0 & 0 & 0 & -1 \\ 0 & 0 & -1 & 0 \\ 0 & 1 & 0 & 0 \\ 1 & 0 & 0 & 0 \end{pmatrix}, \quad T_y = \frac{1}{2}\begin{pmatrix} 0 & -1 & 0 & 0 \\ 1 & 0 & 0 & 0 \\ 0 & 0 & 0 & -1 \\ 0 & 0 & 1 & 0 \end{pmatrix},$$

$$T_z = \frac{1}{2}\begin{pmatrix} 0 & 0 & -1 & 0 \\ 0 & 0 & 0 & 1 \\ 1 & 0 & 0 & 0 \\ 0 & -1 & 0 & 0 \end{pmatrix}.$$

They satisfy the commutation relations

$$[T_x, T_y] = T_z, \quad [T_y, T_z] = T_x, \quad [T_z, T_x] = T_y, \tag{7.39}$$

so that the map

$$\bar{\sigma}_x \leftrightarrow T_x, \quad \bar{\sigma}_y \leftrightarrow T_y, \quad \bar{\sigma}_z \leftrightarrow T_z$$

is a Lie algebra isomorphism. The cost $J(u_x, u_y)$ in (7.37) can be written, denoting by x_f the $2n$-dimensional real vector corresponding to $\vec{\psi}_f$,

$$J(u_x, u_y) = -x^T(T)x_f + \eta \int_0^T u_x^2(t) + u_y^2(t)dt. \tag{7.40}$$

We now apply the maximum principle to determine the form of the optimal control in this case. Notice that the integrand in (7.40) does not depend on x and therefore the adjoint equations for the costate λ are the same as the equations for x, i.e.,

$$\dot{\lambda} = (T_z u_z + T_y u_y(t) + T_x u_x(t))\lambda. \tag{7.41}$$

The optimal control Hamiltonian $h(\lambda, x, u_x, u_y)$ is given by

$$h(\lambda, x, u_x, u_y) = \lambda^T(T_z u_z + T_y u_y + T_x u_x)x - \eta(u_x^2 + u_y^2).$$

Maximization with respect to u_x and u_y gives

$$u_x = \frac{1}{2\eta}\lambda^T T_x x, \tag{7.42}$$

$$u_y = \frac{1}{2\eta}\lambda^T T_y x. \tag{7.43}$$

At this point, we can place (7.42) and (7.43) into (7.38) and (7.41) and try to solve the associated two point boundary value problem. A better alternative, in this case, is to differentiate (7.42) and (7.43). Using (7.38) and (7.41) and the commutation relations (7.39), we obtain

$$\dot{u}_x = \frac{1}{2\eta}(\lambda^T T_z x u_y - u_z u_y), \tag{7.44}$$

$$\dot{u}_y = \frac{1}{2\eta}(-\lambda^T T_z x u_x + u_z u_x). \tag{7.45}$$

A similar calculation gives

$$\frac{d}{dt}\lambda^T T_z x \equiv 0,$$

so that (7.44) and (7.45) can be rewritten as

$$\dot{u}_x = ku_y,$$

$$\dot{u}_y = -ku_x,$$

where k is the constant $k := \frac{1}{2\eta}(\lambda^T T_z x - u_z)$. This shows that the optimal controls are trigonometric functions of the form

$$u_x(t) = M\cos(\omega t + \gamma), \tag{7.46}$$

$$u_y(t) = M\sin(\omega t + \gamma). \tag{7.47}$$

By replacing these in the equations (7.38) or (7.36) we can solve the corresponding differential equation with ω, M and γ as free parameters and then determine the parameters ω, γ and M which minimize the cost.

As we can see from this example, the Pontryagin maximum principle allows us to transform an infinite-dimensional optimization problem (the search over a set of functions) into a finite-dimensional optimization problem (the search over a set of parameters).

7.4 Time Optimal Control of Quantum Systems

Consider the problem of steering the state between two values in minimum time. The standard Filippov existence theorems (see, e.g., [110]) guarantee that the time optimal control exists, under the assumptions of controllability, i.e., existence of a control performing the desired task, Lebesgue measurable control functions with values in a compact set, and smoothness and growth conditions on the functions describing the dynamics. These smoothness and growth conditions are satisfied for bilinear quantum control models described in Chapter 2.

The problem of finding the time optimal control is related to the problem of describing the reachable sets $\mathcal{R}(t)$, $\mathcal{R}(\leq t)$, $\mathcal{R}$ starting from a given state, as defined in Chapter 3 (see discussion before Formula (3.2)). In fact we have the following fact

Proposition 7.4.1 Consider system (7.1) with an initial condition x_0. If u is the minimum time control to reach the target x_f, T the minimum time and $x = x(t)$ the corresponding trajectory, then for every $t \in (0, T)$, $x(t)$ belongs to the boundary of the reachable set $\mathcal{R}(\leq t)$, starting from x_0.

Proof. The idea of the proof is illustrated in Figure 7.3. Fix $t \in (0, T)$. If $x(t)$ belongs to the interior of $\mathcal{R}(\leq t)$ then there exists an $\epsilon > 0$ such that $x(t + \epsilon)$ also belongs to $\mathcal{R}(\leq t)$. Therefore, using another control and an alternative trajectory x_1, it is possible to reach $x(t + \epsilon)$ in time $\leq t$. From $x(t + \epsilon)$ we can then follow the trajectory x, for time $T - t - \epsilon$ and reach x_f in time $\leq T - \epsilon$ which contradicts the minimality of T. $\qquad\square$

A natural question is whether x_f in the above theorem is also in the boundary of $\mathcal{R}(T)$. Under assumptions typically verified for quantum control systems (bilinear systems on $SU(n)$ with the control u with values in a compact set) reachable sets are compact and vary continuously with respect to the Hausdorff metric (see, e.g., [29], [218]). This guarantees that x_f is in the boundary of $\mathcal{R}(T)$.

For time optimal control problems, we will give a version of the PMP different from Theorem 7.2.1 in order to single out a set of candidate controls, the *extremals*. From these we need to select the ones that give the minimum control. The explicit expression of the time optimal controls can be obtained only in special cases and typically for low-dimensional problems [43]. For some problems, there are relevant properties that simplify the search for the time optimal control (see, e.g., [42] and the references therein). An example of a class of systems will be described in the next subsection.

In the case where the controls are not required to be bounded in magnitude, the time optimal control does not exist in general for quantum control systems, and one has to look for a control which gives the infimum time. In some situations [161], such controls can be found using the Cartan decompositions of Lie groups and the Cartan classification of symmetric spaces described in the previous chapter for the case of $SU(n)$.

The following subsection deals with the case of bounded control by describing the necessary conditions of optimality and some examples of problems that can be solved explicitly. Subsection 7.4.2 deals with the case of unbounded control and the search for a control which makes the time of transfer arbitrarily close to an infimum.

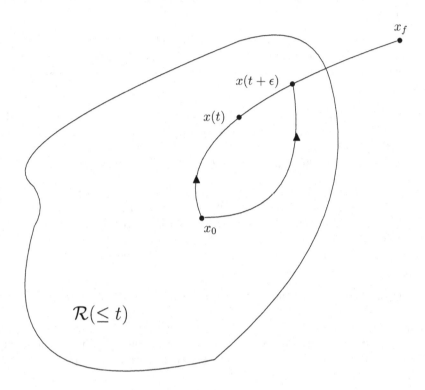

FIGURE 7.3:   Illustration of the proof of Proposition 7.4.1.

7.4.1   The time optimal control problem; bounded control

For time optimal control problems, we have the following version of the PMP which gives conditions analogous to the ones of Theorem 7.2.1, with the only difference that the condition (7.18) on the final value of the costate is now replaced by $x(T) = x_f$.[5] Recall that, for a time optimal control problem, we have fixed initial and final conditions for the state x and let the final time free and to be minimized.

Theorem 7.4.2 *Consider the problem of driving the state x of system (7.1) from an initial condition x_0 to a final condition x_f in minimum time. Consider a Lebesgue integrable control function $u = u(t)$ such that, for almost every t, $u(t)$ belongs to a compact set $\mathcal{U} \subseteq \mathbf{R}^m$. If u is the time optimal*

---

[5]In fact, a general version of the PMP can be given which includes as special cases Theorem 7.2.1 and Theorem 7.4.2 (see, e.g., [110]).

control in the interval $[0, T]$ and x the associated trajectory solution of (7.1) in the interval $[0, T]$, then there exists an n-dimensional vector of functions λ called the costate, not identically zero and satisfying the adjoint equation

$$\dot{\lambda}^T = -\lambda^T f_x(x, u).$$

Define the optimal control Hamiltonian,

$$h(\lambda, x, u) := \lambda^T f(x, u).$$

Then, for every t in $[0, T]$,

$$h(\lambda(t), x(t), u(t)) \geq h(\lambda(t), x(t), v),$$

for every $v \in \mathcal{U}$. Moreover, $h(\lambda(t), x(t), u(t))$ is constant along an optimal trajectory.

For quantum control systems, as described in the previous section, the state x and the costate λ satisfy the same differential equation (cf. (7.34) where $\tilde{L}_x \equiv 0$).

7.4.1.1 Example of a time optimal quantum control problem

For some classes of systems, properties can be proved that simplify the calculation of the time optimal control. We consider as an example the class of quantum control systems (7.8) having the Hamiltonian $H(u)$ of the form

$$H(u_1, u_2, v_1, v_2) := \begin{pmatrix} E_1 & u_1 + iu_2 & 0 \\ u_1 - iu_2 & E_2 & v_1 + iv_2 \\ 0 & v_1 - iv_2 & E_3 \end{pmatrix}. \tag{7.48}$$

Here the controls are u_1, u_2, v_1, v_2. The real constants E_1, E_2, E_3 represent the natural energy levels of the quantum system in the absence of the controls. This class of models describes the dynamics of certain three-level systems coupled with two laser fields, which are the controls $u := (u_1, u_2)^T$ and $v := (v_1, v_2)^T$. The laser field u (v) activates the coupling between levels 1 and 2 (2 and 3). A straightforward calculation of the dynamical Lie algebra for this system shows that it is operator controllable. This class of models was studied in a series of papers (see [42] and references therein). It can be generalized to an n-level problem and it has several properties which are useful for the calculation of the time optimal control. The following proposition holds.[6]

---

[6]The properties in Proposition 7.4.3 are valid for more general systems and costs. In particular, the existence of a minimizer in resonance is a more general property (it also holds for the problem considered in section 7.3). We refer to [42] and the references therein. The three-level problem for the evolution operator was treated in [14], [87].

Proposition 7.4.3 ([42]) Consider the problem of driving the state $\vec{\psi}$ of (7.8) from an initial state $\vec{\psi}_0$ to a final state $\vec{\psi}_f$ in minimum time, with controls $||u_1 + iu_2|| \leq 1$, and $||v_1 + iv_2|| \leq 1$. Then there always exists a minimizer in resonance, namely

$$(u_1 + iu_2)(t) := u(t)e^{i[(E_2 - E_1)t + \xi_1]}, \tag{7.49}$$

$$(v_1 + iv_2)(t) := v(t)e^{i[(E_3 - E_2)t + \xi_2]}, \tag{7.50}$$

with real functions $u(t)$ and $v(t)$ and real phase parameters ξ_1 and ξ_2.

Let us consider the problem of steering the state $\vec{\psi}$ from a state of the form $\vec{\psi}_0 = [e^{i\eta_1}, 0, 0]^T$, corresponding to an energy level E_1 (i.e., an eigenstate $|E_1\rangle$), to a state of the form $\vec{\psi}_f = [0, 0, e^{i\eta_2}]^T$, corresponding to an energy level E_3 (i.e., an eigenstate $|E_3\rangle$). Here η_1 and η_2 are arbitrary real parameters. It is convenient to perform a change of coordinates

$$\vec{r} = DU(t)\vec{\psi}, \tag{7.51}$$

with $U(t) := \text{diag}(iE_1t, iE_2t, iE_3t)$ and D an appropriate diagonal, constant, unitary matrix. This transforms system (7.8), (7.48) into the system

$$\frac{d}{dt}\vec{r} = -S_z u(t)\vec{r} - S_x v(t)\vec{r}, \tag{7.52}$$

where $S_{x,y,z}$ are defined in (6.8) and satisfy the commutation relations (6.9). In deriving (7.52), we have used Proposition 7.4.3 and in particular we have restricted ourselves to controls in resonance of the form (7.49) (7.50). In particular u and v are the amplitudes appearing in (7.49), (7.50). The change of coordinates does not change the structure of the initial state and of the final state, and we can choose them real as the phase factors η_1 and η_2 are arbitrary. In particular, we can assume the initial value $\vec{r}_0 = [1, 0, 0]^T$ and the desired final value $\vec{r}_f := [0, 0, 1]^T$. Using the result on the existence of a minimizer in resonance, the problem is therefore reduced to a real problem of time optimal control on the two-dimensional sphere.[7] The problem has low dimension and one can apply the techniques of optimal control on two-dimensional manifolds of [43]. The final result is summarized in the following theorem which is proved in [42].

Theorem 7.4.4 *The optimal control to steer the state $\vec{r}$ of system (7.52) between two points is a concatenation of controls (u,v) of the type $(u,v) = (\pm 1, \pm 1)$, $(u,v) = (\pm 1, 0)$, and $(u,v) = (0, \pm 1)$. In particular, the time optimal control steering from $\vec{r}_0 = [1,0,0]^T$ to $\vec{r}_f = [0,0,1]^T$ is given by $u \equiv 1$, $v \equiv 1$ in the interval $[0, \frac{\pi}{\sqrt{2}}]$ (cf. Figures 7.4, 7.5).*

---

[7]As discussed in subsection 7.1.2, considering a real problem is always possible. However, in this case, we have a reduction of the dimension of the problem which is crucial in the actual solution.

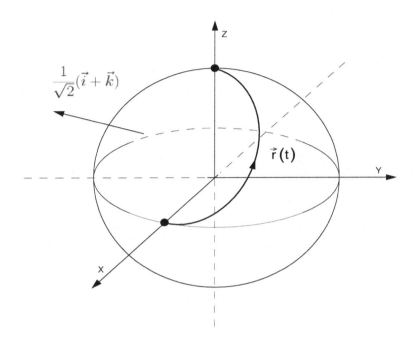

FIGURE 7.4: Time optimal trajectory from the point $\vec{r}_0 = [1,0,0]^T$ to $\vec{r}_f = [0,0,1]^T$ according to Theorem 7.4.4. It is a rotation about the axis $\frac{1}{\sqrt{2}}(\vec{i} + \vec{k})$.

We shall not report the complete proof of this theorem here but we shall add some considerations to show how the PMP determines the nature of the optimal controls. In particular the arguments that follow prove the first sentence of the theorem. The optimal control Hamiltonian of Theorem 7.4.2, $h(\lambda, \vec{r}, u, v)$, reads in this case as

$$h(\lambda, \vec{r}, u, v) = -u\lambda^T S_z \vec{r} - v\lambda^T S_x \vec{r},$$

and it is constant along the optimal trajectory. The adjoint equations for λ read as

$$\dot{\lambda} = -u S_z \lambda - v S_x \lambda. \tag{7.53}$$

The procedure to find the time optimal control consists of calculating all the controls satisfying the PMP, i.e., the extremals, and then selecting, among them, the one which gives minimum time. To understand the nature of the extremals, notice that extremals which are identically zero, i.e., $u \equiv 0$ and $v \equiv 0$, in an interval of nonzero measure, must be excluded because they are clearly not time optimal. In fact the trajectory would 'stop' in some point for some time. From the PMP, we obtain that in intervals where $\lambda^T S_z \vec{r} < 0$ (> 0), $u \equiv 1$ ($u \equiv -1$), while in intervals where $\lambda^T S_x \vec{r} < 0$ (> 0), $v \equiv 1$ ($v \equiv -1$). We cannot have intervals where

$$\lambda^T S_z \vec{r} \equiv 0 \tag{7.54}$$

and

$$\lambda^T S_x \vec{r} \equiv 0, \tag{7.55}$$

because, in that case, by differentiating these expressions, using (7.52) and (7.53), and the fact that we have excluded extremals with $u \equiv 0$ and $v \equiv 0$, we obtain

$$\lambda^T S_y \vec{r} \equiv 0. \tag{7.56}$$

This together with (7.55) and (7.54) would give $\lambda \equiv 0$, which contradicts the PMP. If (7.54) is verified in some interval, differentiating (7.54) with (7.52) and (7.53) gives

$$v \lambda^T S_y \vec{r} \equiv 0,$$

and since (7.55) cannot be verified in the same interval, as just shown, we must have $v \neq 0$ which implies (7.56). Differentiating (7.56), we obtain

$$-\lambda^T S_x \vec{r} u + \lambda^T S_z \vec{r} v \equiv 0,$$

which, with (7.54), and since $\lambda^T S_x \vec{r} \neq 0$, gives $u \equiv 0$. Analogously one proves that, when (7.55) is verified, $v \equiv 0$.

7.4.2 Minimum time control with unbounded control; Riemannian symmetric spaces

We consider now the problem of minimum time control with no a priori bound on the controls for the *Schrodinger operator equation*. The problem is to drive the *unitary evolution operator* to a desired target in minimum time. In several cases the problem can be solved explicitly using the theory of Riemmanian symmetric spaces, some of which was discussed in 6.3, and concepts of Lie group theory and homogeneous spaces.

We set up the problem in general, namely, we look at a differential system on a general compact matrix Lie group $e^{\mathcal{L}}$, with semisimple Lie algebra $\mathcal{L}$, of the form

$$\dot{X} = AX + \sum_{j=1}^{m} B_j X u_j, \qquad X(0) = \mathbf{1}, \tag{7.57}$$

where $\mathbf{1}$ is the identity matrix in the group, u_j, $j = 1, ..., m$, are the controls and the matrices A and B_j, $j = 1, ..., m$ generate the Lie algebra $\mathcal{L}$. The

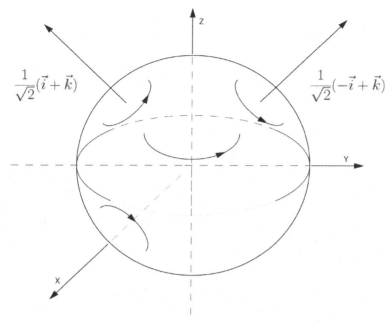

FIGURE 7.5: According to Theorem 7.4.4, every optimal trajectory is the concatenation of the four types of trajectories depicted in this picture. These are rotations about the x and z axes and about the directions of $\frac{1}{\sqrt{2}}(\vec{i} + \vec{k})$ and $\frac{1}{\sqrt{2}}(-\vec{i} + \vec{k})$ (all possibly in both directions).

problem is to steer X from the identity to a target X_f in minimum time and there is no a priori bound on the control. Since the minimum time does not exist in general for this problem, we shall look for the infimum time. In formulas, the infimum time $t_f^* := t_f^*(X_f)$ is defined as

$$t_f^*(X_f) := \inf\{T \geq 0 | X_f \in \bar{\mathcal{R}}(\leq T)\},$$

with the reachable set $\mathcal{R}(\leq T)$ defined in Chapter 3 (cf. the discussion before formula (3.2)) and $\bar{\mathcal{R}}$ here means the closure of the set $\mathcal{R}$. In words, for every $\epsilon > 0$ it is always possible to drive the state arbitrarily close to X_f in time less than $t_f^* + \epsilon$ and t_f^* is the smallest time for which this is true. The problem here is to find an expression for the function $t_f^*(X_f)$ and then to find a control which will drive X from the identity arbitrarily close to X_f in time arbitrarily close to $t_f^*(X_f)$.

7.4.2.1 The control subgroup

An important role, in this problem, is played by the Lie subgroup of $e^{\mathcal{L}}$ corresponding to the Lie subalgebra of $\mathcal{L}$ generated by the control matrices $B_1, ..., B_m$, i.e., $\mathcal{K} := \{B_1, ..., B_m\}_{\mathcal{L}}$. This subgroup is called the *control* *subgroup* and the Lie algebra $\{B_1, ..., B_m\}_\mathcal{L}$ is called the *control subalgebra*. It has the feature that, for every $K \in e^\mathcal{K}$, $t_f^*(K) = 0$. In order to see this, we show that, for every $K \in e^\mathcal{K}$, there exists a control steering the evolution operator X of (7.57) arbitrarily close to K in arbitrarily short time. Consider the system corresponding to (7.57) but without the drift term, i.e.,

$$\dot{X}_c = \sum_{j=1}^m B_j X_c u_j, \qquad X_c(0) = \mathbb{1}. \tag{7.58}$$

Every element K of $e^\mathcal{K}$ can be reached by an appropriate control.[8]

Consider the control $u = u_j(t)$, $j = 1, ..., m$, steering X_c from the identity to K, in time T. Then the control $M u_j(Mt)$ steers to K in time $\frac{T}{M}$. In fact, it is easy to verify that the solution of (7.58) with this control in the interval $[0, \frac{T}{M}]$ is $X_c(Mt)$, if X_c denotes the solution corresponding to the control $u_j(t)$ in $[0, T]$. Consider now the control $M u_j(Mt)$ for system (7.57) and for system (7.58) and subtract one from the other to obtain

$$\dot{X} - \dot{X}_c = AX + \left(\sum_{j=1}^m B_j M u_j(Mt) \right)(X - X_c).$$

Integrating this in the interval $[0, \frac{T}{M}]$, we obtain

$$X\left(\frac{T}{M}\right) - X_c\left(\frac{T}{M}\right)$$

$$= \int_0^{\frac{T}{M}} AX(t)\, dt + \int_0^{\frac{T}{M}} \left(\sum_{j=1}^m B_j M u_j(Mt) \right)(X(t) - X_c(t))\, dt.$$

Let P be a uniform bound on $\|AX\|$ (which exists because the group $e^\mathcal{L}$ is assumed to be compact) and let N be a uniform bound on $\|\sum_{j=1}^m B_j u_j(Mt)\|$. Then we have

$$\left\| X\left(\frac{T}{M}\right) - X_c\left(\frac{T}{M}\right) \right\| < P\frac{T}{M} + MN \int_0^{\frac{T}{M}} \|X(t) - X_c(t)\| dt.$$

By applying the Gronwall-Bellman inequality[9] (cf., e.g., [110], Appendix A), we obtain

$$\|X(\frac{T}{M}) - X_c(\frac{T}{M})\| \leq P\frac{T}{M} + PTN \int_0^{\frac{T}{M}} e^{MN(\frac{T}{M} - r)} dr,$$

---

[8]Since system (7.58) is driftless one can apply results for driftless systems [154] to obtain that the reachable set is $e^\mathcal{K}$ independently of whether $e^\mathcal{K}$ is compact or not (cf. Theorem 3.2.1).

[9]Assume $k \geq 0$ and nonnegative functions m and h. Then

$$0 \leq m(t) \leq k + \int_0^t h(s)m(s)ds \rightarrow m(t) \leq k + \int_0^t kh(s)e^{\int_s^t h(r)dr} ds \tag{7.59}$$

which gives

$$\left\| X\left(\frac{T}{M}\right) - X_c\left(\frac{T}{M}\right) \right\| \leq P\frac{T}{M} + PT\frac{e^{TN}}{M}(1 - e^{-NT}).$$

As $X_c(\frac{T}{M}) = K$ for every M, letting $M \to \infty$ we obtain the desired result. With $K \in e^{\mathcal{K}}$, for every $X_f \in e^{\mathcal{L}}$, we have

$$t_f^*(X_f) = t_f^*(KX_f) = t_f^*(X_fK), \tag{7.60}$$

because after, or before, having driven to X_f we can produce K in arbitrarily small time. We summarize this discussion in terms of coset spaces (cf. the definition in subsection 3.3.1.2) in the following proposition.

Proposition 7.4.5 Let π be the natural projection (cf. Definition (3.19))

$$\pi \ : \ e^{\mathcal{L}} \to e^{\mathcal{L}}/e^{\mathcal{K}}.$$

If $\pi(X_1) = \pi(X_2)$, then $t_f^*(X_1) = t_f^*(X_2)$. In particular, if $\pi(X_1) = \pi(1)$, i.e., X_1 belongs to $e^{\mathcal{K}}$, then $t_f^*(X_1) = 0$.

7.4.2.2 Motion in the coset space $e^{\mathcal{L}}/e^{\mathcal{K}}$

For a control u_j, $j = 1, ..., m$, and the corresponding trajectory $X = X(t)$ solution of (7.57), we can consider the projection of such a trajectory onto the coset space $e^{\mathcal{L}}/e^{\mathcal{K}}$, $\pi(X)$. We shall focus in particular on right cosets. If X moves for time $T > 0$ inside a coset, i.e., $\pi(X)$ is constant, we can always find another control and trajectory which will give (approximately) the same resulting state transfer in time less than T. This is because in the transfer $X_f \to KX_f$, $K \in e^{\mathcal{K}}$, can be obtained in approximately no time according to the discussion in the previous subsection. To look for the minimum time control we have to look for the fastest way to go from the initial coset $\{e^{\mathcal{K}}1\}$ to the final coset $\{e^{\mathcal{K}}X_f\}$ (cf. Figure 7.6). So we look at the minimum time problem for the associated system on $e^{\mathcal{L}}/e^{\mathcal{K}}$. To study the motion of $\pi(X)$, i.e., the motion in the coset space $e^{\mathcal{L}}/e^{\mathcal{K}}$, we write the solution $X = X(t)$ of (7.57) as

$$X(t) := X_c(t)U(t), \tag{7.61}$$

where X_c is the solution of (7.58) and $U(t)$ is the solution of the *auxiliary system*:

$$\dot{U} = (X_c^{-1}AX_c)U, \qquad U(0) = \mathbf{1}. \tag{7.62}$$

In fact, by differentiating the right-hand side of (7.61) using (7.62) and (7.58), one obtains that $X_c(t)U(t)$ satisfies (7.57), and by uniqueness of the solution of a differential equation, it is equal to X. In the factorization (7.61), $U(t)$ describes how $X(t)$ moves from one coset to the other, while X_c only contributes motion *inside* a right coset.

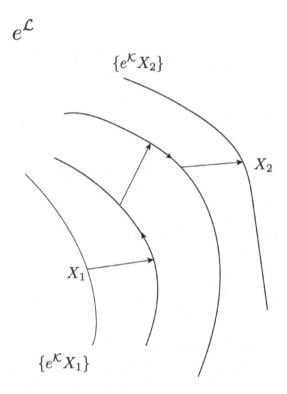

$e^{\mathcal{L}}$

$\{e^{\mathcal{K}} X_2\}$

X_2

X_1

$\{e^{\mathcal{K}} X_1\}$

FIGURE 7.6: Motion in the Lie Group $e^{\mathcal{L}}$: The curves represent cosets. In going from a coset $\{e^{\mathcal{K}} X_1\}$ to a coset $\{e^{\mathcal{K}} X_2\}$, motion along the curves is arbitrarily fast while motion along the straight arrows is slow. It represents going from one coset to the other.

One way to approach the problem of infimum time to X_f for system (7.57) is to study the minimum time problem for the auxiliary system (7.62) to reach the coset $\{e^{\mathcal{K}} X_f\}$. In this problem, the control is the set of matrix functions with values in the *adjoint orbit* $\mathcal{O}_{\mathcal{K}}(A)$ defined by

$$\mathcal{O}_{\mathcal{K}}(A) := \{X_c^{-1} A X_c | X_c \in e^{\mathcal{K}}\}. \tag{7.63}$$

For this problem, we denote the infimum time to reach the coset $\{e^{\mathcal{K}} X_f\}$ by $t_A^*(\{e^{\mathcal{K}} X_f\})$, analogously to how we have defined the infimum time $t_f^*(X_f)$ for the original problem. Denote by $\mathcal{B}$, $\mathcal{B}(T)$ and $\mathcal{B}(\leq T)$ the reachable sets of the system (7.62), i.e., $\mathcal{B}(\leq T) := \cup_{0 \leq t \leq T} \mathcal{B}(t)$ and $\mathcal{B} := \cup_{t \geq 0} \mathcal{B}(t)$, with $\mathcal{B}(t)$ the set of points reachable at time t. With $\bar{\mathcal{B}}$ denoting the closure of a set $\mathcal{B}$, we have

$$t_A^*(\{e^{\mathcal{K}} X_f\}) := \inf\{t \geq 0 | \exists K \in e^{\mathcal{K}}, K X_f \in \bar{\mathcal{B}}(\leq t)\}.$$

In words, $t_A^*(\{e^{\mathcal{K}} X_f\})$ is the infimum time to drive the state of (7.62) from the identity to arbitrarily close to a coset containing X_f by using controls with values in $\mathcal{O}_{\mathcal{K}}(A)$. The calculation of $t_f^*(X_f)$ is equivalent to the calculation of $t_A^*(\{e^{\mathcal{K}} X_f\})$, as the following equivalence result holds.

Proposition 7.4.6

$$t_f^*(X_f) = t_A^*(\{e^{\mathcal{K}} X_f\}).$$

Proof. (Sketch) By definition of $t_A^*(\{e^{\mathcal{K}} X_f\})$, given $\epsilon > 0$, there must exist a $K \in e^{\mathcal{K}}$, a $T \le t_A^*(\{e^{\mathcal{K}} X_f\} + \epsilon$ and a piecewise continuous (control) trajectory, $X_c = X_c(t)$, such that the corresponding solution $U(t)$ of (7.62) is arbitrarily close to $K X_f$ at time T. From results in geometric control theory (see, e.g., Theorem 3.2. in [171]) we can find a control $u(t)$ such that the *trajectory* of (7.58) is arbitrarily close (in an appropriate topology) to $X_c = X_c(t)$ and therefore the corresponding trajectory of (7.62) is arbitrarily close to $K X_f$ at time T. The solution of the original system (7.57) is then arbitrarily close to $K_1 X_f$ for some $K_1 \in e^{\mathcal{K}}$, and therefore

$$t_f^*(\{K_1 X_f\}) = t_f^*(X_f) \le t_A^*(\{e^{\mathcal{K}} X_f\}).$$

To show

$$t_f^*(X_f) \ge t_A^*(\{e^{\mathcal{K}} X_f\}), \tag{7.64}$$

notice that if there exists a control steering the solution X of (7.57) arbitrarily close to X_f in time $T \le t_f^*(X_f) + \epsilon$, then, with the same control, the solution of (7.62) will be arbitrarily close to $K X_f$ for some $K \in e^{\mathcal{K}}$. This gives the inequality (7.64). □

Proposition 7.4.6 gives a way of studying the problem of minimum time control for system (7.57). This means to study the minimum time problem to steer the state of the auxiliary system (7.62) between two right cosets using controls which are constrained to be in the adjoint orbit (7.63).

7.4.2.3 Determination of the minimum time in the case where $e^{\mathcal{L}}/e^{\mathcal{K}}$ is a symmetric space

We now examine how the problem of minimum time control for system (7.62) specializes in the case where $e^{\mathcal{L}}/e^{\mathcal{K}}$ is a symmetric space. Let $\mathcal{L}$ be semisimple and have the Cartan decomposition

$$\mathcal{L} = \mathcal{K} \oplus \mathcal{P},$$

where $\mathcal{K}$ and $\mathcal{P} = \mathcal{K}^{\perp}$ satisfy the commutation relations (6.24), (6.25), (6.26). Consider A in (7.57) and assume $A \in \mathcal{P}$. Let $\mathcal{A}$ be a Cartan subalgebra containing A (it always exists because of Theorems 6.3.4, 6.3.5) and define the *Weyl orbit* $\mathcal{W}(A)$ to be the set of elements in the adjoint orbit $\mathcal{O}_{\mathcal{K}}(A)$ that also belong to $\mathcal{A}$, i.e.,

$$\mathcal{W}(A) := \mathcal{A} \cap \mathcal{O}_{\mathcal{K}}(A).$$

The *positive span* of $\mathcal{W}(A)$, $\mathrm{span}_{\geq}(\mathcal{W}(A))$, is defined as

$$\mathrm{span}_{\geq}(\mathcal{W}(A)) := \left\{ \sum_j \beta_j A_j \,\middle|\, A_j \in \mathcal{W}(A), \beta_j \geq 0 \right\}.$$

We make the following assumption

Assumption 1:

$$\mathrm{span}_{\geq}(\mathcal{W}(A)) = \mathcal{A}.$$

This assumption is certainly satisfied if $\tilde{A} \in \mathcal{W}(A)$ implies $-\tilde{A} \in \mathcal{W}(A)$ and $\mathrm{span}(\mathcal{W}(A)) = \mathcal{A}$, as it will be the case for the one and two spin examples below.

If Assumption 1 is verified, then, from the Cartan decomposition (6.31) of Theorem 6.3.7, we have that, for every element X_f, there exist elements K_1 and K_2 in $e^{\mathcal{K}}$ and commuting elements in the adjoint orbit $\mathcal{O}_\mathcal{K}$, A_j, $j = 1, 2, ..., p$, along with positive coefficients $\beta_j = 1, ..., p$ so that

$$X_f = K_1 e^{\sum_{j=1}^{p} \beta_j A_j} K_2. \tag{7.65}$$

The following theorem and remark solve the problem of calculating $t_f^*(X_f)$ and giving the minimum time control in the case of a Riemannian symmetric space when Assumption 1 is verified. The proof of the theorem is given in [160], [161].

Theorem 7.4.7 *Under the above assumptions, $t_f^*(X_f)$ is the minimum value of $\sum_{j=1}^{p} \beta_j$ for which it is possible to write (7.65), with commuting $A_j \in \mathcal{O}_\mathcal{K}(A)$.*

Remark 7.4.8 Since $A_j = X_j^{-1} A X_j$, $j = 1, ..., p$, for some $X_j \in e^{\mathcal{K}}$, a time minimum (infimum) control and trajectory can be obtained by driving X first to K_2 with a very large, very short, pulse, then to X_1 with the same method, then setting the control equal to zero for a time β_1, so as to produce $e^{A\beta_1}$. Then a new large and short pulse is used to drive the solution X of (7.57) from the identity to arbitrarily close to $X_2 X_1^{-1} \in e^{\mathcal{K}}$. Then the control is set equal to zero for time β_2 and so on. The last large and short pulse is used to generate K_1 in (7.65). This control achieves the transfer of the solution X of (7.57) from the identity $\mathbf{1}$ to arbitrarily close to X_f in time arbitrarily close to $t_f^*(X_f)$ given in Theorem 7.4.7.

The main idea behind Theorem 7.4.7 is as follows. Given a decomposition (7.65) of X_f, we have from (7.60) that $t_f^*(X_f) = t_f^*(e^{\sum_{j=1}^{p} \beta_j A_j})$, and by the equivalence result of Proposition 7.4.6 we can calculate the infimum time

$t_A^*(\{e^{\mathcal{K}} e^{\sum_{j=1}^p \beta_j A_j}\})$ for the auxiliary system (7.62) to reach the coset containing $e^{\sum_{j=1}^p \beta_j A_j}$. We can, without loss of generality, assume for the system (7.62) 'controls' $X_c^{-1} A X_c$, piecewise constant. Moreover we need to have all the control matrices $X_c^{-1} A X_c$ commuting for the trajectory to be time optimal. In fact, if two of them, say A_k and A_l, do not commute, then write, using (for example) the exponential formula (E.5) of Appendix E,

$$e^{A_k \frac{t}{n}} e^{A_l \frac{t}{n}} = e^{\frac{t}{n}(A_k + A_l) + \frac{t^2}{2n^2}[A_k, A_l] + o(\frac{1}{n^3})}.$$

We now use the assumption that $e^{\mathcal{L}}/e^{\mathcal{K}}$ is a Riemannian symmetric space. From (6.26), we obtain that $[A_k, A_l] \in \mathcal{K}$ and therefore there is a component of the motion in $e^{\mathcal{K}}$ which we could have obtained in a much quicker way by using large and short pulses. Therefore, in this case, the control cannot be optimal. The candidate optimal final points for system (7.62) are all of the form $e^{\sum_{j=1}^p \beta_j A_j}$, with commuting A_j's, satisfying (7.65), and the corresponding time is $\sum_{j=1}^p \beta_j$. Among those, the infimizing time is the minimum $\sum_{j=1}^p \beta_j$ so that (7.65) holds.

7.4.2.4 One and two spin example

Consider now the model of a spin-$\frac{1}{2}$ particle in a magnetic field studied in Chapter 2, subsection 2.4.2, in section 6.1.4 of the previous chapter, and in section 7.3. In particular, let us assume that we have a constant magnetic field in the y direction and a varying control magnetic field in the z direction, without any a priori bound on its magnitude. After a re-scaling of the control and of time we can write the Schrödinger operator equation (cf. (6.19)) as

$$\dot{X} = (\bar{\sigma}_y + u(t)\bar{\sigma}_z)X,$$

with $\bar{\sigma}_{x,y,z}$ defined in (6.5). The Cartan decomposition of the dynamical Lie algebra $\mathcal{L} = su(2)$ is $\mathcal{L} = \mathcal{K} \oplus \mathcal{P}$ with the Lie algebra $\mathcal{K} := \text{span}\{\bar{\sigma}_z\}$ and $\mathcal{P} := \text{span}\{\bar{\sigma}_y, \bar{\sigma}_x\}$. It induces a decomposition of the Lie group $SU(2)$ which is the Euler decomposition of subsection 6.1.2. The rank of the corresponding symmetric space is one. The Weyl orbit $\mathcal{W}(\bar{\sigma}_y)$ is given by

$$\mathcal{W}(\bar{\sigma}_y) = \{\bar{\sigma}_y, -\bar{\sigma}_y\}. \tag{7.66}$$

Applying Theorem 7.4.7, the minimum time to reach a final state X_f is the smallest nonnegative t_2 (Euler angle) so that we can write (cf. (6.13))

$$X_f = e^{\bar{\sigma}_z t_3} e^{\bar{\sigma}_y t_2} e^{\bar{\sigma}_z t_1}.$$

Methods to calculate the Euler angles were discussed in subsection 6.1.3. It follows from the discussion in that subsection that the infimum time $t_f^*(X_f)$ is to be taken in the interval $[0, 2\pi]$.

The situation is only slightly more complicated in the case of two spin-$\frac{1}{2}$ particles interacting through Ising interaction (cf. subsection 2.4.2). We assume that it is possible to address each one of the two spins separately with magnetic fields u and v in independent directions so that the model is written as

$$\dot{X} = (i\sigma_z \otimes \sigma_z)X + \left(i \left(\sum_{j=x,y,z} u_j \sigma_j \right) \otimes 1 + i1 \otimes \left(\sum_{j=x,y,z} v_j \sigma_j \right) \right) X,$$

where $\sigma_{x,y,z}$ are the Pauli matrices (1.20). In this case, the relevant Cartan decomposition is the one described in Example 6.3.10 and the rank of the associate symmetric space is 3. The Weyl orbit $\mathcal{W}(i\sigma_z \otimes \sigma_z)$ is given by

$$\mathcal{W}(i\sigma_z \otimes \sigma_z) = \{\pm i\sigma_z \otimes \sigma_z, \pm i\sigma_x \otimes \sigma_x, \pm i\sigma_y \otimes \sigma_y\}. \qquad (7.67)$$

Theorem 7.4.7 applies and it says that $t_f^*(X_f)$ is the minimum value of $\beta_x + \beta_y + \beta_z$ for which we can write

$$X_f = L_1 e^{\sum_{j=x,y,z} i\beta_j \sigma_j \otimes \sigma_j} L_2,$$

with $L_{1,2}$ of the form $L_{1,2} = K_{1,2} \otimes R_{1,2}$ with $K_{1,2}$ and $R_{1,2}$ in $SU(2)$. Methods to calculate such a decomposition were discussed in subsection 6.3.6.

7.5 Numerical Methods for Optimal Control of Quantum Systems

There exists a wealth of numerical methods to solve optimal control problems. Some of them directly aim at finding the optimal control in a specific problem, some others are useful in the solution of auxiliary problems, as for example the integration of the adjoint equations of the maximum principle. We summarize in the following subsections some of the main ideas with emphasis on application to quantum systems. The following two sections describe, respectively, methods to find the optimal control by discretization and by iterative algorithms. These iterative methods generate a sequence of control functions converging to the optimal control or to an optimal candidate. Subsection 7.5.3 deals with the numerical solution of two points boundary value problems arising in the application of the PMP.

7.5.1 Methods using discretization

Given the (realification of a) quantum system

$$\dot{x} = \tilde{H}(u)x, \qquad (7.68)$$

and a cost of the Bolza type

$$J(u) = \phi(x(T)) + \int_0^T L(x(t), u(t))dt,$$

we can choose a time step Δt and a positive integer N so that $N\Delta t = T$. Δt has to be chosen small if the dynamics of (7.68) is fast. Defining $u_k := u(k\Delta t)$, $x_k := x(k\Delta t)$, $k = 0, ..., N$, we write

$$x_{k+1} \approx x_k + (\Delta t)\tilde{H}(u_k)x_k, \tag{7.69}$$

$$J(u) \approx \phi(x_N) + \sum_{k=0}^{N-1} L(x_k, u_k)\Delta t. \tag{7.70}$$

If n is the dimension of the state vector x and m the dimension of the control vector u, there are $Nn + Nm$ variables, $x_1, ..., x_N$, $u_0, ..., u_{N-1}$, in (7.69), (7.70). We wish to find the values that minimize the function on the right-hand side of (7.70) subject to the constraint in (7.69). This constraint can be eliminated and the problem transformed into an unconstrained optimization problem by defining $G(u_k) := I + (\Delta t)\tilde{H}(u_k)$ and setting, from (7.69),

$$x_k = G(u_{k-1})G(u_{k-2}) \cdots G(u_1)G(u_0)x_0.$$

Placing this into the function in (7.70) we only have a function of $u_0, ..., u_{N-1}$ to be minimized. Standard numerical methods of unconstrained optimization, such as *steepest descent method* (see, e.g., [88]) can be used for this problem.

7.5.2 Iterative methods

Iterative methods for optimal control generate sequences of control laws that converge to extremals satisfying the necessary conditions of optimality (the PMP). Several proposals for such algorithms have been put forward and theoretical and numerical studies exist on the convergence properties of such algorithms (see, e.g., [122], [188], [246], [309], [310]).

Let us consider the case of the cost (7.14) and let us assume the common situation in molecular control where the available control is only one laser field. The dynamics can be written as

$$\frac{d}{dt}\vec{\psi} = (A + uB)\vec{\psi},$$

where $\vec{\psi}$ is the pure state, u the control field and A and B are matrices in $su(n)$. Transforming the problem into a real one (cf. subsection (7.1.2)), the cost takes the form

$$J(u) = \frac{1}{2}x^T\tilde{O}x + \frac{k}{2}\int_0^T u^2(t)dt,$$

where x is defined in (7.10), and $\tilde{O}$ is the symmetric negative definite matrix

$$\tilde{O} = \begin{pmatrix} O_R & -O_I \\ O_I & O_R \end{pmatrix},$$

with $O = O_R + iO_I$, $O_R = O_R^T$, $O_I^T = -O_I$. The dynamics take the form

$$\dot{x} = \tilde{A}x + \tilde{B}xu, \qquad x(0) = x_0 \tag{7.71}$$

with (cf. (7.11))

$$\tilde{A} = \begin{pmatrix} A_R & -A_I \\ A_I & A_R \end{pmatrix}, \tilde{B} = \begin{pmatrix} B_R & -B_I \\ B_I & B_R \end{pmatrix}, A = A_R + iA_I, B = B_R + iB_I,$$

and x_0, the appropriate initial condition. The necessary conditions of sub-section 7.2.2 give, in this case, that there exists a nonzero costate vector λ satisfying

$$\dot{\lambda} = \tilde{A}\lambda + \tilde{B}\lambda u, \tag{7.72}$$

with terminal condition

$$\lambda(T) = -\tilde{O}x(T), \tag{7.73}$$

and so that, given $x = x(t)$ and $\lambda = \lambda(t)$,

$$\lambda^T(\tilde{A}x + \tilde{B}xu) - \frac{k}{2}u^2 \geq \lambda^T(\tilde{A}x + \tilde{B}xv) - \frac{k}{2}v^2,$$

for every admissible value v. If we do not assume any bound on the value of the control u, we can calculate the maximum by taking the derivative of the right-hand side with respect to v and setting it equal to zero. This gives

$$u(t) = \frac{1}{k}\lambda^T(t)\tilde{B}x(t). \tag{7.74}$$

Iterative methods generate sequences of functions $\{u^k(t), x^k(t), \lambda^k(t)\}$ which converge to a triple $\{u(t), x(t), \lambda(t)\}$ satisfying the equations of the maximum principle, namely (7.71), and (7.72) with terminal condition given by (7.73).

The most natural and simple iterative method is as follows: Starting with a control $u^0 = u^0(t)$, one integrates (7.71) forward to obtain $x^1 = x^1(t)$. Then using x^1 instead of x in the expression of the terminal condition (7.73), one integrates (7.72) backwards to obtain $\lambda^1 = \lambda^1(t)$. Then x^1 and λ^1 are used in (7.74) to calculate the new estimate for the control $u^1 = \frac{1}{k}\lambda^{1T}\tilde{B}x^1$. This process is then iterated. Unfortunately, such a simple and intuitive iterative scheme does not converge in several situations. This has stimulated much research to design iterative algorithms which are guaranteed to converge. The algorithm given in [188] contains as special cases the Krotov algorithm [268] and its modification, the Zhu-Rabitz algorithm [309]. We describe it here as an example of an iterative algorithm for quantum optimal control. For applications and results of numerical simulations we refer the reader to the original papers.

Algorithm III ([188])

The iteration formulas are as follows. Choose two constants $\delta, \eta \in [0, 2]$,

$$\dot{x}^k = \tilde{A}x^k + \tilde{B}x^k u^k, \qquad x^k(0) = x_0, \tag{7.75}$$

$$u^k = (1 - \delta)v^{k-1} + \frac{\delta}{k}\lambda^{k-1T}\tilde{B}x^k, \tag{7.76}$$

$$\dot{\lambda}^k = (\tilde{A} + \tilde{B}v^k)\lambda^k, \qquad \lambda^k(T) = -\tilde{O}x^k(T), \tag{7.77}$$

$$v^k = (1 - \eta)u^k + \frac{\eta}{k}(\lambda^{kT}\tilde{B}x^k). \tag{7.78}$$

One starts with a guess for an auxiliary control v^0 and a guess for the costate λ^0. Using (7.76) in (7.75) and integrating forward the resulting nonlinear equation, one obtains x^1, which in turn, placed in (7.76), gives u^1. Now, u^1 and x^1, with (7.78), are used to integrate backwards (7.77). This produces λ^1 and using (7.78) v^1, which are then used in place of λ^0 and v^0 to run the next iteration, and the procedure is iterated several times. This scheme has the following convergence property [188]:

Theorem 7.5.1 *Algorithm III is such that*

$$J(u^{k+1}) \leq J(u^k).$$

In particular, since $J(u)$ is bounded from below, $J(u^k)$ will converge. Notice that, in principle, there is no guarantee that the sequence of control laws will converge to a *global* minimum. Experience has however shown that in many cases the result of a numerical optimal control algorithm turns out to be the actual optimal control [223] (also cf. next section).

7.5.3 Numerical methods for two points boundary value problems

As we have observed in subsection 7.2.1, the application of the PMP requires the solution of a system of differential equations with conditions at the initial time as well as at the final time. In general, we would like to have conditions at the initial time only so that we can integrate the relevant differential equations forward. The search for an appropriate initial condition can be cast in the following class of problems:

Problem 1 *Given a differential equation*

$$\dot{y} = f(y(t), t), \tag{7.79}$$

find an initial condition $y(0) = y_0$ so that, for the solution at the final time,
$y(T, y_0)$, we have

$$\tilde{F}(y_0) := F(y(T, y_0), y_0) = 0, \qquad (7.80)$$

where F has the same dimension as y.

In particular for the system (7.29), (7.30) with initial condition $x(0) = x_0$ and final condition $\lambda^T(T) = -\phi_x(x(T))$ we can set the state variable $y := (x^T, \lambda^T)^T$ and the conditions (7.80) can be written as

$$y_1(0) - x_{0,1} = 0, \; y_2(0) - x_{0,2} = 0, \; ..., \; y_n(0) - x_{0,n} = 0,$$

$$y_{n+1}(T, y_0) + (\phi_x^T)_1 \, (y_1(T, y_0), ..., y_n(T, y_0)) = 0,$$

$$..., \; y_{2n}(T, y_0) + (\phi_x^T)_n \, (y_1(T, y_0), ..., y_n(T, y_0)) = 0.$$

In this notation, the subscript gives the component of the vector except for 0 which indicates the initial condition so that $y_0 = y(0)$ so that, in particular $y_{0,j} = y_j(0)$.

The algebraic equation (7.80) can be solved numerically. For example, a Newton iteration (see, e.g., [88]) gives

$$y_0^{k+1} = y_0^k - \left[\frac{\partial \tilde{F}}{\partial y_0} \Big|_{y_0 = y_0^k} \right]^{-1} \tilde{F}(y_0^k). \qquad (7.81)$$

To obtain $\frac{\partial \tilde{F}}{\partial y_0}\big|_{y_0 = y_0^k}$, we calculate with (7.80)

$$\frac{\partial \tilde{F}}{\partial y_0} \Big|_{y_0 = y_0^k} = \frac{\partial F}{\partial y_0} \Big|_{y_0 = y_0^k} + \frac{\partial F}{\partial y} \Big|_{y = y(T, y_0^k)} \frac{\partial y(T, y_0)}{\partial y_0} \Big|_{y_0 = y_0^k}. \qquad (7.82)$$

In this formula $\frac{\partial y(t)}{\partial y_0}\big|_{y_0 = y_0^k}$ describes the dependence of the solution $y(t)$ of (7.79) on the initial condition when the initial condition is y_0^k. It is the solution of the *variational equation* associated with (7.79)

$$\frac{d}{dt} \frac{\partial y(t, y_0)}{\partial y_0} \Big|_{y_0 = y_0^k} = f_y(y(t, y_0^k), t) \frac{\partial y(t, y_0)}{\partial y_0} \Big|_{y_0 = y_0^k}. \qquad (7.83)$$

This suggests the following algorithm to find the vector y_0 solving Problem 1, i.e., satisfying (7.80).

Algorithm IV

- Step 1: Choose the initial guess y_0^0 for y_0 and set $k = 0$.

- Step 2: Integrate (7.79) with the current value of y_0.

- Step 3: Check (7.80). If it is satisfied, up to some a priori tolerance STOP. Otherwise go to Step 4.

- Step 4: Integrate the variational equation (7.83).

- Step 5: Use $\frac{\partial y(T)}{\partial y_0}\big|_{y_0=y_0^k}$ calculated in the previous step to update the estimate of y_0 according to (7.81) and (7.82). Also, increment k by 1 and go to Step 2.

7.6 Quantum Optimal Control Landscape

As we have pointed out already, an optimal control problem consists of the minimization of a functional over an appropriate space of control functions. The functional map which associates to any given control function the corresponding value of the cost is called the corresponding **optimal control landscape.** An, at least qualitative, knowledge of the landscape is important for both theoretical and practical reasons. In particular, such knowledge reveals the presence of *local minimums* and *local maximums* where a numerical search algorithm for the control can get stopped and lead to wrong results. In the vicinity of saddle points, it is of interest to see how steep the direction of descent is and how many directions of descent there are. The problem can be set up at different levels of generality and for different types of systems and costs, as well as for different classes of control functions. It has generated, in the last few decades, a large literature and evolved into a large area of quantum control and nonlinear control, in general, for various applications (see, e.g., [225] and references therein).

We shall limit ourselves to considering the minimization of the expectation value of an observable O

$$J = Tr(O\rho(T)), \qquad (7.84)$$

on a finite interval $[0, T]$, in a density matrix formulation. We shall have fairly general assumptions on the set of control functions which are assumed for example piecewise continuous, so as we can apply the controllability theorems of Chapter 2. Our goal is to illustrate, in a simplified setting, the empirical evidence that finding quantum optimal control in numerical experiments has proved surprisingly easy. This was also the original motivation of the first papers in this area [134], [143], [224] . The following are common assumptions in this setting and we shall assume them verified.

1. *Controllability:* The system is operator controllable, i.e., the dynamical Lie algebra is given by $su(n)$.

2. *Regularity:* Let T be the final time, and u a control to reach the unitary evolution U_f in time T. Then for every (small) $A \in su(n)$ there exists a (perturbed) control driving the unitary evolution to $e^A U_f$ in time T.

The regularity assumption is verified when the time of control T is larger than the critical time T_c defined in Theorem 3.2.8. This ensures us that, assuming that a certain control leading to a final unitary operation U_f is optimal, we can apply a variation to U_f itself. That is, we can perturb U_f as $U_f \to e^A U_f$, for an arbitrary $A \in su(n)$ and be sure that there exists a (perturbed) control leading to $e^A U_f$, in the same time T. In looking for a characterization of the extrema, we can avoid referring to the control and use a perturbation of this type on the final evolution operator. This approach is sometimes referred to as *kinematic*. In this case, the actual form of the Liouville equation (1.51) for the quantum dynamics plays no role, except for the determination of the critical time T_c. The role of the control is hidden in the final value of the evolution operator. This approach was used for example in [143]. Another method is to consider the control explicitly and to calculate the variation of the final cost due to a variation on the control (for example, strong or weak variations) similarly to what was done in the derivation of the maximum principle in Remark 7.2.3. Such an approach can be called *dynamic* and was for example explored in [134]. We shall limit ourselves to a kinematic approach.

Let ρ_0 denote the initial condition for the density matrix. If the optimal control drives the evolution operator from the identity to U_f, we must have, for any $A \in su(n)$, from (7.84),

$$\frac{d}{d\epsilon}|_{\epsilon=0} J(\epsilon) := \frac{d}{d\epsilon}|_{\epsilon=0} Tr\left(O e^{A\epsilon} U_f \rho_0 U_f^\dagger e^{-A\epsilon} \right) = \qquad (7.85)$$

$$Tr(O[A, \rho(T)]) = Tr(A[\rho(T), O]).$$

Since this has to hold for any $A \in su(n)$, it implies

$$[\rho(T), O] = 0. \qquad (7.86)$$

Commutativity of O and $\rho(T)$, implies that they can be simultaneously diagonalized, that is, there exists a matrix $K \in SU(n)$ such that

$$K\rho(T)K^\dagger = \Lambda^\rho, \qquad KOK^\dagger = \Lambda^O, \qquad (7.87)$$

with Λ^O and Λ^ρ, diagonal, displaying the eigenvalues of O and $\rho(T)$, respectively. We remark that the eigenvalues of $\rho(T)$ are the same as the eigenvalues of ρ_0, and therefore Λ^ρ contains the eigenvalues of ρ_0 on the diagonal. For example, if ρ_0 is a pure state, the diagonal of Λ^ρ contains one 1 and all zeros. Replacing (7.87) in (7.84), we obtain for the cost J,

$$J = Tr(\Lambda^O \Lambda^\rho).$$

These are the possible values of the cost J at the stationary points. There are $n!$ (possibly coinciding) values, each corresponding to a permutation of the n eigenvalues of $\rho(T)$ once we fix an order for the eigenvalues of O, or vice versa. The maximum corresponds to the case where the eigenvalues of O and $\rho(T)$

are ordered in the same fashion (i.e., both ascending or both descending). The minimum corresponds to the case where they are ordered in opposite fashion. Stationary points, which are not maxima or minima correspond to the intermediate cases. Therefore, at stationary points of the optimal control landscape, the cost J can only take a finite number (at most $n!$) of values, and the minimum and maximum and all intermediate values are easily determined by considering all the permutations of the eigenvalues of O and $\rho(T)$, which are the same as the eigenvalues of ρ_0.

We have defined the optimal control landscape as the map $\mathcal{U} \to \mathbf{R}$, with $\mathcal{U}$ the set of possible control functions (defined in $[0, T]$), where each control function is mapped to the corresponding value of the cost J in $\mathbf{R}$. In our analysis, we have decomposed the landscape map as $\mathcal{U} \to SU(n) \to \mathbf{R}$ and focused on the part of the map $SU(n) \to \mathbf{R}$, for which we found the values at the stationary points (kinematic approach). Notice that since the (partial) map $\mathcal{U} \to SU(n)$ is surjective (controllability at time T assumption), the stationary values (and in particular the maximum and minimum) we found are also stationary values for the full (dynamical) landscape map $\mathcal{U} \to SU(n) \to \mathbf{R}$. To describe the values of $U_f \in SU(n)$ corresponding to stationary points, from (7.86) we have

$$[U_f \rho_0 U_f^\dagger, O] = 0, \qquad (7.88)$$

which, given ρ_0 and O, defines a closed manifold in $SU(n)$. Every control steering to a point in such a manifold steers to a stationary point of the landscape map. There are an infinite number of such controls. Let R be a matrix in $SU(n)$ diagonalizing ρ_0, i.e., $\rho_0 = R\Lambda^{\rho_0} R^\dagger$, with Λ^{ρ_0} diagonal. Then the first one of (7.87) reads as

$$KU_f R\Lambda^{\rho_0} R^\dagger U_f^\dagger K^\dagger = \Lambda^\rho. \qquad (7.89)$$

Since $\rho(T)$ and ρ_0 are similar, Λ^ρ and Λ^{ρ_0} are diagonal matrices containing the same elements on the diagonal. Therefore, the matrix $KU_f R$ acts by conjugation as a *permutation* matrix. Once K and R are fixed (from O and ρ_0) the value of this permutation determines which value of the stationary point we have.

We now want to determine the nature of the stationary points by deriving a quadratic form in the $A \in su(n)$ matrix used in (7.85). In order to do this, we reconsider, the perturbation on values $U_f \in SU(n)$ used in (7.85). Defining $\rho := \rho(T)$ we have

$$J(\epsilon) = Tr(Oe^{A\epsilon}\rho e^{-A\epsilon}) = Tr\left(O\left(\sum_{k=0}^{\infty} \frac{\epsilon^k}{k!} ad_A^k(\rho)\right)\right) =$$

$$Tr(O\rho) + \epsilon Tr(O[A, \rho]) + \frac{\epsilon^2}{2} Tr(O[A, [A, \rho]]) + h.o.t.,$$

where we used the BCH formula (E.1) of Appendix E. Using property (7.86) at a stationary point, the term linear in ϵ give zero. Moreover, using for the term quadratic in ϵ the cyclic property of the Killing trace form,[10] we obtain

$$J(\epsilon) = Tr(O\rho) - \frac{\epsilon^2}{2} Tr\left([A, \rho][A, O]\right) + h.o.t. \tag{7.90}$$

Consider now the matrix K used in (7.87) which simultaneously diagonalizes ρ and O. We have

$$J(A) = Tr(O\rho) - \frac{\epsilon^2}{2} Tr\left([KAK^\dagger, \Lambda^\rho][KAK^\dagger, \Lambda^O]\right). \tag{7.91}$$

Let us denote by $\alpha_{j,k}$ the entries of $KAK^\dagger$ (with $\alpha_{k,j} = -\bar{\alpha}_{j,k}$), and by $\{\lambda_1, ..., \lambda_n\}$, the eigenvalues of the diagonal matrix Λ^ρ, and by, $\{\mu_1, ..., \mu_n\}$, the eigenvalues of the diagonal matrix Λ^O. We have

$$[KAK^\dagger, \Lambda^\rho]_{j,k} = \alpha_{j,k}(\lambda_k - \lambda_j), \qquad [KAK^\dagger, \Lambda^O]_{j,k} = \alpha_{j,k}(\mu_k - \mu_j).$$

Therefore, up to higher order terms in ϵ, $J(\epsilon)$ in (7.91) becomes (using $\alpha_{k,j} = -\bar{\alpha}_{j,k}$)

$$J(A) = Tr(O\rho) - \frac{\epsilon^2}{2} \sum_j \sum_k \alpha_{j,k}\alpha_{k,j}(\lambda_k - \lambda_j)(\mu_j - \mu_k) = \tag{7.92}$$

$$Tr(O\rho) + \frac{\epsilon^2}{2} \sum_j \sum_k |\alpha_{j,k}|^2(\lambda_k - \lambda_j)(\mu_j - \mu_k) =$$

$$Tr(O\rho) - \frac{\epsilon^2}{2} \sum_j \sum_k |\alpha_{j,k}|^2(\lambda_k - \lambda_j)(\mu_k - \mu_j).$$

In formula (7.92), the $\alpha_{j,k}$'s represent 'directions' where we can move in $SU(n)$ and therefore in the landscape. If we are at a global minimum, the λ_j and μ_j, eigenvalues of ρ_0 and O respectively, will be ordered in opposite fashion (one ascending and one descending or vice versa) and therefore the second term in (7.92) will be nonnegative for each value of the $\alpha_{j,k}$. On the other hand, if we are at a global maximum, the eigenvalues will be ordered in the same fashion and therefore the second term in the above formula is nonpositive. The most interesting situation for us happens at stationary points which are not global maximums nor global minimums. In particular, assume we are looking for the global minimum. At these points, directions (j, k) where $(\lambda_k - \lambda_j)(\mu_k - \mu_j) > 0$ will be such that moving along them will decrease the cost. Thus, they are 'escape directions' in our search for the optimal control. The existence of such escape direction implies that local stationary points are not local maximums nor local minimum and therefore, in principle, they cannot act as traps in our optimal numerical search. Notice that, in particular in the case where O

---

[10]$Tr(A[B, C]) = Tr(B[C, A])$.

is non degenerate, i.e., $\mu_k \neq \mu_j$ for $k \neq j$, such directions always exist (as we are assuming that we are not in the trivial case where ρ_0 is a perfectly mixed state). One can further examine formula (7.92) to determine, using the eigenvalues of ρ_0 and O, how steep the descent is at any stationary point. This formula gives the sought for justification of the apparent ease in finding optimal quantum control in numerical experiments.

7.7 Notes and References

Optimal control of quantum mechanical systems has been considered in many papers, especially in connection with applications to molecular dynamics. Early work was done by A. P. Peirce, M. A. Dahleh, and H. Rabitz in [212], and many studies have been presented since then. The books [233], [254] give several examples of applications to molecular control, and the paper [116] is an extensive review on quantum optimal control with many references. There exist more general versions of the maximum principle which include, as special cases, Theorems 7.2.1 and 7.4.2. These can be found in [110] or other advanced books on optimal control such as [177], [181], [214]. The problem of optimal control for two-level systems was treated in [83], and several generalizations have been given (see, e.g., [40]). A general result of [40] states that minimizers are in resonance for a class of systems, which includes as a special case the one treated in subsection 7.4.1. Moreover this is true also for the control problem of minimizing the energy in a given interval subject to a terminal condition on the state as well as for the problem of minimizing the time of transfer subject to an energy constraint. The geometric treatment of minimum time problems for quantum systems with no a priori bound on the magnitude of the control in subsection 7.4.2 follows [160], [161]. Similar ideas were used for the treatment of systems other than the one and two spin case, for example for a system with three spin-$\frac{1}{2}$ particles in [163]. It is interesting to note that the $P - K$ Cartan type of decomposition assumed in these papers for the dynamics also enters another class of quantum optimal control problems where the roles of the P and K parts of the decomposition are reversed. The P part is the one that contains the control while the drift term belongs to the K part of the decomposition. Such problems, which were called $K - P$ problems, have been extensively studied (see, e.g., [11], [14], [41], [42], [87] and references therein) as they are, in many cases, amenable of explicit solution.

There are many papers on the subject of iterative numerical algorithms for optimal control of quantum systems, and the subject still poses many theoretical and practical questions. In particular, there are several unsolved mathematical problems concerning not only the convergence of the given algorithms, but also the analysis of the speed of convergence, as well as the properties of the sets of limit points. For the subject of iterative numerical algorithms for quantum control we refer to [95], [122], [188], [223], [233], [246], [268], [309], [310]. For the discussion of numerical methods for two point boundary value problems of subsection 7.5.3, we followed [166].

There are many review papers on quantum optimal control landscapes. Some of them are [60][225]. For our mathematical treatment in section 7.6 we mostly elaborated upon the content of [134], [143]. A mathematical study of the critical sub-manifolds defined in (7.88) is given in [301]. As we mentioned, the optimal control landscape can be studied for costs different from (7.84). If the cost is the distance of the final evolution operator from a desired one, we talk about *gate control landscape*. A study in this setting is [224].

7.8 Exercises

Exercise 7.1 Consider the three-level problem

$$\frac{d}{dt}\vec{\psi} = A\vec{\psi} + Bu\vec{\psi},$$

with

$$A := \begin{pmatrix} i & 0 & 0 \\ 0 & -i & 0 \\ 0 & 0 & 0 \end{pmatrix} \qquad B = \begin{pmatrix} 0 & 1 & 1 \\ -1 & 0 & 0 \\ -1 & 0 & 0 \end{pmatrix},$$

with cost

$$J(u) = ||\vec{\psi}(T) - \vec{\psi}_f||^2 + \int_0^T u^2(t)dt,$$

with $T = 2$ and $\vec{\psi}_f = \frac{1}{\sqrt{2}}\begin{pmatrix} 1 \\ 1 \\ 0 \end{pmatrix}$ and initial condition $\vec{\psi}(0) = \begin{pmatrix} 1 \\ 0 \\ 0 \end{pmatrix}$.

a) Transform this problem into an equivalent real problem of Mayer.

b) Write the associate adjoint equation for the six-dimensional costate λ and use the maximum principle to obtain an expression of the optimal control in terms of state and costate.

c) Specialize Algorithm III to this problem.

Exercise 7.2 Consider the example of section 7.3 and notice that the form of the controls, (7.46), (7.47), does not depend on the nonintegral part of the cost in (7.37). Therefore, we can replace the cost with

$$J'(u_x, u_y) = -\|\vec{\psi}^{\,\dagger}(T)\vec{\psi}_f\| + \eta \int_0^T u_x^2(t) + u_y^2(t)dt. \qquad (7.93)$$

Set the desired state $\vec{\psi}_f = \begin{pmatrix} 1 \\ 0 \end{pmatrix}$ and the initial state $\vec{\psi}(0) = \begin{pmatrix} 0 \\ 1 \end{pmatrix}$. Notice that the cost does not change if we replace $\vec{\psi}_f$ with $\vec{\psi}_f e^{ir}$ for some real value r. By plugging the controls (7.46),(7.47) into the differential equations (7.36), the latter equations can be integrated explicitly. Find the parameters ω γ and M in (7.46), (7.47) which minimize the cost (7.93). Discuss the dependence of the minimum cost on T and η.

Exercise 7.3 Verify formulas (7.66) and (7.67). How does the treatment of the two spin case go if we replace Ising interaction with Heisenberg interaction (2.64)?

Exercise 7.4 Prove that the map $\phi : u(n) \to so(2n) \cap sp(n)$, defined by

$$\phi(R + iI) := \begin{pmatrix} R & -I \\ I & R \end{pmatrix},$$

where I and R are real matrices is a Lie algebra isomorphism.

Exercise 7.5 How does Proposition 7.4.1 modify in the case of infimum (rather than minimum) time?

Exercise 7.6 This exercise deals with the passage from complex state $\vec{\psi}$ and costate λ to real and vice versa.

1. Verify that, with the given definitions and notations, the optimal control Hamiltonian (7.35) coincides with the one in (7.33).

2. Redo the treatment of the example of section 7.3 without resorting to transformation of the state vector and costate vector to the corresponding real ones but using the (state and costate) equations and the equations of the PMP with complex vectors.

Exercise 7.7 Consider the one-dimensional system

$$\dot{x} = x + u, \qquad x(0) = 2,$$

and assume that in time $T = 1$ we want to find the optimal control $u \in [-1, 1]$ to minimize the cost $\phi = x^3(1) + x(1)$. Write the equations of the PMP namely the adjoint equations for the costate λ and the form of the optimal control, so as to obtain the system (7.29), (7.30), with appropriate initial and terminal conditions. Set up the condition (7.80), the derivative (7.82) and the variational equation (7.83) for application of Algorithm IV of subsection 7.5.3.

Exercise 7.8 For a given $U_f \in SU(n)$, the Cayley transform is defined as

$$U_f \to (1 + A)(1 - A)^{-1} U_f, \tag{7.94}$$

with $A \in su(n)$.

1. Prove that this transforms unitary matrices to unitary matrices.

2. If U_f is *special* unitary, that is in $SU(n)$, is its Cayley transform special unitary as well? Prove or disprove this fact.

3. Expanding in Taylor series $(1 - A)^{-1} = 1 + A + A^2 + h.o.t.$ in (7.94), derive formula (7.90) using the Cayley transform of U_f as a perturbation of U_f.

Exercise 7.9 Consider the Hamiltonian $H = \sigma_z(1 - u) + \sigma_y u$, where $\sigma_{x,y,z}$ are the Pauli matrices and u is the control in $[0, 1]$ for a two-level quantum system, with initial condition given by the density matrix $\rho_0 = \begin{pmatrix} \frac{1}{2} & \frac{1}{6} \\ \frac{1}{6} & \frac{1}{2} \end{pmatrix}$. Consider the problem of minimizing the expectation value of $Tr(O\rho)$ with $O = \sigma_z$. Use the techniques of Chapter 6 to determine an upper bound for the critical time T_c of Theorem 3.2.8. Consider the optimal control problem on a finite time interval $[0, T]$ for $T \geq T_c$. Find the value of the maximum and minimum for such optimal control problem. Are there critical points besides these ones?

Chapter 8

More Tools for Quantum Control

In this chapter, we discuss some more tools for the control of quantum systems. Some of these are commonly considered in the physics and chemistry literature. Some others, like the Lyapunov control method, have been very well studied in the control literature and have then been extended to quantum systems.

To every level of energy of a quantum system there corresponds an eigenstate (or an eigenspace). Transferring population between two levels means transferring the state from one eigenstate to the other or, more generally, increasing the probability of one value of energy as compared to the other. We start by explaining the fact that, to induce a population transfer between two levels, we need to use a field (approximately) at the frequency corresponding to the energy difference between the two levels. This way, selective population transfer is obtained between two energy levels. This is discussed in section 8.1. In section 8.2, we briefly review *time-varying perturbation theory*, which is a powerful tool to analyze quantum dynamics. Time-varying perturbation theory also gives an alternative illustration of the above described behavior of the solution of the Schrödinger equation in terms of the frequency of the control. In section 8.3, we study the technique of *adiabatic control* and the *adiabatic approximation* on which it is based. The main idea here is that, if the Hamiltonian varies very slowly, the evolution of the state vector will approximately follow an eigenstate of the Hamiltonian. Both the ideas of selective population transfer via frequency tuning and adiabatic control are used in the technique called STIRAP (*STImulated Raman scattering involving Adiabatic Passage*), which can be used to induce population transfer between two levels which are not directly coupled. This is discussed in section 8.4. Section 8.5 summarizes the technique of Lyapunov control as applied to quantum control systems.

8.1 Selective Population Transfer via Frequency Tuning

A common technique in the control of quantum systems is to transfer population between two energy levels using a field whose frequency is tuned at the

DOI: 10.1201/9781003051268-8

energy difference between the selected levels. This idea is at the foundation of quantum optics and the theory of absorption and emission of radiation by atoms (see, e.g., [185]). When one considers the electromagnetic field itself as a quantum system, its energy can assume a discrete set of values. For a monochromatic field the possible energy values differ by integer multiples of a constant quantity called a *photon* which is a quantum of available energy. It classically corresponds (modulo the Plank constant $\hbar$) to the frequency of the field. Population transfer happens because the controlled quantum system absorbs one (or more) quantum of energy. However, the system can only absorb (or emit) quantities of energy corresponding to the differences between its allowed energy levels. Therefore, there must be a correspondence between the frequency of the controlling field and the energy gap between the two levels considered. In a semiclassical treatment, a control at a frequency equal to the energy difference between two levels only affects the population in these two levels and leaves untouched populations in other levels separated by very different energy gaps. This can also be seen as a justification of the multi-level approximation (cf. Chapter 2). In the rest of the section we illustrate and mathematically justify more rigorously such an idea in the semiclassical setting.

For simplicity of notation, we work in units where $\hbar = 1$ so that we shall deal with controlled Schrödinger equation in the bilinear form with one control,[1]

$$\frac{d}{dt}\vec{\psi} = (-iH_0 - iH_1 u(t))\vec{\psi}. \tag{8.1}$$

We shall assume H_0 diagonal,

$$H_0 := \mathrm{diag}(E_1, E_2, \ldots, E_n),$$

and, for simplicity, we shall assume it to be non degenerate, so that $E_j \neq E_k$ if $j \neq k$. We assume we want to control within the subspace identified by the first two energy levels, E_1 and E_2, so that, ideally, the remaining portion of the n-dimensional Hilbert space associated with the system is not affected by the dynamics. The matrix H_1 has zeros on the diagonal, that is,

$$H_1 = \{h_{jk}\}, \qquad h_{jk} = h_{jk}^*, \qquad h_{jj} = 0, \qquad j, k = 1, 2, ..., n.$$

The constants h_{jk} represent the coupling between the j-th and the k-th energy level (eigenstate).

The dynamics of quantum control systems may be represented by diagrams of the type in Figure 8.1 where every horizontal segment represents an energy level, and levels j and k are connected if h_{jk} is different from zero. One example of a possible configuration is a V-system which is a three-level system

---

[1] Only notational changes are needed to handle the case of more than one control.

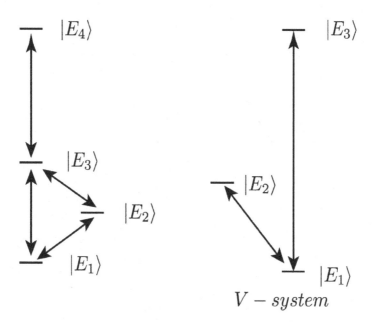

$V - system$

FIGURE 8.1: Level representation of quantum systems for control. The configuration on the right represents a V system.

where the lowest energy level is coupled to the remaining two levels but these are not coupled to each other. Another typical three-level configuration is a Λ system, where the coupling is only between the highest energy level and the remaining two levels (cf. Figure 8.2).

We shall assume a control field $u = u(t)$ with the functional dependence

$$u(t) = v(t)\sin(\omega_L t), \qquad \omega_L \geq 0, \qquad (8.2)$$

where $v = v(t)$ is a smooth *slow-varying envelope* which we shall refer to as the **pulse**, and which has the property

$$\lim_{t \to 0} v(t) = \lim_{t \to +\infty} v(t) = 0. \qquad (8.3)$$

The magnitudes of the various components of the state $\vec{\psi}$ are called the **populations** in the various levels of energy. Going to the interaction picture by defining (cf. subsection 1.3.2) $\vec{\phi} = e^{iH_0 t}\vec{\psi}$ which does not modify the populations, we obtain the differential equation for $\vec{\phi}$

$$\frac{d}{dt}\vec{\phi} = -i\tilde{H}_1(t)\vec{\phi}, \qquad (8.4)$$

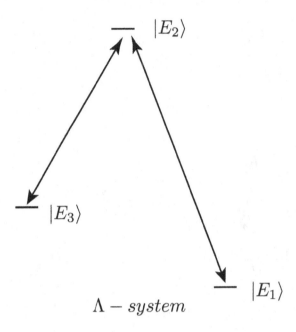

FIGURE 8.2:   Level representation of a Λ system.

where $\tilde{H}_1$ is defined by

$$(\tilde{H}_1)_{jk} := h_{jk}u(t)e^{i\Delta_{jk}t}, \qquad j < k,$$

where the difference in energy levels j and k, Δ_{jk}, is defined by $\Delta_{jk} := E_j - E_k$.

Using expression (8.2) for the control u and separating the terms containing a *difference* of frequencies from those containing a *sum*, we can write $\tilde{H}_1$ as

$$\tilde{H}_1(t) = \tilde{D}_1(t) + \tilde{A}_1(t). \qquad (8.5)$$

Here $\tilde{D}_1$ and $\tilde{A}_1$ are both Hermitian and, for $j < k$,

$$(\tilde{D}_1)_{jk} := \pm v(t)\frac{h_{jk}}{2i}e^{\pm i(\omega_L - |\Delta_{jk}|)t}, \quad (\tilde{A}_1)_{jk} := \mp v(t)\frac{h_{jk}}{2i}e^{\mp i(\omega_L + |\Delta_{jk}|)t},$$

where the $+$ $(-)$ sign for $\tilde{D}_1$ and the $-$ $(+)$ sign for $\tilde{A}_1$ is used if $\Delta_{jk} \leq 0$ $(\Delta_{jk} > 0)$.

Our goal is now to show that terms of the type $v(t)\mu e^{i\omega t}$, for a complex constant μ and a real frequency ω, when ω is large, and slow pulse v, do not significantly affect the solution of the differential equation (8.4). The intuitive idea is that fast oscillating terms average out to zero when integrated. Consider $\tilde{H}_1$ in (8.4), and write it as $\tilde{H}_1(t) := v(t)S(t) + v(t)F(t)$, where $F(t)$

is zero everywhere except in entries (j, k) and (k, j), for one selected pair (j, k), which are of the form $\mu e^{i\omega t}$ and $\mu^* e^{-i\omega t}$, respectively. We assume $j = 1$, $k = 2$, to have a simpler notation. In practice, we pull out of the Hamiltonian $\tilde{H}_1$ a single fast-varying term. We rewrite (8.4) as

$$\vec{\phi}(t) = \vec{\phi}(0) + \int_0^t -iv(s)S(s)\vec{\phi}(s)\,ds + \int_0^t -iv(s)F(s)\vec{\phi}(s)\,ds. \qquad (8.6)$$

Because of the fast dependence of F on s the last integral gives a small contribution. Consider the diagonal matrix

$$G_\omega := \mathrm{diag}(i\omega, -i\omega, 1, ..., 1).$$

We can write the last integral in (8.6), using integration by parts, as

$$\int_0^t -iv(s)F(s)\vec{\phi}(s)\,ds = G_\omega^{-1} \int_0^t -iv(s)\frac{dF(s)}{ds}\vec{\phi}(s)\,ds =$$

$$-iG_\omega^{-1} \left(\left[F(s)v(s)\vec{\phi}(s) \right]_0^t - \int_0^t F(s)\frac{d}{ds}(v(s)\vec{\phi}(s))\,ds \right).$$

Because of the shape of the function v, if t is sufficiently large,

$$\left[F(s)v(s)\vec{\phi}(s) \right]_0^t \approx 0,$$

so that we can write (8.6) as

$$\vec{\phi}(t) = \vec{\phi}(0) - i \int_0^t v(s)S(s)\vec{\phi}(s)\,ds \qquad (8.7)$$

$$+iG_\omega^{-1} \left(\int_0^t \frac{dv(s)}{ds}F(s)\vec{\phi}(s)\,ds + i\int_0^t v^2(s)(F(s)S(s) + F^2(s))\vec{\phi}(s)\,ds \right).$$

Notice that $F^2(s)$ is a constant matrix, independent of s (and ω). Furthermore, in (8.7, only the first two components of the vector multiplying G_ω^{-1} are possibly different from zero. Assuming that $\frac{dv}{ds}$ is bounded (in fact, we assume it to be small), all the quantities inside the integrals in (8.7) multiplying G_ω^{-1} can be bounded independently of ω. Assuming further that $v(s)$ and $\frac{dv(s)}{ds}$ go to zero sufficiently fast,[2] we find an upper bound to the whole term multiplying G_ω^{-1} in (8.7)

By comparing the solution of (8.4) with the one of the ideal case

$$\frac{d}{dt}\vec{\phi}_d = -iS(t)v(t)\vec{\phi}_d, \qquad \vec{\phi}_d(0) = \vec{\phi}(0),$$

---

[2]This is done only to simplify notations. Otherwise we would write $m = m(t)$ below.

we obtain

$$||\vec{\phi}(t) - \vec{\phi}_d(t)|| \leq \int_0^t ||v(s)S(s)||||\vec{\phi}(s) - \vec{\phi}_d(s)|| \, ds + \frac{1}{\omega}m.$$

We have denoted by m the uniform bound for $|| \int_0^t \dot{v}F\vec{\phi} \, ds - i \int_0^t Fv^2(S + F)\vec{\phi} \, ds||$, which we assume exists. We can obtain an explicit bound by using the *Gronwall-Bellman inequality* (see (7.59) of Chapter 7). We have

$$||\vec{\phi}(t) - \vec{\phi}_d(t)|| \leq \frac{m}{\omega} + \int_0^t \frac{m}{\omega}||v(s)S(s)||e^{\int_s^t ||v(r)S(r)|| dr} \, ds,$$

which goes to zero as ω goes to infinity.

We may carry out this procedure for other frequencies in $S(t)$ and obtain various estimates for the error resulting from neglecting the high frequency components, in the Schrödinger equation. In particular, by considering the decomposition of $\tilde{H}_1$ (8.5), we can neglect all the elements in $\tilde{A}_1$, as these contain sums of the energy gaps and the laser frequency ω_L which give a fast-varying contribution. This is the **rotating wave approximation**. Moreover, elements corresponding to energy gaps which are very different from the frequency of the laser ω_L can also be neglected. In an ideal situation where H_0 has eigenvalues very much spaced from each other, a laser frequency ω_L close to the energy gap between two particular levels is such that the system can be effectively considered a two-level system. This idea is used to selectively induce population transfer between two given levels. The method used in this section, which is based on the Gronwall-Bellman inequality, can be used to estimate the error.

8.2   Time-Dependent Perturbation Theory

Our discussion of **time-dependent perturbation theory** will focus on this theory for the evolution operator. Let us write the Schrödinger equation in the interaction picture (8.4) for the evolution operator $X - X(t)$.

$$\dot{X} = -i\tilde{H}_1(t)X(t), \qquad X(0) = 1. \tag{8.8}$$

Written as an integral equation, this becomes

$$X(t) = 1 - i \int_0^t \tilde{H}_1(t_1)X(t_1)dt_1. \tag{8.9}$$

By plugging (8.9) back into (8.9), we obtain

$$X(t) = 1 - i \int_0^t \tilde{H}_1(t_1)dt_1 + (-i)^2 \int_0^t \int_0^{t_1} \tilde{H}_1(t_1)\tilde{H}_1(t_2)X(t_2)dt_2 dt_1.$$

Continuing this way, we obtain the *Dyson series* expansion of the solution of the Schrödinger operator equation in the interaction picture (8.8)

$$X(t) = 1 + \sum_{n=1}^{\infty} D_n(t), \qquad (8.10)$$

with

$$D_n(t) := (-i)^n \int_0^t \int_0^{t_1} \cdots \int_0^{t_{n-1}} \tilde{H}_1(t_1)\tilde{H}_1(t_2) \cdots \tilde{H}_1(t_n) dt_n \cdots dt_1.$$

For finite t and finite-dimensional (bounded) $\tilde{H}_1$ this series converges and the error truncating the series (8.10) after m terms is given by

$$\left\| X(t) - \left(1 + \sum_{n=1}^{m} D_n(t)\right) \right\| =$$

$$\left\| \int_0^t \int_0^{t_1} \cdots \int_0^{t_m} \tilde{H}_1(t_1)\tilde{H}_1(t_2) \cdots \tilde{H}_1(t_{m+1}) X(t_{m+1}) dt_{m+1} dt_m \cdots dt_1 \right\|$$

$$\leq \frac{t^{m+1}}{(m+1)!} \|\tilde{H}_1\|_{[0,t]}^{m+1},$$

where we have denoted by $\|\tilde{H}_1\|_{[0,t]}$ the infinity norm of $\tilde{H}_1$ in the interval $[0,t]$.

In time-varying perturbation theory, one replaces the solution of Schrödinger operator equation with the first m terms of the Dyson expansion, for some m. This is called *m-th order perturbation theory*. It gives a good approximation when the norm of $\tilde{H}_1$ is small, e.g., for a controlled Schrödinger equation (8.1), when the control pulse has small amplitude. The most common situation is first order perturbation theory where one considers

$$X(t) \approx 1 - i \int_0^t \tilde{H}_1(t_1) dt_1. \qquad (8.11)$$

This first order expansion is used to analyze dynamics in several contexts (see, e.g., [245]). First order perturbation theory and formula (8.11) can also be used to give another, elementary, proof that the high frequency terms can be neglected in the dynamics. In particular, if $A_1 = A_1(t)$ is a high frequency term in $\tilde{H}_1$, i.e., having a 2×2 block of the form $\begin{pmatrix} 0 & \mu e^{i\omega t} \\ \mu^* e^{-i\omega t} & 0 \end{pmatrix}$, then integration of A_1 gives

$$\int_0^t A_1(t_1) dt_1 = F_\omega^{-1}(A(t) - 1),$$

with F_ω^{-1} going to zero as $\omega \to \infty$. For $t \to \infty$ a treatment of perturbation theory based on residue calculus is given in [254] (section 2.2).

There has been a lot of study on the solutions of linear time-varying equations such as (8.8). An alternative method of approximating the solution is the *Magnus expansion formula* [189] where one expresses the solution X as

$$X(t) = e^{\Omega(t)}. \tag{8.12}$$

Here Ω is given by a series expansion

$$\Omega(t) = \sum_{k=0}^{\infty} \Omega_k(t), \tag{8.13}$$

where $\Omega_k(t)$ are multiple integrals of nested commutators of $-i\tilde{H}_1$. For example,

$$\Omega_1 = \int_0^t -i\tilde{H}_1(t_1)dt_1, \qquad \Omega_2(t) = (-i)^2 \int_0^t \int_0^{t_1} [\tilde{H}_1(t_1), \tilde{H}_1(t_2)]dt_2 dt_1.$$

The study of Magnus expansion (existence, computation and properties of the Magnus series (8.13)) is more complicated than the study of the Dyson series (8.10). However, the main advantage is that, since all the elements Ω_k (8.13) are in the Lie algebra $u(n)$, so are their sums. Truncation at any order of the series (8.13) gives a matrix which is in $u(n)$ and the resulting approximation of $X(t)$, according to (8.12) is unitary, a property which is not shared by the Dyson expansion. For studies comparing the two expansions, we refer to [211] and the references therein.

8.3   Adiabatic Control

Consider a finite-dimensional, quantum control system in the general (not necessarily bilinear) form

$$\frac{d}{dt}\vec{\psi} = -iH(u(t))\vec{\psi}, \qquad \vec{\psi}(0) = \vec{\psi}_0. \tag{8.14}$$

At every time t the Hamiltonian can be diagonalized via a unitary matrix $T = T(t)$, i.e.,

$$T(t)H(u(t))T^\dagger(t) = \Lambda(t), \tag{8.15}$$

with $\Lambda(t)$ diagonal , and assume that $T(t)$ is at least of class C^1 in the interval of control. Define the vector

$$\vec{\phi}(t) := T(t)\vec{\psi}(t).$$

Then the differential equation for $\vec{\phi}$ reads as, using (8.15),

$$\frac{d}{dt}\vec{\phi} = \frac{d}{dt}(T\vec{\psi}) = (\dot{T}T^\dagger - i\Lambda)\vec{\phi}, \qquad \vec{\phi}(0) = T(0)\vec{\psi}_0. \qquad (8.16)$$

Notice that the norm of $\dot{T}T^\dagger$ depends only on $\dot{T}$, since $T^\dagger$ is unitary. If $\dot{T}$ is small, which means that the control $u = u(t)$ is *slow*, we may neglect the term $\dot{T}T^\dagger$ in equation (8.16) and write the solution of (8.16) as

$$\vec{\phi}(t) = e^{-i\int_0^t \Lambda(r)dr}T(0)\vec{\psi}_0,$$

and the solution of (8.14) as

$$\vec{\psi}_a(t) = T^\dagger(t)e^{-i\int_0^t \Lambda(r)dr}T(0)\vec{\psi}_0. \qquad (8.17)$$

This is the **adiabatic approximation**, whose validity depends on the fact that the control, and therefore the Hamiltonian, varies slowly with time. In particular, assume $\vec{\psi}_0$ is an eigenstate of $H(u(0))$ corresponding to a nondegenerate eigenvalue λ_0, and assume that the time evolution of $H(u(t))$ is such that the eigenvalue function $\tilde{\lambda} = \tilde{\lambda}(t)$, with $\tilde{\lambda}(0) = \lambda_0$ remains a nondegenerate eigenvalue of $H(u(t))$, for all times t. Then the corresponding eigenvector $\vec{v} = \vec{v}(t)$ can be taken as a column of $T^\dagger(t)$ and the *adiabatic solution* $\vec{\psi}_a$ in (8.17) is equal to $\vec{v}$ modulo a phase factor, i.e.,

$$\vec{\psi}_a(t) = e^{-i\int_0^t \tilde{\lambda}(r)dr}\vec{v}(t).$$

We shall assume, from now on, to be in the above mentioned scenario, namely that the initial state is an eigenvector corresponding to a nondegenerate eigenvalue and that this (non degeneracy) situation is preserved during the time evolution.

The relevant question is how much the actual solution $\vec{\psi}(t)$ differs from the adiabatic solution $\vec{\psi}_a(t)$, and therefore to obtain *estimates* of the norm of the difference $\vec{\psi}(t) - \vec{\psi}_a(t)$ in terms of the norm of $\dot{T}$. This topic has been the subject of many papers which have proved such estimates in the so called *adiabatic theorems* (see, e.g., [28], [126], [150], [152], [156], [205], [232]).[3] Common wisdom in quantum mechanics states that the adiabatic approximation is good when the *eigenvalue gap*, $\gamma = \gamma(t)$, is large. The eigenvalue gap is defined as the minimum distance between the eigenvalue $\tilde{\lambda}$ and the other eigenvalues of $H = H(u(t))$, i.e., denoting by $\lambda_j = \lambda_j(t)$, the other eigenvalues, the eigenvalue gap is defined as

$$\gamma(t) := \min_j |\lambda_j(t) - \tilde{\lambda}(t)|. \qquad (8.18)$$

---

[3]The subject of adiabatic theorems is in fact much more general, in particular dealing with linear operators in infinite-dimensional spaces.

To make this statement precise, we shall illustrate below a theorem of [150] which gives simple estimates.

To study the norm of the difference $||\vec{\psi} - e^{ir}\vec{\psi}_a||$ for every real r, and, in particular, the minimum over real r, we equivalently study the norm $||\vec{\psi} - \vec{\psi}_a \vec{\psi}_a^\dagger \vec{\psi}||$, i.e., the norm[4] of the difference between the true solution $\vec{\psi}$ and its orthogonal projection onto the one-dimensional subspace spanned by $\vec{\psi}_a$ (cf. Exercise 8.6). Defining $P_a(t) := \vec{\psi}_a(t)\vec{\psi}_a^\dagger(t)$, we look for estimates of the quantity $||(1 - P_a(t))\vec{\psi}(t)||$. We introduce a scaled time $s := \frac{t}{T}$ where T is some characteristic time scale of the Hamiltonian. Defining in general for a (matrix) function $f = f(t)$ a function of s, $\hat{f}(s) = f(Ts)$, we can rewrite the Shrödinger equation (8.14) in terms of the scaled time s. This gives

$$\frac{d}{ds}\hat{\vec{\psi}} = -iT\hat{H}(s)\hat{\vec{\psi}}, \qquad \hat{\vec{\psi}}(0) = \hat{\vec{\psi}}_0, \tag{8.19}$$

where we used the shorthand notation $\hat{H} = H(\hat{u})$. We shall look for estimates of the quantity $||(1 - \hat{P}_a(s))\hat{\vec{\psi}}(s)||$. We have the following adiabatic Theorem (cf. Theorem 3 in [150]) .[5]

Theorem 8.3.1 *With the above notations concerning the scaled time $s := \frac{t}{T}$, assume that the Hamiltonian $\hat{H} = \hat{H}(s)$ is of class at least C^2 in the interval $[0, 1]$. Then, for every $s \in [0, 1]$, we have* [6]

$$||(1 - \hat{P}_a(s))\hat{\vec{\psi}}(s)|| \leq \frac{||\hat{H}'(0)||}{T\hat{\gamma}^2(0)} + \frac{||\hat{H}'(s)||}{T\hat{\gamma}^2(s)} + \int_0^s \frac{||\hat{H}''(r)||}{T\hat{\gamma}^2(r)} + 7\frac{||\hat{H}'(r)||^2}{T\hat{\gamma}^3(r)} dr. \tag{8.20}$$

Remark 8.3.2 Formula (8.20) is formula (6) of [150] written in the special case where we follow a one-dimensional eigenspace and the system is finite-dimensional. It is a typical adiabatic theorem in that it gives estimates of the error which depend on three quantities: The interval of control $[0, T]$, which we would like to be large; the minimum eigenvalue gap γ as defined in (8.18); and the derivatives of the Hamiltonian with respect to the normalized time, which give a rate of change of the Hamiltonian as compared to the interval of control. Formula (8.20), assuming bounded derivatives of the Hamiltonians

---

[4]Recall we are using the Euclidean norm $||\vec{\psi}|| = \sqrt{\vec{\psi}^\dagger \vec{\psi}}$.

[5]The operator norm used in this theorem is the induced norm, i.e., $||A|| = \max_{||\vec{\psi}||=1} ||A\vec{\psi}||$, which depends on the norm chosen for the vectors. In the case of the Euclidean vector norm $||\vec{\psi}|| := \sqrt{\vec{\psi}^\dagger \vec{\psi}}$, this is equal to the maximum singular value of A, namely the largest nonnegative square root of the eigenvalues of $A^\dagger A$.

[6]Consistently with the introduced notation $\hat{\gamma}(r) = \gamma(Tr)$ where γ is the eigenvalue gap.

and eigenvalue gap bounded away from zero implies the adiabatic theorem in the form

$$\lim_{T \to \infty} ||(\mathbf{1} - \hat{P}_a(1))\hat{\tilde{\psi}}(1)|| = \lim_{T \to \infty} \min_{r \in \mathbf{R}} ||\hat{\tilde{\psi}}(1) - e^{ir}\hat{\tilde{\psi}}_a(1)|| = 0.$$

Remark 8.3.3 In **adiabatic quantum computation** (see, e.g., [106], [107], [248], [281]), the solution of a computational problem is encoded in an eigenstate (typically the ground state) of a Hamiltonian H_1. As this eigenstate is difficult to prepare, one lets a quantum system evolve according to an *interpolating Hamiltonian*

$$(1 - u(s))H_0 + u(s)H_1, \qquad s \in [0,1], \tag{8.21}$$

where $u(s)$ is a slowly varying control function with $u(0) = 0$, $u(1) = 1$. H_0 is a Hamiltonian whose ground state is known and easy to prepare. According to the adiabatic theorem, under adiabatic evolution, the state of the quantum system evolving according to the Hamiltonian (8.21), starting at the ground state of H_0, will be close to the (unknown) ground state of H_1 at time $s = 1$. From this state it is then possible to extract the solution of the original computational problem. The paper [106] contains several examples of this idea applied to quantum computations of interest in applications.

8.4 STIRAP

A population transfer between two levels which are not directly coupled can be obtained with a technique called **STImulated Raman scattering involving Adiabatic Passage** (**STIRAP**), which combines control via frequency tuned pulses (section 8.1) with adiabatic control (section 8.3). The method is best explained in the simplest case of a Λ type system, although extensions exist to different types of configurations (see, e.g., [35], [69], [233], [254]). Let a system be in the configuration depicted in Figure 8.2. There is a direct coupling between the eigenstates $|E_1\rangle$ and $|E_2\rangle$ corresponding to energy level E_1 and E_2, respectively, and between the eigenstates $|E_2\rangle$ and $|E_3\rangle$ corresponding to the energy levels E_2 and E_3, respectively. There is no direct coupling between the eigenstates $|E_1\rangle$ and $|E_3\rangle$ and the problem consists of transferring population from $|E_1\rangle$ to $|E_3\rangle$.

Two laser pulses are used to couple, respectively, levels $|E_1\rangle$ and $|E_2\rangle$ and levels $|E_2\rangle$ and $|E_3\rangle$. We denote them by

$$u_{12}(t) := 2v_{12}(t)\cos(\omega_{12}t), \qquad u_{23}(t) := 2v_{23}(t)\cos(\omega_{23}t), \tag{8.22}$$

where v_{12} and v_{23} are slow-varying pulse functions and ω_{12} and ω_{23} are the corresponding laser frequencies. The pulses v_{12} and v_{23} are called *pump pulse* and *Stokes pulse*, respectively. The Hamiltonian writes as

$$H = H_0 + H_1(t),$$

with

$$H_0 := \operatorname{diag}(E_1, E_2, E_3), \qquad H_1(t) = \begin{pmatrix} 0 & h_{12}u_{12}(t) & 0 \\ h_{12}^* u_{12}(t) & 0 & h_{23}u_{23}(t) \\ 0 & h_{23}^* u_{23}(t) & 0 \end{pmatrix},$$

where h_{12} and h_{23} are constants coupling the levels 1 and 2 and 2 and 3, respectively. The initial state is $\vec{\psi}(0) = [1, 0, 0]^T$, i.e., all the initial population is assumed to be in level $|E_1\rangle$. Let D be given by

$$D := \operatorname{diag}(E_2 - \omega_{12}, E_2, E_2 - \omega_{23}).$$

The Schrödinger equation

$$\frac{d}{dt}\vec{\psi} = -i(H_0 + H_1(t))\vec{\psi},$$

is transformed, using the transformation of coordinates $\vec{\phi} = e^{iDt}\vec{\psi}$, into

$$\frac{d}{dt}\vec{\phi} = -i\tilde{H}_1(t)\vec{\phi}, \tag{8.23}$$

where

$$\tilde{H}_1(t) := -i \begin{pmatrix} -(E_2 - E_1 - \omega_{12}) & h_{12}u_{12}e^{-i\omega_{12}t} & 0 \\ h_{12}^* u_{12}e^{i\omega_{12}t} & 0 & h_{23}u_{23}e^{i\omega_{23}t} \\ 0 & h_{23}^* u_{23}e^{-i\omega_{23}t} & -(E_2 - E_3 - \omega_{23}) \end{pmatrix}.$$

By using the expression of u_{12} and u_{23} in (8.22) along with a rotating wave approximation (cf. section 8.1) and defining

$$\Delta_{12} := E_2 - E_1 - \omega_{12}, \qquad \Delta_{23} := E_2 - E_3 - \omega_{23},$$

we replace $\tilde{H}_1$ by

$$\tilde{H}_2 := \tilde{H}_2(t) = \begin{pmatrix} -\Delta_{12} & h_{12}v_{12}(t) & 0 \\ h_{12}^* v_{12}(t) & 0 & h_{23}v_{23}(t) \\ 0 & h_{23}^* v_{23}(t) & -\Delta_{23} \end{pmatrix}. \tag{8.24}$$

In order to transfer population from level $|E_1\rangle$ to level $|E_3\rangle$, the *intuitive strategy* would be to use the pump pulse first at (nearly) the resonance frequency, $E_2 - E_1$, i.e., with $\Delta_{12} = 0$, so as to couple level $|E_1\rangle$ and $|E_2\rangle$ and transfer population from level $|E_1\rangle$ to level $|E_2\rangle$. Then, one would use a Stokes pulse at (nearly) the resonance frequency $E_2 - E_3$, i.e., with $\Delta_{23} = 0$, to couple levels $|E_2\rangle$ and $|E_3\rangle$ to transfer population to level $|E_3\rangle$. This intuitive strategy has the drawback of populating the intermediate level $|E_2\rangle$.

The interaction with the external environment, i.e., with energy levels which are not included in the model, makes some of the population transfer from level $|E_2\rangle$ to these levels. In probabilistic terms, this means that there is an increasing probability to find the system at an energy level lower than E_2. This phenomenon of *spontaneous decay* is particularly important as $|E_2\rangle$ is a higher energy level and open systems tend to the lowest energy configuration. In this setting one could pose an interesting optimal control problem and try to shape the pulses v_{12} and v_{23} so as to minimize $\int_{-\infty}^{+\infty} |\psi_2(t)|^2 dt$, where $|\psi_2|$ represents the population in level $|E_2\rangle$.

The STIRAP method uses a *counterintuitive pulse sequence* in which the Stokes pulse v_{23} *precedes* the pump pulse v_{12}. The two pulses overlap for some time in which the magnitude of the Stokes pulse slowly decreases while the magnitude of the pump pulse slowly increases. The typical situation is described in Figure 8.3. If the pulses are slow enough, the resulting evolution can be considered adiabatic and the resulting trajectory analyzed using the ideas in section 8.3.

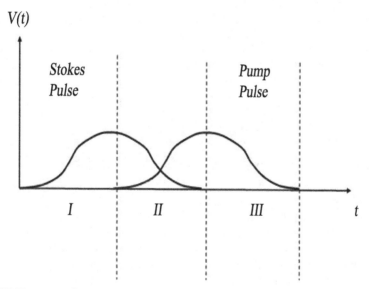

FIGURE 8.3: Counterintuitive pulse sequence in the STIRAP control. In the given notation the Stokes pulse is v_{23} and the Pump pulse is v_{12}.

Assume

$$\Delta_{12} = \Delta_{23} := \Delta.$$

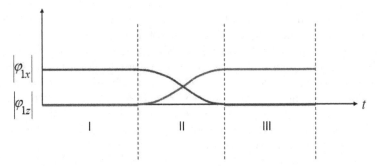

FIGURE 8.4: Time dependence of the x and z component (φ_{1x} and φ_{1z}, respectively) of the state vector $\vec{\phi}_1$ in adiabatic approximation for the counterintuitive pulse sequence in the STIRAP control method.

The eigenvalues of $\tilde{H}_2$ in (8.24) are

$$\lambda_1 = -\Delta,$$

$$\lambda_2(t) = -\frac{\Delta}{2} + \frac{1}{2}\sqrt{\Delta^2 + 4(|h_{23}|^2 v_{23}^2(t) + |h_{12}|^2 v_{12}^2(t))},$$

$$\lambda_3(t) = -\frac{\Delta}{2} - \frac{1}{2}\sqrt{\Delta^2 + 4(|h_{23}|^2 v_{23}^2(t) + |h_{12}|^2 v_{12}^2(t))}.$$

Of particular interest to us is the normalized eigenvector corresponding to the eigenvalue λ_1, i.e.,

$$\vec{\phi}_1 := \frac{1}{\sqrt{|h_{12}v_{12}|^2 + |h_{23}v_{23}|^2}} \begin{pmatrix} -h_{23}v_{23} \\ 0 \\ h_{12}^* v_{12} \end{pmatrix}. \tag{8.25}$$

We now follow the trajectory of the state $\vec{\phi}$ solution of (8.23), assuming that we start with an initial state $\vec{\phi}(0) = [1, 0, 0]^T$. During the interval I in Figure 8.3, such a state is not changed except for an irrelevant phase factor, since $v_{12} \equiv 0$ in this interval. For the whole interval, the state coincides with the eigenstate $\vec{\phi}_1$ (8.25) corresponding to λ_1. In the interval II, assuming adiabatic evolution, the trajectory follows $\vec{\phi}_1(t)$ in (8.25). As the magnitude of v_{23} slowly decreases while the one of v_{12} slowly increases, $\vec{\phi}_1$ and therefore $\vec{\phi}$ moves from a state parallel to $[1, 0, 0]^T$ to a state parallel to $[0, 0, 1]^T$. The magnitude of the pump pulse v_{23} is then slowly decreased to zero in the interval III, and this does not modify the final state. Notice that since $\vec{\phi}_1$ does not have any component along the state $|E_2\rangle$, the process ideally does not place any population in $|E_2\rangle$ so as to avoid the problems of the intuitive strategy. The behavior of the adiabatic vector $\vec{\phi}_1$ during the interval of control $I + II + III$ is summarized in Figure 8.4.

The errors of the adiabatic approximation can be evaluated using one of the adiabatic theorems quoted in the previous section. In particular (cf. Theorem

8.3.1) the error is smaller if the gap (8.18) between the eigenvalues is large. In our case, the gap is given by

$$\gamma(t) := \min\{|\lambda_1 - \lambda_2|, |\lambda_1 - \lambda_3|\} = \frac{1}{2}\left|\sqrt{\Delta^2 + 4(|h_{12}|^2 v_{12}^2 + |h_{23}|^2 v_{23}^2)} - |\Delta|\right|.$$

In the interval of interest, where the actual state transfer occurs, i.e., the interval II, the quantity measuring the strength of the coupling with the laser field, namely $|h_{12}|^2 v_{12}^2 + |h_{23}|^2 v_{23}^2$, is always larger than a positive lower bound. If the strength of the coupling with the laser field is sufficiently large, γ is also large, which makes the adiabatic scheme valid. A *resonant* STIRAP algorithm is such that $\Delta = 0$. However, resonance is not necessary for the validity of the scheme.

8.5 Lyapunov Control of Quantum Systems

In Lyapunov control of quantum systems,[7] one specifies a function of the state and designs the control so that the value of the function decreases to a desired value. Such a function is called a **Lyapunov function**. The Lyapunov method of control is a powerful tool in nonlinear control systems design [159].

8.5.1 Quantum control problems in terms of a Lyapunov function

The main form of a Lyapunov function we shall consider, written in terms of the density matrix ρ, is

$$V(\rho) := \text{Tr}(P\rho), \tag{8.26}$$

where P is a Hermitian matrix, which, without loss of generality, we can assume positive semi-definite, so as to satisfy the standard requirement for a Lyapunov function, $V(\rho) \geq 0$.[8] This has the meaning of the expectation value of the observable P (cf. 1.39). Several quantum control problems can be formulated in terms of finding a control function u so that the solution of

$$\dot{\rho} = -i[H_0 + H_1 u, \rho], \qquad \rho(0) = \rho_0, \tag{8.27}$$

---

[7]During the computations in this section, we often use the identity $\text{Tr}(A[B, C]) = \text{Tr}(B[C, A])$.

[8]We can always replace P with $P + k\mathbf{1}$, with k sufficiently large. This will just shift the value of V by a quantity k, having no effect on the following analysis.

converges to the minimum of $V(\rho)$ in (8.26).[9] In the case where the system is operator controllable the state ρ which minimizes $V(\rho)$ has to be sought among the states which are unitarily equivalent to ρ_0, or, equivalently, the ones that have the same spectrum as ρ_0, that is matrices in the orbit

$$\mathcal{O}_{u(n)}(\rho_0) := \{X\rho_0 X^\dagger | X \in U(n)\}. \tag{8.28}$$

We shall assume operator controllability in the following.

Example 8.5.1 Consider the problem of driving the pure state of (8.27) to a state collinear to $\vec{\psi}_f := \begin{pmatrix} 1 \\ 0 \\ 0 \end{pmatrix}$. The problem is equivalent to driving the state ρ to the minimum of the Lyapunov function (8.26) with P given by $P := \mathrm{diag}(0,1,1)$. The minimum is obtained for $\mathrm{Tr}(P\rho) = 0$. This condition is equivalent to $\rho = \vec{\psi}_f \vec{\psi}_f^\dagger$[10] and therefore the problem can be posed as driving the state ρ to the minimum of $V = V(\rho)$.

Both the minimum and maximum of the function $V = V(\rho)$ in the set (8.28) correspond to matrices that commute with P. In fact, by adapting the computation we did to obtain (7.86) in Chapter 7, we have:

Proposition 8.5.2 For any stationary point ρ of $V(\rho)$ (which includes maximum and minimum value)
$$[\rho, P] = 0.$$
Conversely, if ρ is such that $[\rho, P] = 0$, then ρ is a stationary point of $V(\rho)$.

From the previous proposition, again using similar considerations to those in section 7.6 we have the following (see Exercise 8.4).

Proposition 8.5.3 Assuming, without loss of generality, that P is diagonal, i.e., of the form

$$P := \mathrm{diag}(p_1 \mathbf{1}_{n_1 \times n_1}, p_2 \mathbf{1}_{n_2 \times n_2}, ..., p_r \mathbf{1}_{n_r \times n_r}),$$

with increasing $p_1 < \cdots < p_r$, and $n_1 + n_2 + ... + n_r = n$, then the *maximum* (*minimum*) has the block diagonal form $\rho = \mathrm{diag}(B_{n_1 \times n_1}, ..., B_{n_r \times n_r})$. The eigenvalues, possibly coinciding, $\lambda_1, \lambda_2, ..., \lambda_n$, of ρ are eigenvalues of the blocks $B_{n_1 \times n_1}, ..., B_{n_r \times n_r}$, in *non decreasing* (*non increasing*) order.

---

[9]We shall treat the case of a general density matrix and then specialize to pure states in Remarks 8.5.4, 8.5.7, 8.5.8. We assume a bilinear quantum control system (8.27) with one control.

[10]The Lyapunov control for a pure state will be discussed in the following. One way to see that the minimum is obtained for $\rho = \vec{\psi}_f \vec{\psi}_f^\dagger$ is to write P as $P = \mathbf{1} - \vec{\psi}_f \vec{\psi}_f^\dagger$ and ρ as $\rho = (\alpha\vec{\psi}_f + \vec{\psi}_o)(\alpha^*\vec{\psi}_f^\dagger + \vec{\psi}_o^\dagger)$, for $\vec{\psi}_o$ a vector perpendicular to $\vec{\psi}_f$. We have $\mathrm{Tr}(P\rho) = \|\vec{\psi}_o\|^2$, which is zero if and only if $\vec{\psi}_o = 0$ and $|\alpha| = 1$. We could also apply Proposition 8.5.3 as in Remark 8.5.4.

Remark 8.5.4 The Lyapunov functions considered in the literature are typically special cases of the general Lyapunov function in (8.26). In particular, for the case of a pure state $\vec{\psi}$,

$$V(\psi) = \vec{\psi}^{\,\dagger} P \vec{\psi} = \mathrm{Tr}(P \vec{\psi} \vec{\psi}^{\,\dagger}) \qquad (8.29)$$

is considered (cf. [123]). By taking

$$P := 1 - \vec{\psi}_f \vec{\psi}_f^{\,\dagger} \qquad (8.30)$$

for some desired final state $\vec{\psi}_f$, one obtains the *Hilbert Schmidt distance*

$$V(\vec{\psi}) = 1 - |\vec{\psi}^{\,\dagger} \vec{\psi}_f|^2, \qquad (8.31)$$

which is used, for example, in [283]. In agreement with Proposition 8.5.3, the minimum of this function is obtained for $\vec{\psi} = e^{ir} \vec{\psi}_f$ for any $r \in \mathbf{R}$, the maximum for any vector perpendicular to $\vec{\psi}_f$. The Lyapunov function $V(\vec{\psi})$ in (8.29) has the advantage that it is independent of a phase factor, which does not have physical meaning. In other words, $V(\vec{\psi}) = V(e^{ir} \vec{\psi})$ for any $r \in \mathbf{R}$. Another Lyapunov function, used for example in [195], is the square norm of the error, i.e.,

$$V_1(\vec{\psi}) = (\vec{\psi} - \vec{\psi}_f)^{\,\dagger} (\vec{\psi} - \vec{\psi}_f) = 2(1 - Re(\vec{\psi}_f^{\,\dagger} \vec{\psi})).$$

The objective of the control is to drive $\vec{\psi}$ to the minimum of $V_1(\vec{\psi})$ up to a phase factor. The relation

$$\min_{r \in \mathbf{R}} V_1(e^{ir} \vec{\psi}) = 2(1 - |\vec{\psi}_f^{\,\dagger} \vec{\psi}|),$$

along with (8.31), establishes the connection with the Hilbert-Schmidt Lyapunov function $V(\vec{\psi})$. The approach of [195] uses the Lyapunov function $V_1(\vec{\psi})$ and treats the phase degree of freedom by adding a fictitious control w in the Schrödinger equation

$$\frac{d}{dt} \vec{\psi} = -i(H_0 + H_1 u) \vec{\psi}, \qquad (8.32)$$

which becomes

$$\frac{d}{dt} \vec{\psi} = -i(H_0 + w + H_1 u) \vec{\psi}. \qquad (8.33)$$

The equivalence between the treatment with the Lyapunov function $V(\vec{\psi})$ with Schrödinger equation (8.32) and the one with Lyapunov function $V_1(\vec{\psi})$ with Schrödinger equation (8.33) is explained in the following FACT, whose proof is left as an exercise (see Exercise 8.5).

FACT: Assume u_d is such that the corresponding solution $\vec{\psi}$ of (8.32) satisfies $\lim_{t \to \infty} V(\vec{\psi}) = 0$, then there exists a function w_d such that the solution

$\vec{\psi}$ of (8.33), with $(u, \omega) = (u_d, \omega_d)$, satisfies $\lim_{t \to \infty} V_1(\vec{\psi}) = 0$. Conversely, if, with an augmented control $(u, \omega) := (u_d, \omega_d)$, the solution (8.33) satisfies $\lim_{t \to \infty} V_1(\vec{\psi}) = 0$, then, with the control $u = u_d$, the solution $\vec{\psi}$ of (8.32) satisfies $\lim_{t \to \infty} V(\vec{\psi}) = 0$.

8.5.2   Determination of the control function

As we wish to minimize the value of the Lyapunov function $V(\rho)$ in (8.26), we choose the control u to make the time derivative of $V(\rho)$, along the corresponding trajectory, negative at all times. We calculate, using (8.27),

$$\frac{d}{dt}V(\rho) = -i\operatorname{Tr}(P[H_0 + H_1 u, \rho]) = -i\operatorname{Tr}(\rho[P, H_0]) - i\operatorname{Tr}(\rho[P, H_1])u.$$

Since $\operatorname{Tr}([P, H_0]) = 0$ and therefore $[P, H_0]$ is not sign defined we assume $[P, H_0] = 0$, and therefore, we assume that both P and H_0 are diagonal. We state this formally.

Assumption 1 The matrices P and H_0 are both diagonal.

With this assumption, we have

$$\frac{d}{dt}V(\rho) = -i\operatorname{Tr}(\rho[P, H_1])u,$$

and the natural choice for the control is

$$u = -iK\operatorname{Tr}(\rho[P, H_1]), \qquad (8.34)$$

for some $K > 0$. With this, we obtain

$$\frac{d}{dt}V(\rho) = -K|\operatorname{Tr}(\rho[P, H_1])|^2 \leq 0.$$

Therefore, the value of $V(\rho)$ will decrease towards a given value.

8.5.3   Study of the asymptotic behavior of the state ρ

With the above control law, the value of the state ρ will tend to a limit set which can be characterized using **La Salle's invariance principle** (see, e.g., [159], Theorem 4.4, p. 128).[11]

Theorem 8.5.5 (*La Salle invariance principle*) *Consider a system of differential equations*

$$\dot{x} = f(x), \qquad (8.35)$$

---

[11]We state the theorem in a weaker form which is sufficient for our purposes.

with smooth f, and $x \in \mathbf{R}^n$. *Given a smooth function* $V : \mathbf{R}^n \to \mathbf{R}$ *such that* $\frac{d}{dt}V = (\frac{d}{dt}V)(x) = (\nabla V \cdot f)(x) \leq 0$, *for every* $x \in \mathbf{R}^n$, *let* M *be the largest invariant set in* $\mathbf{R}^n$ *where* $(\frac{dV}{dt})(x) = 0$. *Then, every solution of (8.35) converges to* M *as* $t \to \infty$.

The set M for the Lyapunov function V in (8.26), with dynamics (8.27) and control (8.34), is the set of all the Hermitian matrices ρ_1 with the same spectrum as ρ_0 which satisfy

$$\mathrm{Tr}(e^{-iH_0 t}\rho_1 e^{iH_0 t}[P, H_1]) = 0, \quad \forall t \in \mathbf{R}, \tag{8.36}$$

i.e.,

$$M := \{\rho_1 \in \mathcal{O}_{u(n)}(\rho_0)|\, \mathrm{Tr}(e^{-iH_0 t}\rho_1 e^{iH_0 t}[P, H_1]) = 0, \quad \forall t \in \mathbf{R}\}.$$

This set is clearly invariant under the differential equation (8.27) with control (8.34). That is, if we start with an initial condition in M, the corresponding trajectory remains in M. Moreover M is included in the set where $(\frac{d}{dt}V)(\rho) = 0$. Furthermore, this is the *largest* invariant set having the property of being included in $\{\rho|(\frac{d}{dt}V)(\rho) = 0\}$. In fact, if $\tilde{\rho}(t)$ is such that $(\frac{d}{dt}V)(\tilde{\rho}(t)) = 0$, then $\tilde{\rho}(t) = e^{-iH_0 t}\tilde{\rho}(0)e^{iH_0 t}$, and therefore, from (8.36), $\tilde{\rho}(0) \in M$.

We now want to characterize the set M of density matrices ρ_1 satisfying (8.36), which we rewrite using Assumption 1,

$$\mathrm{Tr}(e^{iH_0 t}H_1 e^{-iH_0 t}[\rho_1, P]) = 0, \quad \forall t \in \mathbf{R}.$$

Taking the n-th derivative with respect to t at time $t = 0$ and recalling the ad notation (3.45), this is equivalent to[12]

$$\mathrm{Tr}((ad_{iH_0}^n H_1)[\rho_1, P]) = 0, \quad n = 0, 1, 2, \dots. \tag{8.37}$$

By writing H_0 as

$$H_0 := \mathrm{diag}(\lambda_1, \dots, \lambda_n),$$

and

$$H_1 := \{h_{jk}\}, \qquad h_{jk} = h_{kj}^*, \qquad j, k = 1, \dots, n,$$

we have

$$(ad_{iH_0}^n H_1)_{lm} = (ad_{iH_0}^n H_1)_{ml}^* = (i)^n \omega_{lm}^n h_{lm}, \qquad l < m,$$

where $\omega_{lm} := \lambda_l - \lambda_m$. Let us denote the (j, k)-th element of $[\rho_1, P]$ by $[\rho_1, P]_{jk} := s_{jk} = -s_{kj}^*$. For a fixed n, in the condition (8.37), s_{jk}, with $j < k$, appears in the trace only two times: $-s_{jk}^*$ multiplies $(i)^n \omega_{jk}^n h_{jk}$ and

---

[12] Alternatively use the BCH formula (E.1) in Appendix E.

s_{jk} multiplies $(-i)^n \omega_{jk}^n h_{jk}^*$. Therefore, for a fixed n, condition (8.37) writes as

$$\sum_{j<k} s_{jk}(-i)^n \omega_{jk}^n h_{jk}^* - s_{jk}^*(i)^n \omega_{jk}^n h_{jk} = 0.$$

For n even, this gives

$$\sum_{j<k} \omega_{jk}^n Im(s_{jk}h_{jk}^*) = 0, \qquad n = 0, 2, 4, \dots \qquad (8.38)$$

For n odd, this gives

$$\sum_{j<k} \omega_{jk}^n Re(s_{jk}h_{jk}^*) = 0, \qquad n = 1, 3, 5, \dots \qquad (8.39)$$

These are the general conditions that have to be satisfied by $[\rho_1, P]$ and, therefore by ρ_1, according to the La Salle invariance principle (8.5.5). As we want ρ_1 to be a minimum for $V(\rho)$, we would like, according to Proposition 8.5.3, $[\rho_1, P] = 0$, i.e., $s_{jk} = 0$ for all $j < k$.[13] This is implied by (8.38) and (8.39) under the following two additional conditions.

Assumption 2 The transition frequencies ω_{jk} of the natural Hamiltonian H_0 are such that

$$\omega_{jk} \neq \omega_{lm}, \qquad j, k \neq l, m.$$

Assumption 3 For every pair j, k, $j < k$, $h_{jk} \neq 0$.

Assumption 2 says that the transition energies between two different levels are clearly identified. In principle, it is possible to use tailored radio-frequency pulses to selectively transfer population between any two levels (cf. section 8.1). Assumption 3 says that all the levels are directly coupled. Under Assumption 2, a Vandermonde determinant argument on the equations (8.38) and (8.39) shows that both $Im(s_{jk}h_{jk}^*)$ and $Re(s_{jk}h_{jk}^*)$ are equal to zero. Therefore, since $s_{jk}h_{jk}^* = 0$, for all $j < k$, Assumption 3 implies that $s_{jk} = 0$. Therefore, we can conclude with the following result

Theorem 8.5.6 *Under Assumptions 1, 2 and 3, the control (8.34) asymptotically drives the state ρ of (8.27) to a stationary point of $V(\rho) = Tr(P\rho)$.*

The nature of the stationary limit point in the above theorem has to be studied for the particular situation at hand. For example, if P has n distinct eigenvalues, that is, all the eigenvalues are nondegenerate, then there exists a finite number of matrices with a given spectrum (the spectrum of the initial condition ρ_0) which commute with P. Call them $\tilde{\rho}_1, \tilde{\rho}_2, \dots, \tilde{\rho}_r, \rho_{min}$, for

---

[13]Notice that, as P is diagonal, all the diagonal elements of $[\rho_1, P] = 0$ are automatically zero.

some integer r and ρ_{min} corresponds to the minimum of $V(\rho)$. Since $V(\rho)$ is decreasing, if $V(\rho(0)) < V(\tilde{\rho}_j)$, $j = 1, 2, ..., r$, then the solution will necessarily converge to ρ_{min}. In other words, the condition $V(\rho(0)) < V(\tilde{\rho}_j)$, $j = 1, 2, ..., r$, determines the *region of attraction* for the given design (cf. Figure 8.5). This observation can be used to shape P in problems where the goal is to drive the state to a given value.

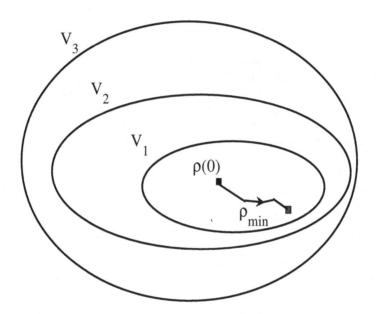

FIGURE 8.5: The matrices commuting with P, $\tilde{\rho}_1$, $\tilde{\rho}_2$,...,$\tilde{\rho}_r$, ρ_{min} determine level lines in the space of density matrices for the Lyapunov function $V = \text{Tr}(P\rho)$. The level lines corresponding to $\tilde{\rho}_1$, $\tilde{\rho}_2$,...,$\tilde{\rho}_r$ are labeled in this example by V_1, V_2 and V_3, with $V_1 < V_2 < V_3$. If $\text{Tr}(P\rho(0)) < V_1$ then the trajectory will converge to ρ_{min}. The innermost curve is the boundary of the region of attraction.

The case of a rank one density matrix ρ, namely of a pure state, is treated in the following remark.

Remark 8.5.7 In the special case where ρ is a pure state, i.e., of the form $\rho = \vec{\psi}\vec{\psi}^\dagger$, and the objective is to drive the state asymptotically to a given eigenstate $\vec{\psi}_f$ of H_0, the Lyapunov function takes the form $V(\vec{\psi})$ in (8.29) with P of the form in (8.30). Assumption 1 does not give any factual constraint since P and H_0 commute if $\vec{\psi}_f$ is an eigenstate of H_0. Under Assumptions 2 and 3, the state will converge to a value $\vec{\psi}_1$ such that $\vec{\psi}_1\vec{\psi}_1^\dagger$ commutes with $\vec{\psi}_f\vec{\psi}_f^\dagger$. Writing this as

$$(\vec{\psi}_f^\dagger\vec{\psi}_1)\vec{\psi}_f\vec{\psi}_1^\dagger = (\vec{\psi}_1^\dagger\vec{\psi}_f)\vec{\psi}_1\vec{\psi}_f^\dagger,$$

and multiplying on the right by $\vec{\psi}_f$, we obtain that the limit state $\vec{\psi}_1$ is collinear with $\vec{\psi}_f$ unless $\vec{\psi}_f^\dagger\vec{\psi}_1 = 0$, which is the situation corresponding to the maximum for $V(\psi)$. If this is not the situation at the beginning it never is since V is always not increasing. Therefore, for almost all initial conditions the state will converge to the desired state parallel to $\vec{\psi}_f$.

In general, if we do not assume Assumptions 2 and 3, all we can say, applying La Salle's invariance principle of Theorem 8.5.5, is that the state ρ will converge to a set of values ρ_1 such that $[\rho_1, P]$ is perpendicular to the space $\mathcal{V}$ spanned by $ad_{iH_0}^n H_1$, $n = 0, 1, \ldots$. In the pure state case, however, Assumptions 2 and 3 can be weakened to a condition equivalent to local controllability of the Schrödinger equation about the final target eigenstate. This is explained in the following remark.

Remark 8.5.8 Consider system (8.33) and a control $\omega \equiv -\lambda$, $u \equiv 0$, starting from a state $\vec{\phi}$. If $\vec{\phi}$ is an eigenvector of H_0 corresponding to the eigenvalue λ, then $\vec{\phi}$ is an equilibrium point for system (8.33). To linearize this system about $\vec{\phi}$, set $\vec{\psi} = \vec{\phi} + \Delta\vec{\psi}$, $\omega := -\lambda + \Delta\omega$, $u = \Delta u$. From (8.33), we obtain

$$\frac{d}{dt}(\Delta\vec{\psi}) = (-iH_0 + i\lambda)\Delta\vec{\psi} - i\vec{\phi}\Delta\omega - iH_1\vec{\phi}\Delta u - i\Delta\omega\Delta\vec{\psi} - iH_1\Delta u\Delta\vec{\psi}.$$

By assuming $\Delta\omega$, Δu and $\Delta\vec{\psi}$ small, we neglect the last two terms. We obtain the linear system

$$\frac{d}{dt}(\Delta\vec{\psi}) = (-iH_0 + i\lambda)\Delta\vec{\psi} - i\vec{\phi}\Delta\omega - iH_1\vec{\phi}\Delta u.$$

According to the Kalman controllability condition (see, e.g., [155]) this system is controllable if and only if the rank of the controllability matrix[14]

$$C := \qquad\qquad\qquad\qquad\qquad\qquad\qquad\qquad\qquad (8.40)$$

$$\left((-iH_0 + i\lambda)^{n-1}(-iH_1\vec{\phi}) \quad \ldots \quad (-iH_0 + i\lambda)(-iH_1\vec{\phi}) \quad (-iH_1\vec{\phi}) \quad -i\vec{\phi} \right),$$

---

[14]Recall the controllability matrix for a linear system $\dot{x} = Ax + Bu$ of dimension n, with A an $n \times n$ matrix and B an $n \times m$ matrix with m the number of controls, is given by $C := [A^{n-1}B, A^{n-2}B, \ldots, AB, B]$. It is an $n \times nm$ matrix. In our case $B = [(-iH_1\vec{\phi}), -i\vec{\phi}]$ and $A = (-iH_0 + i\lambda)$. Since $(-iH_0 + i\lambda)\vec{\phi} = 0$ all the columns of C in (8.40) corresponding to $(-iH_0 + i\lambda)^k(-i\vec{\phi})$, with $k \geq 1$, will be zero. We have omitted these columns in (8.40).

is equal to n. This condition is independent of the coordinates. So we can assume $\vec{\phi} := (1, 0, ..., 0)$ and $H_0 := \operatorname{diag}(0, \lambda_2 - \lambda, ..., \lambda_n - \lambda)$. The controllability matrix [155] takes the form

$$
C = \begin{pmatrix}
0 & \cdot & \cdot & \cdot & 0 & -ih_{11} & -i \\
(-i)^{n-1}(\lambda_2 - \lambda)^{n-1}h_{12}^* & \cdot & \cdot & \cdot & -i(\lambda_2 - \lambda)h_{12}^* & 0 & 0 \\
(-i)^{n-1}(\lambda_3 - \lambda)^{n-1}h_{13}^* & \cdot & \cdot & \cdot & -i(\lambda_3 - \lambda)h_{13}^* & 0 & 0 \\
\cdot & \cdot & \cdot & \cdot & \cdot & \cdot & \cdot \\
\cdot & & \cdot & \cdot & \cdot & \cdot & \cdot \\
\cdot & & & \cdot & \cdot & \cdot & \cdot \\
(-i)^{n-1}(\lambda_n - \lambda)^{n-1}h_{1n}^* & \cdot & \cdot & \cdot & -i(\lambda_n - \lambda)h_{1n}^* & 0 & 0
\end{pmatrix}.
$$

If Assumptions 2 and 3 are verified, then all the h_{1j}, $j = 2, ..., n$ are different from zero and the factors $\lambda_n - \lambda$ are all different from each other. Therefore, a Vandermonde determinant argument shows that the rank of C is equal to n. Conversely, it is easily seen that if controllability holds for *all* the eigenvectors, then Assumptions 2 and 3 are verified.

In [195], only the pure state case is considered and convergence is proved under conditions weaker than Assumption 2 and 3. In particular, these conditions are that $\omega_{1j} \neq \omega_{1k}$, for $j \neq k$, and that the level corresponding to the target eigenstate[15] is directly coupled to all the others. Here the index 1 refers to the target eigenstate. This condition is equivalent to controllability of the linearized system about the target eigenstate. In our treatment, we have considered the more general case of density matrix control, which led to the stronger Assumptions 2 and 3.

In the general case of density matrices control, as treated here, Assumptions 2 and 3 guarantee convergence to a stationary point of the Lyapunov function $V(\rho) = \operatorname{Tr}(P\rho)$ with the control (8.34), and the value of such a function will be decreasing along the resulting trajectory. If the state reached is not the desired minimum, then in principle one could try to switch the control to a function which moves the state away from the reached stationary point so that the value of $V(\rho)$ decreases. Then one switches again to the Lyapunov control to reach the next stationary point. Methods to design such a strategy are an interesting research problem. In the case of a pure state, as discussed in Remarks 8.5.7 and 8.5.8, assumptions weaker than 2 and 3 and equivalent to local controllability about a given eigenstate guarantee the convergence to a desired eigenstate of H_0, for almost all the initial conditions. There have been several attempts in the literature to overcome the limitations of such assumptions. The technique used in [195] uses an adiabatic trajectory which starts at the desired eigenstate $\vec{\psi}_f$ and terminates at the same eigenstate. The control is designed via a Lyapunov function so as to track such an adiabatic trajectory. More precisely, one first considers a *slow* control

---

[15]instead of all levels

$u_T(t)$ in an interval $[0, T]$, with $u_T(0) = u_T(T) = 0$. Given the *Hamiltonian* $H(t) = H_0 + u_T(t)H_1$, denote by $\vec{\psi}_T := \vec{\psi}_T(t)$ the eigenvector of $H(t)$ corresponding to initial condition $\vec{\psi}_T(0) = \vec{\psi}_T(T) = \vec{\psi}_f$. If $\vec{\psi}_T$ was constant, we would use a control corresponding to (8.34) for the pure state case (cf. Remark 8.5.7)

$$u = -iK \operatorname{Tr}(\vec{\psi}\vec{\psi}^{\,\dagger}[(\mathbf{1} - \vec{\psi}_f\vec{\psi}_f^{\,\dagger}), H_1])$$

$$= iK \operatorname{Tr}(\vec{\psi}\vec{\psi}^{\,\dagger}[\vec{\psi}_f\vec{\psi}_f^{\,\dagger}, H_1]) = iK \operatorname{Tr}(\vec{\psi}\vec{\psi}^{\,\dagger}[\vec{\psi}_T\vec{\psi}_T^{\,\dagger}, H_1]).$$

The fact that $\vec{\psi}_T$ varies slowly suggests using the control

$$u = u_T + iK \operatorname{Tr}(\vec{\psi}\vec{\psi}^{\,\dagger}[\vec{\psi}_T\vec{\psi}_T^{\,\dagger}, H_1]),$$

i.e., the sum of a term inducing the adiabatic trajectory and a term which is the one suggested by the Lyapunov design. For this control, assume there exists a time $\bar{t} \in (0, T)$ such that the system (8.33) linearized about the eigenvector $\vec{\psi}_T(\bar{t})$ and the controls $u = u_T(\bar{t})$ and $\omega = -\lambda(\bar{t})$ (with $\lambda(\bar{t})$ the corresponding eigenvalue of $H_0 + u_T(\bar{t})H_1$) is controllable. Then, given an initial distance between the initial state $\vec{\psi}(0)$ and the desired state $\vec{\psi}_f$ and a desired (small) final distance ϵ between the two states, there exists a $\bar{T}$ such that

$$1 - |\vec{\psi}^{\,\dagger}(T)\vec{\psi}_f|^2 < \epsilon,$$

for every $T > \bar{T}$. The proof of this result is given in [195], and it is based on an adiabatic theorem given in [103].

8.6   Notes and References

We have surveyed some more tools to achieve the control of finite-dimensional quantum systems. The list of the techniques indicated here along with the ones of the previous two chapters is not complete. The books [233] and [254], for example, describe some more techniques in the setting of molecular control. One can combine the techniques described here and the ones in the previous two chapters. For example, in [280] a Lie Group decomposition was used to decompose a control problem into lower-dimensional problems and each of them was treated using a Lyapunov approach.

Our main references for the adiabatic theorems have been [150] and [232], which can be used as a starting point to explore the literature on adiabatic theorems. It should be mentioned that Theorem 8.3.1 is much more general in [150] than in the version we have given here. In particular, the authors do not consider the adiabatic trajectory corresponding to a single eigenvector but one corresponding to an element in a given eigenspace with spectrum separated by a gap from the rest of the spectrum of the time-varying Hamiltonian.

The discussion of the STIRAP technique in section 8.4 is focused on the case of Λ systems. The technique can be extended to other configurations. We refer to the books [233] and [254] for these extensions as well as for several practical applications of STIRAP.

The main references in the treatment of Lyapunov control have been [123], [195] and [283]. Although these papers treat the case of pure state, we have presented the treatment in the general case of density matrix dynamics and then considered the special case of pure states in Remarks 8.5.7, 8.5.8. For example, Proposition 8.5.2 is an extension of Lemma 1 in [123] to the density matrix setting. We believe that the treatment with the density matrix is not only more general but also more compact. However, convergence is proved at the price of stronger assumptions (Assumptions 2 and 3) which can be weakened for the pure state case to conditions equivalent to the controllability of the system linearized about the target eigenstate (cf. Remark 8.5.8).

Further analysis of the Lyapunov control methods for quantum systems can be found in [287] and [288].

All the Lyapunov based controls are in feedback form, i.e., the control is a function of the state (cf. (8.34)). For quantum systems the state is not available for measurement and, in fact, it follows from the measurement postulate that a measurement will modify the state and therefore (abruptly) change the dynamics of the system. Therefore, such closed loop controls have to be first simulated. From the simulation one obtains a control function which is then, in practice, applied in open loop. In the setting of quantum feedback, one considers stochastic dynamics for the quantum state and the Lyapunov function approach, in the stochastic version, has proved a useful tool to design control laws for these models (see, e.g., [282]).

8.7 Exercises

Exercise 8.1 Consider an interpolating $1-$qubit Hamiltonian of the form

$$H_T(t) := (1 - \frac{t}{T})H_0 + \frac{t}{T}H_p,$$

with $H_0 = \frac{1}{2} + \frac{1}{2}\sigma_z$ (cf. (1.20)).

a) Give necessary and sufficient conditions on H_p so that it is possible to analyze the dynamics in terms of adiabatic evolution in the interval $[0, T]$, i.e., no level crossing occurs in this interval ($\gamma_{min} \neq 0$).

b) Pick one such H_p Hamiltonian and estimate the error of the adiabatic approximation $\|(\mathbf{1} - P_a(t))\vec{\psi}(t)\|$ using Theorem 8.3.1, for $T = 10$, $T = 100$ and $T = 1000$.

Exercise 8.2 Consider two spin $\frac{1}{2}$ particles interacting via Ising interaction (cf. subsection 7.4.2). Assume the interaction is weak. Consider two constant magnetic fields in the z direction different from each other, and a common control field in the y-direction. The Hamiltonian may be written, after some normalization, as

$$H = J\sigma_z \otimes \sigma_z + (u_{z1}\sigma_z \otimes 1 + u_{z2}1 \otimes \sigma_z) + u_y(1 \otimes \sigma_y + \sigma_y \otimes 1),$$

where J is the *small* coupling constant. By making u_{z1} very much different from u_{z2}, we can create two different energy gaps and, this way, we can effectively address the two spins separately.

a) Using this fact and the strategy of section 8.1 devise a methodology to address the two spins separately.

b) Combine this, with the methodology of section 7.4.2, to obtain a (almost) minimum time control.

c) Consider the transition from the state $|00\rangle$ (both spin down), to the 'entangled' state $\frac{1}{2}(|00\rangle + |11\rangle)$. Use this strategy to find a control to approximately perform this state transfer. Use the methods of section 8.1 and 8.2 to obtain estimates for the error.

Exercise 8.3 In section 8.1 we have assumed that the diagonal elements of H_1 in (8.1) are all zero. How does the treatment modify if this is not the case?

Exercise 8.4 Prove Proposition 8.5.3.

Exercise 8.5 Prove the **FACT** of Remark 8.5.4.

Exercise 8.6 Prove that

$$\min_{r \in \mathbf{R}} \|\vec{\psi} - e^{ir}\vec{\psi}_a\| = \sqrt{2(1 - |\vec{\psi}^\dagger\vec{\psi}_a|)},$$

and

$$\|\vec{\psi} - \vec{\psi}_a\vec{\psi}_a^\dagger\vec{\psi}\| = \sqrt{1 - |\vec{\psi}_a^\dagger\vec{\psi}|^2}.$$

Take these two functions as measures of error in the adiabatic approximation. Deduce that these two measures are equivalent in the sense that one is an increasing function of the other.

Exercise 8.7 How can Propositions 8.5.2 and 8.5.3 be generalized if we replace the full orbit (8.28) with $\mathcal{O}_{e^{\mathcal{L}}}(\rho_0) := \{X\rho_0 X^\dagger | X \in e^{\mathcal{L}}\}$, where $e^{\mathcal{L}}$ is a Lie subgroup of $SU(n)$ corresponding to the Lie subalgebra $\mathcal{L}$ of $su(n)$?

Chapter 9

Analysis of Quantum Evolutions; Entanglement, Entanglement Measures and Dynamics

Entanglement of quantum systems refers to the situation where two or more systems interact and the state of the whole system cannot be seen as a superposition of states of the single systems. This feature of quantum dynamics is crucial in many algorithms of quantum communication and quantum information processing and it is a resource for many fascinating applications. Moreover entanglement, as a feature of quantum mechanics, has stimulated in-depth investigation on the value of quantum mechanics itself as a physical theory.

In their famous paper [102] on the *EPR paradox*, Einstein, Podolski and Rosen proposed an experiment to argue that quantum mechanics is not a complete physical theory. In the Bohm simplified version of this experiment, a pair of spin-$\frac{1}{2}$ particles are placed in a *singlet state* $|\psi_s\rangle$, defined as

$$|\psi_s\rangle := \frac{1}{\sqrt{2}}(|10\rangle - |01\rangle),\tag{9.1}$$

where we have denoted by $|0\rangle$ ($|1\rangle$) the spin down (up) state. This state is **entangled** in that it cannot be written in the form $|a\rangle \otimes |b\rangle$, for two states $|a\rangle$ and $|b\rangle$ of the two systems. The two particles are sent to two different locations, A and B. If a measurement is performed on the spin of the particle at location A and the result is a positive value of the spin angular momentum (say in the z direction), then the state of the system of two spins will collapse, according to the measurement postulate, into $|\psi\rangle := |10\rangle$. This means that a measurement of the spin at location B will, with certainty, give a negative value of the spin (corresponding to the $|0\rangle$ eigenvector). It appears therefore that the result of an experiment at one site has immediately affected the result of an experiment at another (possibly very distant) site. This appears to contradict a principle of *locality*, i.e., that experiments at two different locations are independent of each other, a fact consistent with the prescription from *special relativity* that information cannot be transmitted at a speed faster than the speed of light. The EPR paradox has generated many theoretical and experimental studies. We shall not be concerned here with the attempts to solve the paradox, whether by providing different interpretations of quantum

DOI: 10.1201/9781003051268-9

mechanics, alternative theories or experimental evidence. Within the standard interpretation of quantum mechanics, we shall be concerned with the mathematical aspects of entanglement, i.e., its precise definition, detection, and quantification, as well as the aspects concerning dynamics, in particular its evolution. We shall see how Lie group decompositions of the type used in Chapter 5 for control purposes can be used to analyze how entanglement evolves.

This chapter is organized in three sections. The first one, section 9.1, is devoted to the definition of entanglement both for pure states and mixed states. We also discuss methods to detect entanglement in subsection 9.1.2 and to quantify it in subsection 9.1.3. The second section, section 9.2, is devoted to the study of the dynamics of entanglement. Several decompositions are given there which allow us to factorize a unitary evolution into factors which produce entanglement and factors that do not. The last section, section 9.3, is devoted to the problem of local equivalence between states, i.e., how to determine when, given two states, it is possible to evolve from one to the other by transformations which act on systems separately (local transformations). This can be seen as a problem of controllability.

9.1   Entanglement of Quantum Systems

9.1.1   Basic definitions and notions

9.1.1.1   Separable and entangled pure states

As discussed in Chapter 1, subsection 1.1.3, the state of a quantum system composed of N subsystems is described by a vector $|\psi\rangle$ in the tensor space

$$\mathcal{H} := \mathcal{H}_1 \otimes \mathcal{H}_2 \otimes \cdots \otimes \mathcal{H}_N,$$

where $\mathcal{H}_j$, $j = 1, 2, ..., N$, is the Hilbert space associated with the j-th system. A state $|\psi\rangle \in \mathcal{H}$ is called **separable** if it can be written as

$$|\psi\rangle = |\psi_1\rangle \otimes |\psi_2\rangle \otimes \cdots \otimes |\psi_N\rangle,$$

where $|\psi_j\rangle$ are states of $\mathcal{H}_j$, $j = 1, ..., N$. States that are not separable are called **entangled states**.

Example 9.1.1 Consider two two-level systems so that the Hilbert space of the total system is four-dimensional. Let $|1\rangle := \begin{pmatrix} 1 \\ 0 \end{pmatrix}$ and $|0\rangle := \begin{pmatrix} 0 \\ 1 \end{pmatrix}$ form a basis for both systems. Therefore, the Hilbert space of the total system is spanned by $|j\rangle \otimes |k\rangle$, $j, k = 0, 1$. The state

$$|1\rangle \otimes |0\rangle := \begin{pmatrix} 0 \\ 1 \\ 0 \\ 0 \end{pmatrix}$$

is separable, while the singlet state (9.1),

$$\frac{1}{\sqrt{2}}(|1\rangle \otimes |0\rangle - |0\rangle \otimes |1\rangle) := \frac{1}{\sqrt{2}} \begin{pmatrix} 0 \\ 1 \\ -1 \\ 0 \end{pmatrix}, \tag{9.2}$$

is entangled. In fact, if we try to express the state in (9.2) as a tensor product $\begin{pmatrix} a \\ b \end{pmatrix} \otimes \begin{pmatrix} c \\ d \end{pmatrix}$, we find that $a \times c$ should be zero, i.e., at least one between a and c should be zero, which contradicts the fact that both $a \times d$ and $b \times c$ should be different from zero.

9.1.1.2 Separable and entangled mixed states

If the quantum system is an ensemble, a density matrix description of the state must be used. In this context, a state ρ is said to be **separable** if

$$\rho := \sum_j \alpha_j \rho_{j1} \otimes \rho_{j2} \otimes \cdots \otimes \rho_{jN}, \tag{9.3}$$

$$\alpha_j > 0, \quad \sum_j \alpha_j = 1,$$

where all the ρ_{jk}, $k = 1, ...N$, are themselves states (i.e., self-adjoint, positive semidefinite, trace 1 operators) for the single subsystems. We call a **product state** a state of the form

$$\rho_{Prod} := \rho_1 \otimes \rho_2 \otimes \cdots \otimes \rho_N,$$

where the ρ_j's, $j = 1, ..., N$, are themselves states. Therefore, we can say that a separable state is a statistical (convex) superposition of product states. Notice that there is no loss of generality in assuming that each factor in the product states in (9.3) is a pure state. In fact, by the definition (1.17) every term ρ_{jk}, $k = 1, ..., N$ in (9.3) has the form

$$\rho_{jk} = \sum_l c_{jk,l} |\psi_{jk,l}\rangle \langle \psi_{jk,l}|, \quad c_{jk,l} > 0, \quad \sum_l c_{jk,l} = 1,$$

and we can re-arrange the sum (9.3) by possibly redefining the coefficients α_j, and define a separable state as a state of the form

$$\rho := \sum_j \alpha_j |\psi_{j1}\rangle \langle \psi_{j1}| \otimes |\psi_{j2}\rangle \langle \psi_{j2}| \otimes \cdots \otimes |\psi_{jN}\rangle \langle \psi_{jN}|, \tag{9.4}$$

with

$$\alpha_j > 0, \quad \sum_j \alpha_j = 1.$$

Notice that each term in the sum (9.4) represents a pure separable state since

$$|\psi_{j1}\rangle\langle\psi_{j1}| \otimes |\psi_{j2}\rangle\langle\psi_{j2}| \otimes \cdots \otimes |\psi_{jN}\rangle\langle\psi_{jN}| =$$

$$(|\psi_{j1}\rangle \otimes |\psi_{j2}\rangle \otimes \cdots |\psi_{jN}\rangle)(\langle\psi_{j1}| \otimes \langle\psi_{j2}| \otimes \cdots \langle\psi_{jN}|).$$

Therefore, we can say that a mixed separable state is the statistical superposition of pure separable states. In particular, rank one mixed separable states are pure separable states.

A mixed state which is not separable is called **entangled**.

Notice that, being a Hermitian operator, ρ can *always* be written in the form (9.3), with general real coefficients α_j and general Hermitian matrices $\rho_{j1},...,\rho_{jN}$. However, only when the α_j satisfy the conditions in (9.3) and the operators $\rho_j,...,\rho_j$ represent states can we say that ρ is a separable state.

9.1.1.3   Partial trace

Given a general mixed state ρ describing the state of a multipartite quantum system, we would like to have a way to extract the state of one of its subsystems, say the first one. 'To extract' means here that when we measure any observable which concerns only the first subsystem, i.e., it is of the type $A \otimes 1 \otimes 1 \otimes \cdots \otimes 1$, we are able to assess the probability distribution of the possible results, just as if we had the state of the first system alone. The appropriate tool is the **partial trace** of the density operator ρ. The definition can be given for a bipartite system first and then extended (see Property 3 in Proposition 9.1.3).

Definition 9.1.2 Consider a bipartite system with Hilbert space[1] $\mathcal{H} := \mathcal{H}_S \otimes \mathcal{H}_P$. Let ρ be a Hermitian operator on the total Hilbert space $\mathcal{H}$. Write ρ in the form $\rho = \sum_l \alpha_l \sigma_l^S \otimes \sigma_l^P$, for some real coefficients α_l and Hermitian operators σ_l^S and σ_l^P on $\mathcal{H}_S$ and $\mathcal{H}_P$, respectively. The partial trace Tr_P of a state ρ with respect to (the second system) P is defined as

$$\rho_S := \mathrm{Tr}_P(\rho) := \sum_l \alpha_l \sigma_l^S \mathrm{Tr}(\sigma_l^P).$$

Analogously one defines the partial trace with respect to the system S.

The main properties of the partial trace are listed in the following proposition whose proof is left as an exercise (see Exercise 9.1).

---

[1]The letters S and P are used here to indicate the main object system S and an auxiliary system P (probe).

Proposition 9.1.3 The partial trace Tr_P has the following properties

1. It is the *unique* linear operator from the space of linear operators on $\mathcal{H}_S \otimes \mathcal{H}_P$ to the space of linear operators on $\mathcal{H}_S$ having the property

$$\text{Tr}_P(A \otimes B) = A\,\text{Tr}(B).$$

2. For general operators F, and unitary operator $G \in U(n_P)$,

$$\text{Tr}_P(F \otimes G\rho F^\dagger \otimes G^\dagger) = F\,\text{Tr}_P(\rho)F^\dagger \tag{9.5}$$

3. Consider a multipartite system on $\mathcal{H} := \mathcal{H}_S \otimes \mathcal{H}_{P_1} \otimes \mathcal{H}_{P_2} \otimes \cdots \otimes \mathcal{H}_{P_{N-1}}$ and denote by $P_1 P_2 \cdots P_{N-1}$ the subsystem corresponding to $\mathcal{H}_{P_1} \otimes \cdots \otimes \mathcal{H}_{P_{N-1}}$. Then

$$\text{Tr}_{P_1 P_2 \ldots P_{N-1}}(\rho) = \text{Tr}_{P_1}(\text{Tr}_{P_2}(\cdots(\text{Tr}_{P_{N-1}}(\rho))\cdots)).$$

4. For every observable M on $\mathcal{H}_S$

$$\text{Tr}(M\,\text{Tr}_P(\rho)) = \text{Tr}(M \otimes \mathbf{1}\rho). \tag{9.6}$$

5. Consider an orthonormal basis of $\mathcal{H}_S$, $\{|s_j\rangle\}$, $j = 1, 2, ..., n_S$, and an orthonormal basis of $\mathcal{H}_P$, $\{|p_k\rangle\}$, $k = 1, 2, ..., n_P$. The corresponding basis of $\mathcal{H}_S \otimes \mathcal{H}_P$ is $\{|s_j, p_k\rangle\}$ and, in this basis, a Hermitian operator A can be written as

$$A = \sum_{jk,lm} a_{jk,lm}|s_j, p_k\rangle\langle s_l, p_m|,$$

with $a^*_{jk,lm} = a_{lm,jk}$. Then

$$\text{Tr}_P(A) = \sum_{j,l} b_{j,l}|s_j\rangle\langle s_l|,$$

with $b_{j,l} := \sum_{k=1}^{n_P} a_{jk,lk}$.

Property (9.6) is the reason to choose the partial trace to describe the state of a subsystem of a quantum system. Given M, an observable on system $\mathcal{H}_S$, the corresponding observable on the overall subsystem is $M \otimes \mathbf{1}$. The state of the system S has to be chosen so that the statistics of M on S is the same as the one of $M \otimes \mathbf{1}$ on the system $S + P$. In particular, the expectation value of M should be the same, which gives equation (9.6). The partial trace is the only linear map which has this property (see, e.g., [207]). In fact, consider any other linear map f from the space of linear Hermitian operators on $\mathcal{H}_S \otimes \mathcal{H}_P$ to the space of linear operators on $\mathcal{H}_S$, satisfying, for every M,

$$\text{Tr}(Mf(\rho)) = \text{Tr}(M \otimes \mathbf{1}\rho).$$

Since this is true for every element M_j of an orthonormal basis in the space of linear operators on $\mathcal{H}_S$, comparing with (9.6) we obtain that the components of $f(\rho)$ and $\mathrm{Tr}_P(\rho)$ along every M_j coincide, which means that $f(\rho) = \mathrm{Tr}_P(\rho)$.

9.1.1.4   Von Neumann entropy

In probability theory, given a probability distribution $\mathcal{P} := \{p_1, ..., p_n\}$, $p_j \geq 0$, $\sum_{j=1}^n p_j = 1$, describing the results of an experiment, one defines **entropy** as a *measure of the uncertainty on the outcome of the experiment*. Entropy, $E(\mathcal{P})$, is defined as

$$E(\mathcal{P}) := -\sum_{j=1}^n p_j \log(p_j),$$

where the log is usually taken in base 2, and it is understood that $0 \log(0) = 0$. $E(\mathcal{P})$ is minimum and equal to zero when only one of the probabilities p_j is different from zero (i.e., one of the events has probability one). It is maximum and equal to $\log(n)$ when all the outcomes have the same probability.

In analogy with probability theory, in quantum mechanics one defines **Von Neumann entropy** as a measure of the uncertainty of a quantum state. In particular, the state of an ensemble ρ can always be written as

$$\rho := \sum_{j=1}^n \lambda_j |\psi_j\rangle\langle\psi_j|,$$

for some orthonormal states $|\psi_j\rangle$, $\lambda_j \geq 0$, $\sum_j \lambda_j = 1$. This can be interpreted as an ensemble of systems where a fraction λ_j is in the pure state $|\psi_j\rangle$. The Von Neumann entropy of the state ρ is defined as

$$\hat{S}(\rho) := -\sum_{j=1}^n \lambda_j \log(\lambda_j) = -\mathrm{Tr}(\rho \log(\rho)),$$

with the understanding that $0 \log(0) = 0$. It is minimum and equal to zero for a pure state. It is maximum and equal to $\log(n)$ for the perfectly mixed state $\rho := \frac{1}{n}\mathbf{1}$. Von Neumann entropy is a fundamental quantity in quantum mechanics and in particular in quantum information theory. We refer to [209] for an exposition on the use of entropy in quantum mechanics. Von Neumann entropy can be used as a test for a quantum state to be pure. We have (Exercise 8.2)

Proposition 9.1.4 The state ρ is a pure state if and only if $\hat{S}(\rho) = 0$.

9.1.1.5   Bipartite and multipartite systems

Entanglement theory is different if we consider bipartite systems and general multipartite systems and much fewer results are known in the second case.

Even for the bipartite case, many problems are still open and only the simplest case of two coupled two-level systems (*qubits*) seems to be sufficiently understood at the moment (see, e.g., [1], [132], [139], [307] and references therein). The theory is constantly evolving and new results are continuously added. The comprehensive survey paper [140] gives an account updated to 2008. .

From an application point of view, the bipartite case is by far the most interesting one as it seems very difficult in practice to entangle (i.e., to create entangled states of) more than two quantum systems. We shall devote our discussion in the following mainly to the bipartite case and then indicate ideas and generalizations for the multipartite case.

9.1.2 Tests of entanglement

A fundamental and practical problem in entanglement theory is to give mathematical criteria to decide whether a quantum state is entangled or not. In the following we shall discuss some of the most powerful criteria for bipartite pure and mixed states, and, more briefly, for multipartite states.

9.1.2.1 Tests of entanglement for bipartite pure states

The partial trace itself provides a test for a pure state of a bipartite system to be entangled or separable. In fact, when the state of system S is entangled with the state of system P, the partial trace will give a mixed state. In information theoretic terms, some information on the state of system S is lost in the interaction with system P, so the Von Neumann entropy of the partial trace will be strictly positive (cf. Proposition 9.1.4). Formally, we have the following test of entanglement for bipartite pure systems.

Proposition 9.1.5 Let $\rho := |\psi\rangle\langle\psi|$ represent a pure state for a bipartite system $S + P$. Then ρ is separable if and only if $\mathrm{Tr}_P(\rho)$ is pure, that is (cf. Proposition 9.1.4) if and only if the entropy of the partial trace is equal to zero, i.e.,

$$\hat{S}(\mathrm{Tr}_P(\rho)) = 0.$$

Notice that there is nothing special about P in the above statement and we could have replaced the partial trace Tr_P with the partial trace Tr_S. This also means that $\mathrm{Tr}_P(\rho)$ is pure if and only if $\mathrm{Tr}_S(\rho)$ is pure.

In the proof of 9.1.5 we use an important and very useful result for bipartite systems, known as **Schmidt decomposition**, given in the following lemma.

Lemma 9.1.6 Let $|\psi\rangle$ be a state of the tensor product space $\mathcal{H}_S \otimes \mathcal{H}_P$ of dimensions $n_S \times n_P$. Assume without loss of generality $n_S \leq n_P$. Then there exists an orthonormal basis $\{|s_j\rangle\}$, $j = 1, ..., n_S$, of $\mathcal{H}_S$ and an orthonormal basis $\{|p_k\rangle\}$, $k = 1, ..., n_P$, of $\mathcal{H}_P$ such that

$$|\psi\rangle = \sum_{l=1}^{n_S} r_l |s_l\rangle \otimes |p_l\rangle. \qquad (9.7)$$

Moreover the coefficients r_l can be chosen real and nonnegative.

Proof. The proof follows from the *singular value decomposition* (see, e.g., [135]). Consider an orthonormal basis $\{|e_j\rangle\}$, $j = 1, ..., n_S$, of $\mathcal{H}_S$ and an orthonormal basis $\{|f_k\rangle\}$, $k = 1, ..., n_P$, of $\mathcal{H}_P$. Then every vector $|\psi\rangle \in \mathcal{H}_S \otimes \mathcal{H}_P$ can be written as

$$|\psi\rangle = \sum_{j=1}^{n_S} \sum_{k=1}^{n_P} a_{jk} |e_j\rangle \otimes |f_k\rangle. \qquad (9.8)$$

The $n_S \times n_P$ matrix $A := \{a_{jk}\}$ has the singular value decomposition $A = U\Lambda V$, where $U := \{u_{jl}\}$ and $V := \{v_{lk}\}$ are unitary matrices of dimensions n_S and n_P, respectively and Λ is an $n_S \times n_P$ matrix, where the last $n_P - n_S$ columns are zero while the first n_S columns are occupied by the diagonal matrix $\mathrm{diag}(\lambda_{11}, ..., \lambda_{n_S n_S})$. Using this, we can write (9.8) as

$$|\psi\rangle := \sum_{j=1}^{n_S} \sum_{k=1}^{n_P} \sum_{l=1}^{n_S} u_{jl} \lambda_{ll} v_{lk} |e_j\rangle \otimes |f_k\rangle = \sum_{l=1}^{n_S} \lambda_{ll} \left(\sum_{j=1}^{n_S} u_{jl} |e_j\rangle \right) \otimes \left(\sum_{k=1}^{n_P} v_{lk} |f_k\rangle \right).$$

This gives

$$|\psi\rangle = \sum_{l=1}^{n_S} \lambda_{ll} |s_l\rangle \otimes |p_l\rangle,$$

if one defines $|s_l\rangle := (\sum_{j=1}^{n_S} u_{jl} |e_j\rangle)$ and $|p_l\rangle := (\sum_{k=1}^{n_P} v_{lk} |f_k\rangle)$. Using the fact that U and V are unitary and that the bases $\{|e_j\rangle\}$ and $\{|f_k\rangle\}$ are orthonormal, it is easily seen that $\{|s_l\rangle\}$ and $\{|p_l\rangle\}$, $l = 1, ..., n_S$ are orthonormal sets. One obtains (9.7) defining $r_l := \lambda_{ll}$ and recalling that, from the properties of the singular value decomposition, $\lambda_{ll} \geq 0$.[2]   □

The main advantage of Schmidt decomposition is that it replaces a double sum (9.8) with a single sum (9.7). This allows us to give a straightforward proof of the entanglement test of Proposition 9.1.5.

Proof of Proposition 9.1.5 It is obvious that if $|\psi\rangle$ is a separable pure state, i.e.,

$$|\psi\rangle = |\psi_S\rangle \otimes |\psi_P\rangle,$$

with $|\psi_S\rangle \in \mathcal{H}_S$ and $|\psi_P\rangle \in \mathcal{H}_P$, then

$$\mathrm{Tr}_P(|\psi\rangle\langle\psi|) = \mathrm{Tr}_P(|\psi_S\rangle\langle\psi_S| \otimes |\psi_P\rangle\langle\psi_P|) = |\psi_S\rangle\langle\psi_S|$$

---

[2]Alternatively by writing $\lambda_{ll} := r_l e^{i\phi_l}$ with $r_l := |\lambda_{ll}|$ and $\phi_l := arg(\lambda_{ll})$ and redefining $e^{i\phi_l} |s_l\rangle \to |s_l\rangle$, one obtains that $r_l \geq 0$.

is a pure state. Conversely, assume that $\text{Tr}_P(|\psi\rangle\langle\psi|)$ is a pure state. Write $|\psi\rangle$ according to the Schmidt decomposition (9.7). We have

$$\text{Tr}_P(|\psi\rangle\langle\psi|) = \text{Tr}_P\Big(\sum_{lj} r_l r_j |s_l\rangle\langle s_j| \otimes |p_l\rangle\langle p_j|\Big) =$$

$$\sum_{lj} r_l r_j |s_l\rangle\langle s_j| \delta_{lj} = \sum_l r_l^2 |s_l\rangle\langle s_l|.$$

As this is a pure state, all the r_l's must be equal to zero except one. The one which is different from zero is equal to one. Therefore, from (9.7), $|\psi\rangle$ is a separable state. $\square$

9.1.2.2 Tests of entanglement for bipartite mixed states

For general mixed states the problem of finding a test to check whether the state is entangled or separable has not received a complete answer and it is one of the unsolved problems in quantum information theory. There have been, however, several important results in this area, some of which are of fundamental value.

A fundamental result is the necessary condition for states to be separable in terms of **positive maps**, which we now describe. Let $\mathcal{H}$ be a finite-dimensional Hilbert space and the set of linear operators on $\mathcal{H}$ be denoted by $\mathcal{B}(\mathcal{H})$. This is also a Hilbert space with the inner product $\langle A, B \rangle := \text{Tr}(B^\dagger A)$. A linear operator $\mathcal{B}(\mathcal{H}) \to \mathcal{B}(\mathcal{H})$ is called a *superoperator* or simply a *map*. We shall use the definition of *positive* and *completely positive* maps which is given in Appendix A. We have the following fundamental result proved in [139].

Theorem 9.1.7 *A density matrix ρ on $\mathcal{H}_S \otimes \mathcal{H}_P$ is separable if and only if for any positive map $\Gamma : \mathcal{B}(\mathcal{H}_S) \to \mathcal{B}(\mathcal{H}_S)$,*

$$(\Gamma \otimes \mathbf{1}_{n_P})(\rho) \geq 0.$$

From the above Theorem, if we had all the positive maps Γ we could in principle test separability of any state ρ. However, finding the structure of all the positive maps is an extremely difficult problem. It is, however, possible that a quorum of maps could be sufficient to discern separable and entangled states. In general if we find a positive map Γ such that

$$\Gamma \otimes \mathbf{1}_{n_P}(\rho) \ngeq 0, \tag{9.9}$$

then we can say that ρ is entangled. Several entanglement criteria for mixed states are of this type. They use specific positive maps Γ, for which condition (9.9) is satisfied for some entangled states ρ. In that case, one says that the map Γ *detects* entanglement in ρ. A suitable map Γ cannot be chosen among the completely positive maps otherwise (9.9) will never be satisfied.

More generally, if a map Γ_0 detects a certain set of entangled states, any map $F + G \circ \Gamma_0$, with F and G completely positive, will recognize a subset of the states recognized by Γ_0. In fact, if ρ is not detected by Γ_0, i.e., $\Gamma_0 \otimes 1_{n_P}(\rho) \geq 0$, then

$$(F \otimes 1_{n_P})(\rho) + (G \otimes 1_{n_P}) \circ (\Gamma_0 \otimes 1_{n_P})(\rho) \geq 0.$$

Notice that if we consider the completely positive map G given by the unitary evolution $G(\rho_S) := U\rho_S U^\dagger$, Γ_0 and $G \circ \Gamma_0$ are able to recognize exactly the same entangled states. The tests based on Γ_0 and $G \circ \Gamma_0$ are therefore equivalent (they are called *unitarily equivalent*).

We present next three important tests of entanglement for bipartite mixed states, in the following three theorems. The first two correspond to the choice of positive maps in Theorem 9.1.7 which are not completely positive.

Theorem 9.1.8 (Peres-Horodecki Positive Partial Transposition criterion [139], [213]. *Let* $\Gamma_T : \mathcal{B}(\mathcal{H}_S) \to \mathcal{B}(\mathcal{H}_S)$ *be given by transposition, i.e.,* $\Gamma_T(A) := A^T$. *Then, if* ρ *is separable,* $(\Gamma_T \otimes 1_{n_P})(\rho) \geq 0$. *Assume* $n_S = 2$ *and* $n_P = 2$ *or* $n_S = 2$ *and* $n_P = 3$. *Then* ρ *is separable if and only if* $\Gamma_T \otimes 1_{n_P}(\rho) \geq 0$.

Example 9.1.9 Consider the state

$$\rho := \frac{1}{4}\begin{pmatrix} 1 & i & 0 & 0 \\ -i & 1 & 0 & 0 \\ 0 & 0 & 2 & 0 \\ 0 & 0 & 0 & 0 \end{pmatrix},$$

which can be written as

$$\rho = \frac{1}{2}\begin{pmatrix} 0 & 0 \\ 0 & 1 \end{pmatrix} \otimes \begin{pmatrix} 1 & 0 \\ 0 & 0 \end{pmatrix} + \frac{1}{4}\begin{pmatrix} 1 & 0 \\ 0 & 0 \end{pmatrix} \otimes \begin{pmatrix} 1 & 0 \\ 0 & 1 \end{pmatrix} + \frac{1}{4}\begin{pmatrix} 1 & 0 \\ 0 & 0 \end{pmatrix} \otimes \begin{pmatrix} 0 & i \\ i & 0 \end{pmatrix}.$$

From this expression, it is clear that $(\Gamma_T \otimes 1_2)(\rho) = \rho \geq 0$ and this along with the fact that the system is such that $n_S = n_P = 2$ shows that the state is separable. In fact, ρ can be written as the convex combination of two product states,

$$\rho = \frac{1}{2}\begin{pmatrix} 1 & 0 \\ 0 & 0 \end{pmatrix} \otimes \begin{pmatrix} \frac{1}{2} & \frac{i}{2} \\ \frac{-i}{2} & \frac{1}{2} \end{pmatrix} + \frac{1}{2}\begin{pmatrix} 0 & 0 \\ 0 & 1 \end{pmatrix} \otimes \begin{pmatrix} 1 & 0 \\ 0 & 0 \end{pmatrix}.$$

Consider now the state

$$\rho = \frac{1}{2}\begin{pmatrix} 0 & 0 & 0 & 0 \\ 0 & 1 & -1 & 0 \\ 0 & -1 & 1 & 0 \\ 0 & 0 & 0 & 0 \end{pmatrix},$$

which can be written as

$$\rho = \frac{1}{2}\begin{pmatrix} 1 & 0 \\ 0 & 0 \end{pmatrix} \otimes \begin{pmatrix} 0 & 0 \\ 0 & 1 \end{pmatrix} + \frac{1}{2}\begin{pmatrix} 0 & 0 \\ 0 & 1 \end{pmatrix} \otimes \begin{pmatrix} 1 & 0 \\ 0 & 0 \end{pmatrix} + \qquad (9.10)$$

$$\frac{1}{2}\begin{pmatrix} 0 & 1 \\ 0 & 0 \end{pmatrix} \otimes \begin{pmatrix} 0 & 0 \\ -1 & 0 \end{pmatrix} + \frac{1}{2}\begin{pmatrix} 0 & 0 \\ 1 & 0 \end{pmatrix} \otimes \begin{pmatrix} 0 & -1 \\ 0 & 0 \end{pmatrix}.$$

From this expression, one can easily check that

$$(\Gamma_T \otimes 1_2)(\rho) = \begin{pmatrix} 0 & 0 & 0 & -1 \\ 0 & 1 & 0 & 0 \\ 0 & 0 & 1 & 0 \\ -1 & 0 & 0 & 0 \end{pmatrix},$$

which has an eigenvalue equal to -1 and therefore is not positive semidefinite. This shows that this state is not separable. It is entangled. In fact this state is the singlet state (9.2).[3] This method of calculation was used in the example.

Although it gives a complete answer only for the low-dimensional cases 2×2 and 2×3, the Positive Partial Transposition (PPT) criterion of Theorem 9.1.7 has proved a very powerful one in that it detects a large class of entangled states. Any interesting new criterion will have to be able to recognize entanglement in states that satisfy the necessary condition of separability of the PPT criterion, which are called *PPT states*. The following two additional criteria are able to recognize some entangled PPT states.

The criterion of [45], [46] is defined in terms of a different positive but not completely positive map. This is done with the help of the time reversal symmetry which is the **AII** Cartan symmetry, $\tilde{\theta}_{II}$ defined in Table 6.1 of Chapter 6 (with $T = 1$). The map $\tilde{\theta}_{II}$ is given, for any element A in $iu(n_S)$, with n_S even, by

$$\tilde{\theta}_{II}(A) = J\bar{A}J^\dagger = JA^T J^\dagger,$$

with the matrix J defined in (3.23). This map is a positive map and, in particular, it is unitarily equivalent to the transposition. Therefore, a criterion based on Theorem 9.1.7 and the map $\Gamma := \tilde{\theta}_{II}$ does not detect any new state other than the ones already detected by the PPT criterion of Theorem 9.1.8. However, let us modify the map $\tilde{\theta}_{II}$ as follows

$$\Phi(A) := \text{Tr}(A)1 - A - \tilde{\theta}_{II}(A). \tag{9.11}$$

Φ is a positive map. To check this, it is enough to check that $\Phi(|\psi\rangle\langle\psi|)$ is positive for any normalized state $|\psi\rangle$. We have, using the definition (9.11),

$$\Phi(|\psi\rangle\langle\psi|) = 1 - \Pi, \quad \Pi := |\psi\rangle\langle\psi| + J|\bar{\psi}\rangle\langle\bar{\psi}|J^\dagger,$$

with $\Pi \geq 0$ as it is an orthogonal projection onto the space spanned by $|\psi\rangle$ and $J|\bar{\psi}\rangle$.[4] Therefore, applying the fundamental Theorem 9.1.7, we have the

---

[3]A simple method to compute the partial transpose of an $n_S n_P \times n_S n_P$ matrix is to divide it in $n_S \times n_S$ blocks of dimension n_P and then transpose the blocks. For example for a $2n_P \times 2n_P$ matrix $\begin{pmatrix} B_{11} & B_{12} \\ B_{21} & B_{22} \end{pmatrix}$, the partial transpose is $\begin{pmatrix} B_{11} & B_{21} \\ B_{12} & B_{22} \end{pmatrix}$.

[4]The main step in the proof that $\Pi^2 = \Pi$ is to recognize that $\langle\psi|J|\bar{\psi}\rangle = 0$. Here $|\bar{\psi}\rangle$ denotes the complex conjugate of $|\psi\rangle$.

following criterion of separability which was proposed by H-P. Breuer in [45], [46].

Theorem 9.1.10 *(Breuer's criterion of separability) If ρ is separable then*

$$(\Phi \otimes 1)(\rho) \geq 0.$$

Therefore, if $(\Phi \otimes 1)(\rho) \ngeq 0$, ρ is entangled. By studying a special family of states, it is possible to show that Breuer's criterion detects some PPT entangled states. We refer for this as well as for several other (optimality) properties of Breuer's criterion to [45], [46]. A related criterion called the *reduction criterion* [138] is discussed in Exercise 9.8.

The **computable cross norm (CCN)**, or **realignment criterion**, [63], [240], [241], [242], involves the calculation of a particular norm of a linear operator $\Gamma : \mathcal{B}(\mathcal{H}_P) \rightarrow \mathcal{B}(\mathcal{H}_S)$. This criterion has several equivalent formulations [241]. We describe one of them. Consider the density matrix ρ as an operator on $\mathcal{H}_S \otimes \mathcal{H}_P$, i.e., as an element of $\mathcal{B}(\mathcal{H}_S \otimes \mathcal{H}_P)$. We can write ρ as

$$\rho = \sum_{j,k=1}^{n_S} \sum_{l,m=1}^{n_P} \rho_{jk,lm} E_{jk} \otimes E_{lm},$$

where $E_{jk} := e_j e_k^T$ ($E_{lm} = e_l e_m^T$), where e_j denotes the element of the standard orthonormal basis in $\mathcal{H}_S$ or $\mathcal{H}_P$. This representation of ρ induces a map $\Gamma_\rho : \mathcal{B}(\mathcal{H}_P) \rightarrow \mathcal{B}(\mathcal{H}_S)$ as follows,

$$\Gamma_\rho(A) := \sum_{j,k=1}^{n_S} \sum_{l,m=1}^{n_P} \rho_{jk,lm} E_{jk} \operatorname{Tr}(E_{lm}^\dagger A),$$

which is clearly a linear map. If T_{Γ_ρ} is the matrix representation of the map Γ_ρ, we can define the *trace class norm* $||T_{\Gamma_\rho}||_1$ as the sum of the magnitudes of the singular values of T_{Γ_ρ}. We have the following criterion.

Theorem 9.1.11 CCN or Realignment Criterion ([63], [240], [241], [242]) *If a state ρ is separable then*

$$||T_{\Gamma_\rho}||_1 \leq 1.$$

It follows that if $||T_{\Gamma_\rho}||_1 > 1$, then ρ is entangled. Contrary to the PPT criterion, the realignment criterion is not necessary and sufficient for 2×2 (two qubits) systems. However, the criterion has proved strong in detecting many PPT entangled states. We refer to [63], [240], [241], [242] for a study of several properties of this criterion. More discussion and generalizations can be found in [306]. We illustrate the calculations involved with a simple two qubits example taken from [240].

Example 9.1.12 Consider the class of 2 qubits states

$$\rho_p := p|00\rangle\langle 00| + \frac{(1-p)}{2}(|01\rangle + |10\rangle)(\langle 01| + \langle 10|), \qquad (9.12)$$

depending on the parameter p, $0 \le p \le 1$. The matrix ρ_p can be written as

$$\rho_p := pE_{00} \otimes E_{00} + \frac{1-p}{2}(E_{00} \otimes E_{11} + E_{11} \otimes E_{00} + E_{10} \otimes E_{01} + E_{01} \otimes E_{10}),$$

where $E_{jk} := |j\rangle\langle k|$. The corresponding linear operator Γ_{ρ_p} is such that

$$\Gamma_{\rho_p}(E_{00}) = pE_{00} + \frac{1-p}{2}E_{11},$$

$$\Gamma_{\rho_p}(E_{01}) = \frac{1-p}{2}E_{10},$$

$$\Gamma_{\rho_p}(E_{10}) = \frac{1-p}{2}E_{01},$$

$$\Gamma_{\rho_p}(E_{11}) = \frac{1-p}{2}E_{00}.$$

Therefore, the matrix representation of Γ_{ρ_p} is

$$T := \begin{pmatrix} p & 0 & 0 & \frac{1-p}{2} \\ 0 & 0 & \frac{1-p}{2} & 0 \\ 0 & \frac{1-p}{2} & 0 & 0 \\ \frac{1-p}{2} & 0 & 0 & 0 \end{pmatrix}.$$

The trace class norm $||T||_1$ is calculated by calculating the singular values of T. The eigenvalues of $T^T T$ are

$$\lambda_{1,2} = \left(\frac{1-p}{2}\right)^2, \quad \lambda_{3,4} := \frac{1}{2}\left(p^2 + \frac{(1-p)^2}{2} \pm \sqrt{p^4 + p^2(1-p)^2}\right).$$

Therefore,

$$||T||_1 = \sum_{l=1}^{4} \sqrt{\lambda_l} = 1 - p + \sqrt{\frac{1}{2}\left(p^2 + \frac{(1-p)^2}{2} + \sqrt{p^4 + p^2(1-p)^2}\right)} +$$

$$\sqrt{\frac{1}{2}\left(p^2 + \frac{(1-p)^2}{2} - \sqrt{p^4 + p^2(1-p)^2}\right)}.$$

From this it follows that

$$||T||_1 \ge 1 - p + p\sqrt{1 + \frac{(1-p)^2}{2p^2}} \ge 1,$$

with equality if and only if $p = 1$. It follows from the criterion that the states (9.12) are entangled for every value of p except for $p = 1$. In this case, clearly ρ_p is a separable state (in fact it is a product state).

9.1.2.3   Tests of entanglement for multipartite systems

For *pure states* of multipartite systems, one may think to naturally extend the partial trace test of Proposition 9.1.5. It is clear in fact that if a pure state is separable, the partial trace with respect to *any* subset of subsystems, taken as a subsystem, must be a pure state. The proof of the converse, however, does not go through in the same way as in the bipartite case because the Schmidt decomposition does not naturally extend to multipartite systems although several generalizations and methods to express states of multipartite systems in a canonical way exist [58]. Nevertheless, this test should detect generic entangled states. More specific results exist for pure quantum states with more structure. For example, tests for separability of pure states on n quantum bits are given in [151].

For mixed states, the tests described above for the bipartite case do not directly extend to the multipartite case. There are some other tests we have not discussed that extend naturally from the bipartite to the multipartite case. One example is given by the numerical tests based on semi-definite programming presented in [98]. In particular in this paper a sequence of tests is proposed that would eventually detect if a state is entangled.

One important concept which extends from the bipartite to the multipartite case is the concept of **entanglement witnesses**. An entanglement witness is an observable W such that $\text{Tr}(W\rho) \geq 0$ for all separable states ρ and $\text{Tr}(W\rho_e) < 0$ for some entangled state ρ_e. If this is the case, one says that W *detects* ρ_e. The existence of a witness which detects an entangled state follows from a classical theorem of convex analysis [234], the *separating hyperplane theorem*, noticing that the set of separable states is a convex and compact set in the space of Hermitian matrices. The separating hyperplane theorem states that for a compact convex set $\mathcal{C}$ in a real inner product space $\mathcal{H}$ and an element $x \notin \mathcal{C}$ there exists a hyperplane separating x from $\mathcal{C}$ (cf. Figure 9.1). This means that there exists a $y := y(x) \in \mathcal{H}$ such that $(y, c) \geq 0$, for all $c \in \mathcal{C}$ and $(y, x) < 0$, where $(\cdot, \cdot)$ denotes the inner product in $\mathcal{H}$. This can be applied to the set of Hermitian matrices with inner product $(A, B) := \text{Tr}(AB)$. Noticing that the set of separable density matrices is compact and convex *both* in the bipartite and in the multipartite case, the theorem says that if a state is entangled, there must exist an entanglement witness which detects it. Therefore, checking whether a state is entangled amounts to checking whether an entanglement witness exists. If a constructive proof of existence is provided, then the witness W also gives, in principle, a method to test entanglement experimentally. In fact $\text{Tr}(W\rho)$ is the expectation value of the observable W. Therefore, a measurement of W over many copies of ρ allows us to detect the entanglement of ρ. Some further information concerning entanglement tests and in particular the use of entanglement witnesses can be found in the survey paper [271].

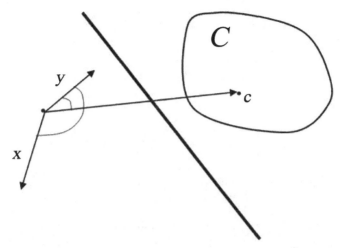

FIGURE 9.1: The separating hyperplane theorem in $\mathbf{R}^2$. The angle between the vector y and the vector x is larger than 90 degrees while for any vector c whose endpoint is in the compact and convex set C the angle is less then or equal to 90 degrees.

9.1.3 Measures of entanglement and concurrence

Measures of entanglement are defined in the *finite regime*, i.e., on a single system, or the *asymptotic regime*, i.e., on a sequence of systems. Their definitions are dictated by *abstract*, i.e., mathematical, or *operational* considerations. In the operational definition one state is more or less entangled if it is more or less utilizable for some task where entanglement is a resource [137]. We are here interested in properties of single states and their evolution; therefore we shall focus on the finite regime.

9.1.3.1 General axioms for entanglement measures

A **measure of entanglement** is a positive real valued function E on the set of density matrices[5] which satisfies the following axioms:

1. E is zero on separable states.

---

[5]Sometimes measures are defined on the set of *all* density matrices so that one can compare for example the entanglement of a four-level system with one of a two-level system. In this setting, it is reasonable to require that, for every measure of entanglement E, $E(\rho \otimes \rho) = 2E(\rho)$, i.e., entanglement does not increase by simply tensoring two states ρ and the amount of entanglement present in two copies of ρ is double the one presented in one copy. In particular, this, along with the convexity property (9.14), implies that the entanglement on a separable state is zero as in property 1. It will be clear from the context whether we are talking about a measure of entanglement defined on the set of all states or for a specific bipartite or multipartite system with a given structure.

2. E does not increase under LOCC operations, i.e., if Φ is an LOCC operation

$$E\left(\frac{1}{\mathrm{Tr}(\Phi(\rho))}\Phi(\rho)\right) \leq E(\rho). \qquad (9.13)$$

3. E is convex, i.e.,

$$E\left(\sum_j p_j \rho_j\right) \leq \sum_j p_j E(\rho_j), \qquad (9.14)$$

for arbitrary states ρ_j and $p_j > 0$, with $\sum_j p_j = 1$, i.e., entanglement does not increase on average under mixing of states.

LOCC operations are *Local Operations along with Classical Communication*. Recall (cf. Appendix A) that an operation Φ is a completely positive linear map from $\mathcal{B}(\mathcal{H}_1)$ (the set of bounded linear operators on $\mathcal{H}_1$, where $\mathcal{H}_1$ is the Hlibert space under consideration) to $\mathcal{B}(\mathcal{H}_2)$ which is trace-nonincreasing, i.e., $\mathrm{Tr}(\Phi(\rho)) \leq \mathrm{Tr}((\rho))$. By the Kraus representation theorem [170] every operation has the *operator sum representation*

$$\Phi(\rho) = \sum_k \Omega_k \rho \Omega_k^\dagger.$$

Local operations are the ones where the Ω_k are of the form

$$\Omega_k := \mathbf{1} \otimes \cdots \otimes \mathbf{1} \otimes \tilde{\Omega}_k \otimes \mathbf{1} \otimes \cdots \otimes \mathbf{1},$$

i.e., they act only on one of the subsystems. LOCC allow for classical communication, i.e., one local operation may depend on the result of another at a different location. Therefore, the composition of two local operations along with classical communication might have the form $\tilde{\Phi}_\alpha \circ \Phi(\rho)$, where α is the result of the operation Φ. The combination of local operations and classical communication has always the form

$$\Phi_{sep}(\rho) = \sum_j \Omega_{1j} \otimes \Omega_{2j} \otimes \cdots \otimes \Omega_{n_r j} \rho \Omega_{1j}^\dagger \otimes \Omega_{2j}^\dagger \otimes \cdots \otimes \Omega_{n_r j}^\dagger,$$

for some operators Ω_{lj}. Operators of the form Φ_{sep} are called *separable* operators. The set of separable operators contains *properly* (cf. [34]) the set of LOCC operators. There is no explicit characterization of LOCC's whose composition can be very complicated as they allow exchange of classical information in various ways as well as for different types of actions following classical communication. Therefore, the set of separable operations, being a super-set of the set of LOCC, is useful in characterizing several properties of LOCC's and, in particular, to study entanglement measures. Notice that the actual state is obtained after *normalization* once an operation is applied.

Remark 9.1.13 It should be mentioned that there is no complete agreement on the requirements that an entanglement measure should satisfy. Several variations of the above axioms are of interest in various situations [137]. For example, the convexity axiom (9.14) is sometimes considered unnecessary (cf. [216]). Requirement 2. is often replaced by the weaker requirement that a measure of entanglement does not increase *on average* under LOCC, namely

$$\sum_j p_j E\left(\frac{\Phi_j(\rho)}{\text{Tr}(\Phi_j(\rho))}\right) \leq E(\rho),$$

where p_j is the probability for the operation Φ_j to occur. We shall, however, assume the above three axioms in the following.

The fact that entanglement measures should not increase under LOCC operations implies that if a state ρ is such that all the states can be obtained from it using LOCC operations then any measure attains a global maximum at ρ, that is ρ is *maximally entangled*. For example, it is possible to prove (see, e.g., [216]) that for a bipartite system $S + P$ with $n_S = n_P := d$, every state (pure or mixed) can be obtained using $LOCC$ starting from the *generalized singlet state*

$$|\psi_d\rangle := \frac{|00\rangle + |11\rangle + \cdots + |d-1, d-1\rangle}{\sqrt{d}},$$

which is therefore maximally entangled. Here the orthogonal bases for the two d-dimensional systems S and P are chosen coinciding and given by $|0\rangle, |1\rangle, ...,$ $|d-1\rangle$. It follows that any measure of entanglement must have its maximum value (which is often normalized to 1) at $|\psi_d\rangle$. For other cases the situation is more complicated. In general, more or less entangled is a *partial* order in that there are states whose entanglements are not comparable.

Local unitary (LU) operations act on single subsystem, i.e., locally, according to $\rho_S \to X\rho_S X^\dagger$, with X unitary. If Φ is LU then Φ is LOCC and invertible and such that Φ^{-1} is also local unitary and therefore LOCC. Therefore, from the axiom (9.13), we obtain

$$E(\rho) \geq E(\Phi(\rho)) \geq E(\Phi^{-1} \circ \Phi(\rho)) = E(\rho).$$

Therefore, any measure of entanglement E must be constant under LU operations.

9.1.3.2   Entanglement measures for bipartite pure states

For bipartite pure states, the Von Neumann entropy of the partial trace, i.e., $E(\rho) = \hat{S}(\text{Tr}_P(\rho))$ is the most widely used measure of entanglement. Notice that it follows directly from Schmidt decomposition, Lemma 9.1.6, that, for a pure state ρ, $\text{Tr}_P(\rho)$ and $\text{Tr}_S(\rho)$ have the same spectrum. Therefore, $\hat{S}(\text{Tr}_P(\rho)) = \hat{S}(\text{Tr}_S(\rho))$. It is possible to show [99], [216] that the Von

Neumann entropy is the only measure of entanglement which satisfies two additional requirements. These are *additivity* on pure states and *asymptotic continuity*. Additivity for a general entanglement measure E means that $E(\rho^{\otimes n}) = nE(\rho)$, for any state ρ and any positive integer n, where $\rho^{\otimes n}$ is the tensor product of n copies of ρ. Asymptotic continuity on pure states means that for every pair of sequences of states $\{|\psi_n\rangle, |\phi_n\rangle\}$ on a sequence of pairs of coinciding Hilbert spaces $\{\mathcal{H}_n, \mathcal{H}_n\}$, if $\lim_{n\to\infty} |||\psi_n\rangle\langle\psi_n| - |\phi_n\rangle\langle\phi_n|||_1 = 0$,[6] then

$$\lim_{n\to\infty} \frac{E(|\psi_n\rangle) - E(\phi_n\rangle)}{1 + \log(\dim(\mathcal{H}_n))} = 0.$$

9.1.3.3 Convex roof extensions to bipartite mixed states; Entanglement of formation

For mixed bipartite states there are two important entanglement measures known as *distillable entanglement* and *entanglement cost* (see [216] for their definitions). These measures are of the operational type, namely they are defined in terms of the amount of operations and resources needed for a given task. Other measures for mixed states can be obtained by **convex roof extension** of measures defined on pure states. This means that if E_p is a measure defined on pure states, then a measure on general mixed states ρ, $E(\rho)$ is defined as

$$E(\rho) = \inf \sum_j p_j E_p(|\psi_j\rangle), \qquad \sum_j p_j = 1, \qquad p_j \geq 0,$$

where the infimum is taken among all the possible sets of $\{p_j, |\psi_j\rangle\}$ such that $\rho = \sum_j p_j |\psi_j\rangle\langle\psi_j|$. If E_p is continuous it is possible to prove [278] that the infimum is in fact a minimum, namely it is reached for a given ensemble $\{p_j, |\psi_j\rangle\}$. Convex roofs are the largest functions E which are convex and give $E(\rho) = E_p(|\psi\rangle)$ if $\rho := |\psi\rangle\langle\psi|$ is a pure state. The **entanglement of formation** E_f is defined as the convex roof associated with the Von Neumann entropy of the partial trace. It is therefore the minimum average entropy of the partial trace among all the possible ensemble realizations of a given state ρ. Its relevance as a measure of entanglement derives from its relation (in some cases proved to be true and in some other cases conjectured, cf. [216]) with operational measures such as the entanglement cost, as well as its own operational meaning [33], [300]. If we have a method to calculate the entanglement of formation we also have a method, alternative to the entanglement tests, to check whether a state is entangled. In fact, it is easily proven that ρ is separable if and only if $E_f(\rho) = 0$.[7] There have been several analytical results

---

[6]Recall that $||A||_1$ denotes the trace norm of the matrix A that is the sum of the singular values.

[7]This is a simple consequence of the continuity of the entropy and the fact that the entropy of the partial trace is zero if and only if the state is separable.

and numerical methods towards the solution of the optimization problem in the definition of E_f. In particular, a result due to A. Uhlmann [277] says that in the infimization problem it is sufficient to consider ensembles with at most r^2 terms where r is the rank of ρ. Estimates have also been found (see, e.g., [46]). However, there is no general method for the explicit calculation.

9.1.3.4 Concurrence

The **concurrence** $C := C(|\psi\rangle)$ is a function of the state which was first defined [132], [299] as instrumental to the calculation of entanglement of formation. It has then become of interest as a measure of entanglement n its own right. Such a quantity has a standard definition only for the case of a pair of qubits and different types of generalizations exist for other classes of systems. We shall consider this next.

Consider two qubits and a pure state $|\psi\rangle$. The concurrence $C(|\psi\rangle)$ is defined in terms of a specific type of symmetry, the time reversal symmetry, which is a Θ_{II} type of symmetry with $T = 1$ (cf. Table 6.1 of Chapter 6), on a single two-level system. In particular, we have the following definition

$$C(|\psi\rangle) := |\langle\psi|\Theta_{II} \otimes \Theta_{II}|\psi\rangle|. \tag{9.15}$$

If $|\psi\rangle$ is a product state $|\psi\rangle := |\psi_1\rangle \otimes |\psi_2\rangle$, we have

$$C(|\psi\rangle) = |\langle\psi_1|\Theta_{II}|\psi_1\rangle\langle\psi_2|\Theta_{II}|\psi_2\rangle| = 0.$$

In fact, if we express $|\psi_j\rangle$, $j = 1, 2$ as a unit vector $\vec{\psi}_j := \begin{pmatrix} x_j \\ y_j \end{pmatrix}$, then, with J defined in (3.23),

$$\Theta_{II}\vec{\psi}_j := J\begin{pmatrix} \bar{x}_j \\ \bar{y}_j \end{pmatrix} = \begin{pmatrix} -\bar{y}_j \\ \bar{x}_j \end{pmatrix}. \tag{9.16}$$

Therefore,

$$\vec{\psi}_j^\dagger \vec{\psi}_j = -\bar{x}_j\bar{y}_j + \bar{y}_j\bar{x}_j = 0.$$

In general, $0 \le C \le 1$. There exists a simple functional relation between the Von Neumann entropy of the partial trace $\hat{S} \circ \mathrm{Tr}_P$ and the concurrence C. It is given by (cf. Exercise 9.4)

$$\hat{S}(\mathrm{Tr}_P(|\psi\rangle\langle\psi|)) = H(C(|\psi\rangle)) := h\left(\frac{1 + \sqrt{1 - C^2(|\psi\rangle)}}{2}\right), \tag{9.17}$$

where $h = h(x)$ is the function[8]

$$h(x) := -x\log(x) - (1 - x)\log(1 - x),$$

where the logarithm is considered base 2 and we have set $0\log(0) = 0$.

---

[8]It is used in the proof of several properties of the concurrence mentioned below that, as a function of C, $H(C) := h(\frac{1+\sqrt{1-C^2}}{2})$ is convex (cf. Figure 9.2).

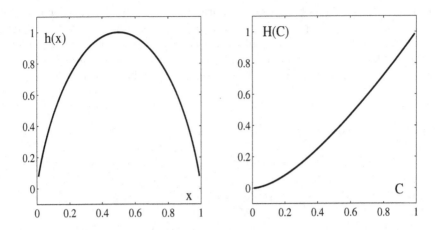

FIGURE 9.2: The function $h(x) := -x\log(x) - (1-x)\log(1-x)$ and the function $H(C) := h(\frac{1+\sqrt{1-C^2}}{2})$ which relate the entanglement of formation to the concurrence.

The entropy of the partial trace is a monotonically increasing function of the concurrence which has its minimum (maximum) at zero (one). Therefore, the entropy and the concurrence agree on which state is more or less entangled and one can take the concurrence itself rather than the entropy as a measure of entanglement.

For the mixed state of two qubits, the concurrence is defined by convex roof extension of the concurrence for pure states, i.e., we have

$$C(\rho) := \inf \sum_k p_k C(|\psi_k\rangle),$$

with the infimum taken over all the possible ensembles $\{p_k, |\psi_k\rangle\}$ with $\rho = \sum_k p_k |\psi_k\rangle\langle\psi_k|$, $p_k \geq 0$, $\sum_k p_k = 1$. It is a remarkable result [299] that (cf. (9.17))

$$E_f(\rho) = H(C(\rho)) := h\left(\frac{1 + \sqrt{1 - \overline{C^2}(\rho)}}{2}\right),$$

and therefore knowing the concurrence of a given state is equivalent to knowing the entanglement of formation. Moreover, there exists an explicit formula for the concurrence of a state ρ given by [299]

$$C(\rho) = \max\{0, \lambda_1 - \lambda_2 - \lambda_3 - \lambda_4\}, \tag{9.18}$$

where $\lambda_{1,2,3,4}$ are, in descending order, the square roots of the eigenvalues of the matrix $\rho\tilde{\theta}_{II} \otimes \tilde{\theta}_{II}(\rho)$ where $\tilde{\theta}_{II}$ is the quantum symmetry associated with Θ_{II} (cf. Table 6.1 and Exercise 9.5).

9.1.3.5 Generalized concurrences

All we have said, from the definition of concurrence on, concerns the case of a system of two qubits. The definition of concurrence (9.15) can be formally extended by replacing the specific quantum symmetry $\Theta_{II} \otimes \Theta_{II}$ with a general quantum symmetry Θ acting on the overall Hilbert space of the total (possibly multipartite) system. One defines a Θ-*concurrence* $C_\Theta(|\phi\rangle)$ as

$$C_\Theta(|\psi\rangle) := |\langle\psi|\Theta|\psi\rangle|, \tag{9.19}$$

and the Θ-concurrence of a mixed state, $C_\Theta(\rho)$ is defined as the convex roof of $C_\Theta(|\psi\rangle)$, i.e., as

$$C_\Theta(\rho) = \inf \sum_k p_k C_\Theta(|\psi_k\rangle), \tag{9.20}$$

with the infimum taken over all the possible pure state decompositions of ρ. The following property proved in [279] directly generalizes (9.18) to the case of a general antiunitary Cartan symmetry Θ, with $\Theta^2 = \mathbf{1}$. In this case, we have

$$C_\Theta(\rho) = \max\{0, \lambda_1 - \sum_{j>1} \lambda_j\}.$$

Here λ_j are the square roots in descending order of the eigenvalues of $\rho\tilde{\theta}(\rho)$, with $\tilde{\theta}$ the quantum symmetry associated with Θ (cf. subsection 6.3.5)). The question arises on whether such generalized concurrences are of interest in the quantification of entanglement. We consider this next.

Dealing with multipartite systems, it is natural to consider quantum symmetries Θ which are tensor products of quantum symmetries on the single subsystems. Moreover, as we want $\tilde{\theta}^2 = \mathbf{1}$ where $\tilde{\theta}(\rho) := \Theta\rho\Theta^{-1}$, we require $\Theta^2 = \pm\mathbf{1}$. Choosing

$$\Theta := \Theta_{II} \otimes \Theta_{II} \otimes \cdots \otimes \Theta_{II}, \tag{9.21}$$

with N factors for a system of N-qubits, where Θ_{II} is as from Table 6.1 with $T = 1$, we obtain the N-concurrence on a system of N qubits, defined in [298], which was proved in [298] to be a viable measure of entanglement for pure states, i.e., a measure satisfying the axioms 1-3 above. This measure is identically zero for odd values of N. However, for N even, it gives another meaningful measure of entanglement which is called the N-*tangle* (cf. [298] and references therein). In the case of a bipartite system $S + P$, with $n_S = 2$ and $n_P \geq 2$ and even, a formula exists relating the entanglement of formation to the generalized concurrences C_Θ. This formula [279] is (cf. (9.17))

$$E_f(\rho) \geq \sup_\Theta h\left(\frac{1 + \sqrt{1 - C_\Theta^2(\rho)}}{2}\right), \tag{9.22}$$

where the sup is taken over the set of all antiunitary quantum symmetries of the form[9] $\Theta := \Theta_S \otimes \Theta_P$, with $\Theta_S^2 := -1$ and $\Theta_P^2 := -1$. This means in particular that, in this case, if there exists a Θ such that $C_\Theta(\rho) > 0$, then the state ρ is entangled. This gives an alternate test of entanglement.[10] This raises the question on whether a certain number of generalized concurrences would be sufficient to detect the entanglement of even more general systems than the bipartite $S + P$, $n_S = 2$, n_P even. In particular one can consider a vector of concurrences. This is the approach used in [27] where the operators Θ are replaced by more general operators Θ_α for some parameters α. The operator Θ_α for a bipartite system on $\mathcal{H}_S \otimes \mathcal{H}_P$ map to zero all the vectors in $\mathcal{H}_S \otimes \mathcal{H}_P$ except the ones which are linear combinations of tensor products of vectors in two two-dimensional subspaces (parametrized by α). On these vectors Θ_α acts as $\Theta_{II} \otimes \Theta_{II}$. The authors of [27] use the generalized concurrences associated with Θ_α to formulate (and conjecture) several separability conditions for bipartite quantum systems.

Another generalization of the concurrence for general (not necessarily two qubits) bipartite pure states is [243]

$$C(|\psi\rangle) := \sqrt{2\,(1 - \text{Tr}[(\text{Tr}_P(|\psi\rangle\langle\psi|))]^2)},$$

which can then be extended by convex roof to mixed states. Estimates for this concurrence on mixed states have been obtained in [46], [62].

9.2   Dynamics of Entanglement

Cartan decompositions of the type introduced in Chapter 6 can be used in the analysis of entanglement dynamics in two ways. First, they relate to generalized Θ-concurrences as defined in the previous subsection, in that they factorize unitary evolutions in a term that modifies concurrence and a term that does not. Secondly, they give expressions of every unitary evolution in terms of local and entangling evolutions. These expressions are useful in the design of evolutions performing a given task. In fact, they allow us to identify exactly the entangling evolutions which are the ones which are usually more difficult to perform in a laboratory. The most complete example of treatment of entanglement dynamics using Lie group decomposition is the two qubits system which we shall describe in the next subsection. Several aspects of the treatment extend to general multipartite systems of arbitrary dimensions, as

---

[9]Notice these antiunitary quantum symmetries can be all parametrized as they are all of the form Θ_{II} of Table 6.1.
[10]The 'sup' in formula (9.22) is, in general, a function of the state ρ under consideration.

we shall see. In section 9.2.3 we shall describe several decompositions of unitary evolutions which recursively use the fundamental Cartan decomposition described in section 6.3 to give decompositions in entangling and local parts for every evolution.

Consider a Θ-concurrence C_Θ as defined in (9.19), (9.20), the Cartan symmetry $\tilde{\theta}$ associated with Θ, the associated Cartan involution θ of $u(\bar{n})^{11}$, and the associated Cartan decomposition of $u(\bar{n})$

$$u(\bar{n}) = \mathcal{K} \oplus \mathcal{P},$$

with $\mathcal{K}$ and $\mathcal{P}$ the $+1$ and -1 eigenspaces of θ in $u(\bar{n})$, respectively. Cartan decompositions and generalized concurrences are related via the following result.

Proposition 9.2.1 If $K \in e^{\mathcal{K}}$, then, for every $|\psi\rangle$,

$$C_\Theta(K|\psi\rangle) = C_\Theta(|\psi\rangle), \tag{9.23}$$

and

$$C_\Theta(K\rho K^\dagger) = C_\Theta(\rho). \tag{9.24}$$

Proof. Formula (9.24) follows from (9.23) (see Exercise 9.6). To show (9.23), notice that since the Lie group $e^{\mathcal{K}}$ is always compact, the exponential map is always surjective (see, e.g., [244]) and we can write every element in $e^{\mathcal{K}}$ as $K = e^A$, with $A \in \mathcal{K}$. To show that $C_\Theta(e^A|\psi\rangle) = C_\Theta(|\psi\rangle)$, we show that (cf. (9.19))

$$e^{-A}\Theta e^A = \Theta,$$

for every $A \in \mathcal{K}$. Equivalently we show

$$\Theta e^A \Theta^{-1} = e^A. \tag{9.25}$$

Since Θ is antiunitary, from the general expression (6.40)

$$\Theta = L \circ Conj,$$

where L denotes a unitary matrix and $Conj$ denotes conjugation of the components of a vector once a basis is fixed, i.e., $Conj(|\psi\rangle) := |\bar{\psi}\rangle$. Θ^{-1} is given by $\Theta^{-1} = Conj \circ L^\dagger$. A straightforward calculation using these definitions gives

$$\Theta e^A \Theta^{-1} = L e^{\bar{A}} L^\dagger = \sum_{k=0}^{\infty} \frac{L \bar{A}^k L^\dagger}{k!}. \tag{9.26}$$

Write A, skew-Hermitian, as $A = iB$, with B Hermitian in the space $i\mathcal{K}$. We have

$$L\bar{A}^k L^\dagger = (-i)^k L \bar{B}^k L^\dagger = (-i)^k \Theta B^k \Theta^{-1} = (-i)^k \tilde{\theta}(B^k) = (-i)^k (-1)^k B^k = A^k,$$

---

[11]$\bar{n}$ denotes the dimension of the overall system, namely, as we are dealing with multipartite systems, $\bar{n} = n_1 \times n_2 \times \cdots \times n_N$, where $n_j, j = 1, ..., r$ is the dimension of the j-th subsystem.

where we have used the fact that B is in the -1 eigenspace of $\tilde{\theta}$ and the definition of B along with (6.43). Plugging this into (9.26), we obtain (9.25).

□

In conclusion, in the factorization (6.29) (or (6.31)) only the factor P (or the factor A) possibly changes the generalized concurrence.

9.2.1 The two qubits example

A special case of the above situation is for the two qubits concurrence (9.15) for which $\Theta := \Theta_{II} \otimes \Theta_{II}$ is defined in (9.16). The associated decomposition is the decomposition of $u(4)$ described in Example 6.3.10,[12] i.e.,

$$u(4) = \mathcal{K} \oplus \mathcal{P}, \tag{9.27}$$

with

$$\mathcal{K} := \operatorname{span}\{i\sigma \otimes \mathbf{1}, i\mathbf{1} \otimes \sigma\}, \qquad \mathcal{P} := \operatorname{span}\{i\sigma \otimes \sigma, i\mathbf{1} \otimes \mathbf{1}\}, \tag{9.28}$$

where σ is one of the Pauli matrices (1.20). If we choose as Cartan subalgebra in $\mathcal{P}$,

$$\mathcal{A} := \operatorname{span}\{i\sigma_x \otimes \sigma_x, i\sigma_y \otimes \sigma_y, i\sigma_z \otimes \sigma_z, i\mathbf{1} \otimes \mathbf{1}\},$$

then every unitary matrix X_f can be written as

$$X_f := e^{i\phi} L_1 \otimes L_2 e^{ic_x \sigma_x \otimes \sigma_x} e^{ic_y \sigma_y \otimes \sigma_y} e^{ic_z \sigma_z \otimes \sigma_z} L_3 \otimes L_4, \tag{9.29}$$

with real parameters c_x, c_y, c_z, ϕ, and matrices $L_1, L_2, L_3, L_4 \in SU(2)$. According to Proposition 9.2.1, the factors $L_1 \otimes L_2$ and $L_3 \otimes L_4$ do not modify the concurrence. Clearly the phase factor $e^{i\phi}$ does not modify the concurrence either. The only terms that create entanglement in (9.29) are the terms $e^{ic_j \sigma_j \otimes \sigma_j}$, $j = x, y, z$. Therefore the *entanglement capability* of a unitary transformation only depends on these terms.

Two unitary evolutions X_1 and X_2 on a system of two qubits are called **locally equivalent** if there exist K_1 and K_2 in $SU(2) \otimes SU(2)$, such that

$$X_1 = K_1 X_2 K_2.$$

This means that it is possible to obtain the evolution X_1 by having a local evolution K_2 followed by the evolution X_2 then followed by another local evolution K_1. It is clear that local equivalence is an equivalence relation and two unitary transformations with the same values of c_x, c_y and c_z in (9.29) are locally equivalent. A *local invariant* on $SU(4)$ is a function $f : SU(4) \to \mathbb{C}$ such that, if X_1 and X_2 are locally equivalent, $f(X_1) = f(X_2)$.

---

[12]This decomposition was also used in the time optimal control problem for two spin-$\frac{1}{2}$ particles in subsection 7.4.2.

A *complete set of local invariants* is a set of local invariants $f_1, f_2, ..., f_m$, such that two matrices in $SU(4)$, X_1 and X_2, are locally equivalent if and only if $f_j(X_1) = f_j(X_2)$, $j = 1, ..., m$.[13] According to a result of [190], [307], once the decomposition of an element in $SU(4)$ is given (i.e., (9.29) is given, with $\phi = 0$), a complete set of invariants is given by the spectrum of A^2 where A is defined as (cf. (9.29))

$$A := e^{ic_x \sigma_x \otimes \sigma_x} e^{ic_y \sigma_y \otimes \sigma_y} e^{ic_z \sigma_z \otimes \sigma_z}. \tag{9.30}$$

In fact, it is possible to parametrize the set of locally equivalent transformations in $SU(4)$ according to the values of the parameters c_x, c_y and c_z [307] as illustrated in Figure 9.3.

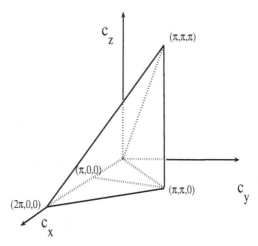

FIGURE 9.3: According to a result of [307], there exists a one to one correspondence between the points of this tetrahedron and the equivalence classes of locally equivalent transformations in $SU(4)$. This correspondence is such that the point (c_x, c_y, c_z) corresponds to the equivalence class containing $e^{ic_x \sigma_x \otimes \sigma_x} e^{ic_y \sigma_y \otimes \sigma_y} e^{ic_z \sigma_z \otimes \sigma_z}$. The only exception concerns points on the base where for each point in the triangle with vertices $(2\pi, 0, 0)$, $(\pi, 0, 0)$, $(\pi, \pi, 0)$, there corresponds an equivalent point in the triangle with vertices $(\pi, 0, 0)$, $(0, 0, 0)$, $(\pi, \pi, 0)$.

For general elements of $U(4)$ (ϕ possibly $\neq 0$ in (9.29)) local equivalence is of interest up to the common phase factor. To check this type of local

---

[13]Similar definitions will be given in the next section for invariants on the set of density matrices.

equivalence for two matrices in $U(4)$, one writes the two matrices in the form (9.29) where $e^{i\phi} = \det(X_f)$ and then checks the local equivalence of the factors in $SU(4)$, according to the mentioned results of [190], [307].

Definition 9.2.2 A unitary evolution $X_f \in U(4)$ is a **perfect entangler** if there exists a separable pure state $|\psi\rangle$, i.e., $C(|\psi\rangle) = 0$, such that $C(X_f|\psi\rangle) = 1$, where C is the concurrence defined in (9.15).[14]

It is clear that with the expression of X_f in (9.29) and A in (9.30), X_f is a perfect entangler if and only if A is a perfect entangler. The entangling capabilities of X_f are the same as those of A. It is possible to characterize perfect entanglers exactly. To do this we need the following definition.

Definition 9.2.3 Given r complex numbers $\lambda_1, ... \lambda_r$, their *convex hull* $\mathcal{C}$ is the set

$$\mathcal{C} := \left\{ \sum_{j=1}^{r} c_j \lambda_j \,\middle|\, c_j \geq 0, \quad j = 1, ..., r, \quad \sum_{j=1}^{r} c_j = 1 \right\}.$$

We notice that all the eigenvalues of A^2 are on the unit circle in the complex plane. We have the following characterization of perfect entanglers in terms of the eigenvalues of A^2. An example of application of this result is given in Figure 9.4.

Proposition 9.2.4 [190] [307] A in (9.30) is a perfect entangler if and only if the convex hull of the eigenvalues of A^2 contains the origin.

Another consequence of the decomposition (9.29) is that only one entangling Hamiltonian (say $\sigma_z \otimes \sigma_z$), along with local transformations, is sufficient to implement every unitary transformation. In fact, every Hamiltonian $\sigma_j \otimes \sigma_k$, $j, k = x, y, z$ (and therefore in particular $\sigma_x \otimes \sigma_x$ and $\sigma_y \otimes \sigma_y$) is similar, via a local similarity transformation, to $\sigma_z \otimes \sigma_z$ (a consequence of the similarity among Pauli matrices).

Notice that not all the nonlocal transformations X_f are entanglers in the sense that there exists a separable state $|\psi\rangle$ with $C(X_f|\psi\rangle) > 0$. For example consider an evolution which is locally equivalent to the *SWAP operator* X_S, which is defined as $X_S(|\psi_1\rangle \otimes |\psi_2\rangle) := |\psi_2\rangle \otimes |\psi_1\rangle$. This type of evolution is not local but it is not an entangler.

In conclusion, every transformation on the system of two qubits can be decomposed, according to Cartan decomposition, into local factors and factors that are responsible for the entanglement of the two qubits. The local factors

---

[14]The definition could have been given equivalently in terms of the entropy of the partial trace.

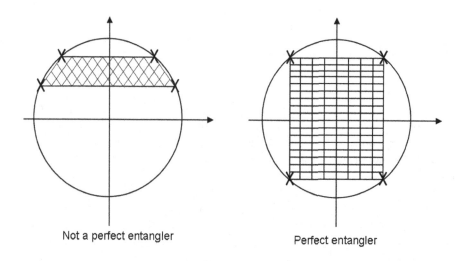

Not a perfect entangler   Perfect entangler

FIGURE 9.4: Illustration of the characterization of perfect entanglers of Proposition 9.2.4. Given that the eigenvalues of A^2 are represented by the symbol $\times$ on the unit sphere, the figure on the left represents an evolution which is **not** a perfect entangler. The figure on the right represents an evolution which is a perfect entangler.

do not modify the concurrence but the entangling factor possibly does. The analysis of the entanglement capability of a given unitary transformation can be performed by considering its entangling part only. In particular, studying the eigenvalues of this factor, it is possible to deduce whether or not the overall evolution is a perfect entangler.

9.2.2 The odd-even decomposition and concurrence dynamics

The Cartan decomposition of $u(4)$ (9.27) (9.28) is an example of a general procedure for a multipartite system to generate a decomposition of the associated unitary Lie algebra in terms of decompositions on the single subsystems [81]. We shall describe this next.

Consider a multipartite quantum system composed of N subsystems of dimensions $n_1,...,n_N$. The Jordan algebra of the Hamiltonians on the j-th subsystem is $iu(n_j)$, $j = 1,...,N$. Assume we perform a Cartan decomposition on each $u(n_j)$ of type **AI** or **AII**,[15] $j = 1, ..., N$ and write

$$u(n_j) := \mathcal{K}_j \oplus \mathcal{P}_j, \tag{9.31}$$

with $\mathcal{K}_j$ a subalgebra (isomorphic to $so(n_j)$ or $sp(\frac{n_j}{2})$) and $\mathcal{P}_j$ its orthogonal complement. To the decomposition (9.31) there corresponds a decomposition of the Jordan algebra $iu(n_j)$,

$$iu(n_j) = i\mathcal{K}_j \oplus i\mathcal{P}_j,$$

a quantum mechanical symmetry Θ_j ($\tilde{\theta}_j$) and the Cartan involution θ_j (cf. section 6.3.5). Consider a (orthonormal) basis of $i\mathcal{K}_j$ and denote its generic element by σ and a (orthonormal) basis of $i\mathcal{P}_j$ and denote its generic element by S. A (orthonormal) basis of $iu(n_1 \times n_2 \times \cdots \times n_N)$ is given by elements of the form

$$F := F_1 \otimes F_2 \otimes \cdots \otimes F_N, \tag{9.32}$$

with F_j, $j = 1, ..., N$, of the type σ or S. Denote now by $\mathcal{I}_o$ the span of elements of the form F in (9.32) with an *odd* number of σ's and by $\mathcal{I}_e$ the span of elements of the form F in (9.32) with an *even* number of σ's. Construct a quantum symmetry Θ_{o-e}, as tensor product of all the symmetries Θ_j, $j = 1, ..., N$,

$$\Theta_{o-e} := \Theta_1 \otimes \Theta_2 \otimes \cdots \otimes \Theta_N,$$

and the associated symmetry on observables $\tilde{\theta}_{o-e}$,

$$\tilde{\theta}_{o-e} := \tilde{\theta}_1 \otimes \tilde{\theta}_2 \otimes \cdots \otimes \tilde{\theta}_N,$$

where, for all $A \in iu(n_j)$, $\tilde{\theta}_j(A) = \Theta_j A \Theta_j^{-1}$.[16] Since $\tilde{\theta}_j(\sigma) = -\sigma$ and $\tilde{\theta}_j(S) = S$, $\tilde{\theta}_{o-e}(A) = A$ if $A \in \mathcal{I}_e$ and $\tilde{\theta}_{o-e}(A) = -A$ if $A \in \mathcal{I}_o$. The involution associated with $\tilde{\theta}_{o-e}$ (cf. (6.42)), θ_{o-e}, is such that, for $A \in iu(n_1 \times n_2 \times \cdots \times n_N)$,

$$\theta_{o-e}(iA) = -i\tilde{\theta}_{o-e}(A),$$

and therefore if $A \in \mathcal{I}_o$, $\theta_{o-e}(iA) = iA$ and if $A \in \mathcal{I}_e$ $\theta_{o-e}(iA) = -iA$. Therefore, θ_{o-e} is a Cartan involution on $u(n_1 \times n_2 \times \cdots \times n_N)$ with $i\mathcal{I}_o$ and $i\mathcal{I}_e$ the $+1$ and -1 eigenspaces, and

$$u(n_1 \times n_2 \times \cdots \times n_N) := i\mathcal{I}_o \oplus i\mathcal{I}_e \tag{9.33}$$

is a Cartan decomposition of $u(n_1 \times n_2 \times \cdots \times n_N)$ with $i\mathcal{I}_o$ and $i\mathcal{I}_e$ being the subalgebra and the complementary orthogonal subspace respectively. This is called the **odd-even decomposition (OED)**. Figure 9.5 summarizes the procedure to obtain this decomposition.

---

[15] Recall that span$\{i\mathbf{1}\}$ is included in $\mathcal{P}_j$ in the decomposition (9.31) (cf. Tables 6.1, 6.2 in Chapter 6).

[16] The definition of the tensor product of two antilinear operators was given in Remark 6.3.11, and it extends naturally to the case of N antilinear operators.

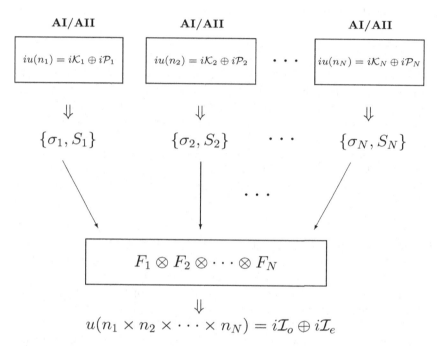

FIGURE 9.5: Scheme for the odd-even decomposition (OED). Decomposition of the types **AI** or **AII** are performed for each subsystem. This determines a set of matrices $\{\sigma_j, S_j\}$, $j = 1, ..., N$, which span orthonormal bases for the associated pairs of subspaces $i\mathcal{K}_j$ and $i\mathcal{P}_j$. By taking the tensor products $F_1 \otimes F_2 \otimes \cdots \otimes F_N$ with F_j of the form σ_j or S_j we obtain the basis of $iu(n_1 \times n_2 \times \cdots \times n_N)$. By splitting this basis in the odd and even part, we obtain the OED decomposition.

Remark 9.2.5 The decomposition (9.27) used in subsection 9.2.1 is a special case of the OED. This special case is obtained when $N = 2$ and both decompositions on $u(n_1) = u(n_2) = u(2)$ are of type **AII**. In fact $sp(1) = su(2)$ and the relevant decomposition of $u(2)$ is $u(2) = \mathcal{K} \oplus \mathcal{P}$, with $\mathcal{K} = su(2) = sp(1)$ and $\mathcal{P} := \text{span}\{i\mathbf{1}\}$. In this case, the decomposition separates local and entangling contributions of every unitary evolution.

The interest of the odd-even decomposition in entanglement theory follows from the fact that such a decomposition of the dynamics is naturally obtained in terms of tensor products of Hamiltonians. Local and entangling Hamiltonians for general dimensions and decompositions are, however, both in the Lie subalgebra part $i\mathcal{I}_o$ and in the complement part $i\mathcal{I}_e$ contrary to what happened for the special case of the previous remark and subsection 9.2.1. For example, the local Hamiltonian $H_{local} := A \otimes \mathbf{1} \otimes \cdots \otimes \mathbf{1}$ belongs to $\mathcal{I}_o$ or $\mathcal{I}_e$ according to whether A belongs to $i\mathcal{K}_1$ or not (cf. (9.31)), respectively. Therefore, consider the associated generalized concurrence $C_{\Theta_{o-e}}$. Proposition 9.2.1 does not guarantee that $C_{\Theta_{o-e}}$ will stay constant for local transformations, i.e., $C_{\Theta_{o-e}}(|\psi\rangle) = C_{\Theta_{o-e}}(L|\psi\rangle)$, for L local, a requirement necessary for any measure of entanglement. Nevertheless, we can obtain invariance if we restrict the local operations to fewer, physically motivated ones. We illustrate this in the following Remark.

Remark 9.2.6 Consider a system of N particles with spin, with variable values of the spin (which can be integer or half-integer). For every spin system of dimension n_j, there exists a three-dimensional Lie subalgebra of $u(n_j)$ isomorphic to $su(2)$, which is given by the infinitesimal rotations. Expressions for the matrices representing these operators in an appropriate basis can be calculated explicitly (see, e.g., [245] section 3.5.). Their exponentials represent rotations of the spins and therefore we call this subalgebra here the *rotation Lie subalgebra*. It is possible to show [10] that, *if n_j is even* (half integer spin), there exists a subalgebra of $u(n_j)$, conjugate to $sp(\frac{n_j}{2})$ and containing as subalgebra the rotation Lie subalgebra. Analogously, it is possible to show that *if n_j is odd* (integer spin) there exists a subalgebra of $u(n_j)$, conjugate to $so(n_j)$ and containing as subalgebra the rotation Lie subalgebra.[17] Let us perform an odd-even decomposition using a decomposition of type **AI** on odd-dimensional systems and of type **AII** on even-dimensional systems and let us use as Lie algebras $\mathcal{K}_j$ (9.31) the ones containing the rotation Lie subalgebra in all cases. Then all the infinitesimal generators of local rotations will be in the $i\mathcal{I}_o$ part of the decomposition (9.33). Therefore, the concurrence $C_{\Theta_{o-e}}$ will be left invariant by local rotations as from Proposition 9.2.1.

Remark 9.2.7 Another special case of the OED is for the case of N qubits when a decomposition of type **AII** is performed on every subsystem. The

---

[17]In fact the converse of these statements also holds. If n_j is even there is no Lie subalgebra of $u(n_j)$ conjugate to $so(n_j)$ and containing the rotation Lie subalgebra. See [10].

resulting decomposition is known as **Concurrence Canonical Decomposition (CCD)** [50] [51].[18] In this case, the local unitary evolutions all belong to $e^{i\mathcal{I}_o}$ since the corresponding Hamiltonians are of the type $\mathbf{1} \otimes \mathbf{1} \otimes \cdots \otimes \sigma \otimes \mathbf{1} \otimes \mathbf{1} \otimes \cdots \otimes \mathbf{1}$, with only one σ which is a linear combination of Pauli matrices. The generalized concurrence $C_{\Theta_{o-e}}$ is the same as the N-concurrence defined after formula (9.21), which is, as we have said, a viable measure of entanglement (which is, however, trivial for odd N).

9.2.2.1 Computation of the OED decomposition

The OED is a Cartan decomposition of $u(n_1 \times n_2 \times \cdots \times n_N)$ and therefore it has to be of the type **AI**, **AII** and **AIII** as defined in subsection 6.3.4. It is of interest to provide a change of coordinates to put the decomposition in the standard basis so as to be able to calculate it using the methods in subsection 6.3.6. The odd-even decomposition cannot be of type **AIII** because the associated quantum symmetry Θ_{o-e} is the tensor product of antiunitary symmetries and therefore also antiunitary (cf. Remark (6.3.11), while for decompositions of the type **AIII**, the associated symmetry is unitary. To discover whether the decomposition is of type **AI** or **AII** we can do a simple dimension count. This is done in [81] and the result is the following.

Proposition 9.2.8 Consider an odd-even decomposition on N subsystems obtained by performing **AII** decompositions on r subsystems and **AI** decompositions on $N - r$ subsystems. Then the resulting decomposition is of type **AII** if r is odd and of type **AI** if r is even.

To put the decomposition in the standard basis, so as to apply computational techniques of subsection 6.3.6, we have to compute the matrix T in Table 6.2 so as to make

$$i\mathcal{I}_o = T so(n_1 \times n_2 \times \cdots \times n_N)T^\dagger,$$

or

$$i\mathcal{I}_o = T sp(\frac{n_1 \times n_2 \times \cdots \times n_N}{2})T^\dagger,$$

according to whether we have an **AI** or **AII** total decomposition. Assume, for simplicity of notation, that all the decompositions are performed in the standard basis. Then we have for all A in $u(n_1 \times n_2 \times \cdots \times n_N)$

$$\theta_{o-e}(A) := i\tilde{\theta}_{o-e}(iA) = (\mathbf{1} \otimes J \otimes \mathbf{1} \cdots \otimes J) \bar{A} \left(\mathbf{1} \otimes J^{-1} \otimes \mathbf{1} \cdots \otimes J^{-1}\right),$$

---

[18]In an independent study [9] the CCD was used to characterize networks of spin-$\frac{1}{2}$ particles which are *input-output equivalent* in the sense that they give the same value of the total average spin when subject to the same (input) magnetic field. The generalization of this result to general spin uses the OED and is provided in [10].

where $\mathbf{1}$ or J, of appropriate dimensions, appear whether we perform a decomposition of the type **AI** or **AII**, respectively on the corresponding subsystem. In the case where the total decomposition is **AI** (but analogous reasoning is valid for the decomposition **AII**) comparing with Table 6.1, we must have

$$(\mathbf{1} \otimes J \otimes \mathbf{1} \cdots \otimes J)\, \bar{A} \left(\mathbf{1} \otimes J^{-1} \otimes \mathbf{1} \cdots \otimes J^{-1}\right) = TT^T \bar{A}(TT^T)^\dagger.$$

As this equation has to hold for every $A \in u(n_1 \times n_2 \times \cdots \times n_N)$ we must have[19]

$$TT^T = \mathbf{1} \otimes J \otimes \mathbf{1} \cdots \otimes J,$$

from which one obtains the unitary T. In this fashion, one can for example obtain the transformation of coordinates given by the magic basis in Example 6.3.10.

9.2.3 Recursive decomposition of dynamics in entangling and local parts

The theorem of Cartan decomposition of section 6.3 can be used recursively to obtain finer decompositions of dynamics in entangling and local parts. There are several proposals in the literature in this direction. We shall summarize some of them in the following.

In [72], M. Dagli, J. D. H. Smith and the author considered the case of N qubits and applied the CCD recursively as follows. First the CCD is applied to write

$$u(2^N) = i\mathcal{I}_o \oplus i\mathcal{I}_e,$$

where $i\mathcal{I}_o$ ($i\mathcal{I}_e$) is the span of elements $i\sigma \otimes \mathbf{1} \otimes \sigma \otimes \cdots \otimes \sigma$ (i.e., tensor products of $\mathbf{1}'s$ and σ's), where σ is a general Pauli matrix, with an odd (even) number of elements σ. Therefore, a general element X of $U(2^N)$ can be written, according to the KAK decomposition (6.31), as

$$X = K_1 A K_2, \tag{9.34}$$

with $K_1, K_2 \in e^{i\mathcal{I}_o}$ and $A \in e^{\mathcal{A}}$ where $\mathcal{A}$ is the Cartan subalgebra associated with the decomposition. From Proposition 9.2.8, according to whether N is even or odd, $i\mathcal{I}_o$ is conjugate to $so(2^N)$ or $sp(2^{N-1})$. In the two cases the Cartan subalgebra[20] has dimension 2^N or 2^{N-1}, for N even or odd, respectively. In the first case, a maximal set of commuting elements of $i\mathcal{I}_e$ can be obtained considering the set

$$\mathcal{S} := \{\sigma_x \otimes \sigma_x, \sigma_y \otimes \sigma_y, \sigma_z \otimes \sigma_z, \mathbf{1} \otimes \mathbf{1}\}. \tag{9.35}$$

---

[19]This argument uses Schur's lemma (see, e.g., [92]). Since $(TT^T)^\dagger \mathbf{1} \otimes J \otimes \mathbf{1} \cdots \otimes J$ commutes with every element in $u(n_1 \times n_2 \times \cdots \times n_N)$ it must be a multiple of the identity.

[20]We are including also multiples of the identity as we are dealing with decompositions of $u(2^N)$.

A Cartan subalgebra is spanned by elements of the form $iA_1 \otimes A_2 \otimes \cdots \otimes A_{\frac{N}{2}}$, where every element A_j, $j = 1, ..., \frac{N}{2}$, belongs to S defined in (9.35). This way one obtains $4^{\frac{N}{2}} = 2^N$ elements, which is the dimension of the Cartan subalgebra. In the second case, a Cartan subalgebra is spanned by elements of the form $iA_1 \otimes A_2 \otimes \cdots \otimes A_{\frac{N-1}{2}} \otimes \mathbf{1}$ where every element A_j, $j = 1, ..., \frac{N-1}{2}$ belongs to S defined in (9.35).

To refine the decomposition (9.34) one needs to decompose the elements K_1 and K_2. This is obtained by performing a Cartan decomposition of $i\mathcal{I}_o$ as

$$i\mathcal{I}_o := \mathcal{K}_o \oplus \mathcal{P}_o,$$

where

$$\mathcal{K}_o = \text{span}\{A \otimes \mathbf{1}, B \otimes \sigma_z | A \in i\mathcal{I}_o^1, B \in i\mathcal{I}_e^1\}, \tag{9.36}$$

$$\mathcal{P}_o := \text{span}\{C \otimes \sigma_{x,y} | C \in i\mathcal{I}_e^1\}, \tag{9.37}$$

where $\mathcal{I}_o^1$ ($\mathcal{I}_e^1$) is the set of tensor products $\sigma \otimes \mathbf{1} \otimes \sigma \otimes \cdots \otimes \mathbf{1}$ with an odd (even) number of σ's and $N - 1$ places.[21] Therefore, K_1 (and analogously K_2) can be decomposed as

$$K_1 = K_{1,1}\tilde{A}K_{1,2},$$

where $K_{1,1}$ and $K_{1,2}$ belong to the Lie group $e^{\mathcal{K}_o}$ and $\tilde{A}$ belongs to $e^{\mathcal{A}_1}$ with $\mathcal{A}_1$ the Cartan subalgebra ($\subseteq \mathcal{P}_o$) associated with the decomposition. This is a decomposition of $so(2^N)$, for N even or a decomposition of $sp(2^{N-1})$ for N odd. The crucial observation to iterate the procedure is to notice that $\mathcal{K}_o$ is isomorphic to $u(2^{N-1})$. In fact the associated decomposition of $i\mathcal{I}_o$ is of type **DIII** and **CI**, respectively (see [130]) and this gives us the information about the dimensions of the associated Cartan subalgebras which are 2^{N-2} and 2^{N-1}, for N even or N odd, respectively. The Cartan subalgebra in the case **DIII** (N even) is spanned by elements of the type $iA_1 \otimes A_2 \otimes \cdots \otimes A_{\frac{N-2}{2}} \otimes \mathbf{1} \otimes \sigma_x$, for A_j, $j = 1, ..., \frac{N-2}{2}$ in the set S defined in (9.35). In the case **CI** (N odd), the Cartan subalgebra is spanned by elements of the type $iA_1 \otimes A_2 \otimes \cdots \otimes A_{\frac{N-1}{2}} \otimes \sigma_x$, where the A_j's, $j = 1, ..., \frac{N-1}{2}$, are defined as above.

The next step is to factorize the elements $K_{1,1}$ and $K_{1,2}$ which belong to $e^{\mathcal{K}_o}$. However, $\mathcal{K}_o$ is the same as $u(2^{N-1})$ and one performs a decomposition equivalent to the CCD on $N - 1$ qubits, given by

$$\mathcal{K}_o := \mathcal{K}_o^1 \oplus \mathcal{P}_o^1,$$

where $\mathcal{K}_o^1 := \text{span}\{A \otimes \mathbf{1} | A \in i\mathcal{I}_o^1\}$ and $\mathcal{P}_o^1 := \text{span}\{B \otimes \sigma_z | B \in i\mathcal{I}_e^1\}$. The procedure repeats itself with $N - 1$ replacing N, and so on. A scheme is summarized in Figure 9.6. At the end of the recursive procedure a decomposition

---

[21]Notice we might as well have replaced σ_z with σ_x (or σ_y) in (9.36) and accordingly $\sigma_{x,y}$ with $\sigma_{y,z}$ (or $\sigma_{x,z}$) in (9.37).

of the unitary evolution in elementary factors is obtained. This decomposition is in terms of tensor products of matrices. We refer to [72] for details and examples.

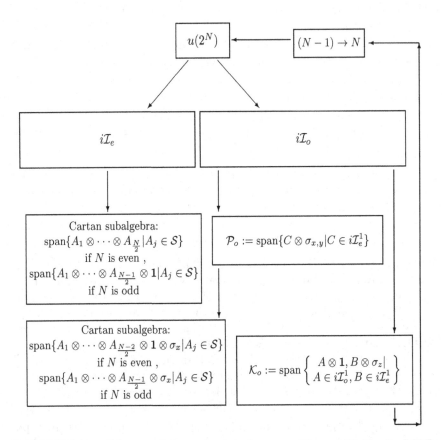

FIGURE 9.6 Scheme of the recursive decomposition of M. Dagli, J.H.D. Smith and the author [72]. The last iteration is with $N = 1$. The set $\mathcal{S}$ is defined in (9.35) and $\mathcal{I}_o^1$ and $\mathcal{I}_e^1$ are the same as $\mathcal{I}_o$ and $\mathcal{I}_e$ but with $N - 1$ factors in the tensor product rather than N. Also, notice that $\mathcal{K}_o$ is isomorphic to $u(2^{N-1})$.

Another recursive decomposition for N qubits was presented by N. Khaneja and S. Glaser in [162]. One starts with $su(2^N)$ and decompose it as

$$su(2^N) = \mathcal{K} \oplus \mathcal{P},$$

where

$$\mathcal{K} := \text{span}\{\sigma_z \otimes A, \mathbf{1} \otimes B, i\sigma_z \otimes \mathbf{1} | A, B \in su(2^{N-1})\},$$

$$\mathcal{P} := \text{span}\{\sigma_{x,y} \otimes A | A \in u(2^{N-1})\}.$$

It is easy to verify that this is a Cartan decomposition of $su(2^N)$. It is in fact an **AIII** type of decomposition with $p = q = 2^{N-1}$ (see subsection 6.3.4 and formula (6.32)) as the subalgebra $\mathcal{K}$ is spanned by all the elements of the form $\begin{pmatrix} C & 0 \\ 0 & D \end{pmatrix}$, $C, D \in u(2^{N-1})$, $\text{Tr}(C) + \text{Tr}(D) = 0$. Therefore, an element $X_f \in SU(2^N)$ can be decomposed in the form

$$X_f = K_1 A K_2, \tag{9.38}$$

with $K_1, K_2 \in e^{\mathcal{K}}$, and $A \in e^{\mathcal{A}}$ with $\mathcal{A}$ a Cartan subalgebra in $\mathcal{P}$. The dimension of $\mathcal{A}$ is the rank of the decomposition which, in this case, is 2^{N-1} ($= \min\{p, q\}$, cf. subsection 6.3.4). A maximal set of linearly independent commuting matrices in $\mathcal{P}$, spanning $\mathcal{A}$ can be obtained as $\sigma_x \otimes F$, where F is in a maximal set of linearly independent commuting matrices in $su(2^{N-1})$, for example diagonal matrices.

The next step of the procedure is to factorize the elements K_1 and K_2 in (9.38). To do this one performs a Cartan decomposition of $\mathcal{K}$,

$$\mathcal{K} := \mathcal{K}_1 \oplus \mathcal{P}_1,$$

where

$$\mathcal{K}_1 = \text{span}\{1 \otimes B | B \in su(2^{N-1})\},$$

$$\mathcal{P}_1 := \text{span}\{\sigma_z \otimes A | A \in u(2^{N-1})\}.$$

This is again a Cartan decomposition where $\mathcal{K}_1$ plays the role of the Lie subalgebra. The element K_1 (and analogously K_2) in (9.38) can be written as

$$K_1 = K_{1,1} \tilde{A} K_{1,2}.$$

Here $K_{1,1}$ and $K_{1,2}$ are in $e^{\mathcal{K}_1}$ and $\tilde{A} \in e^{\mathcal{A}_1}$, with $\mathcal{A}_1$ a maximal Abelian subalgebra in $\mathcal{P}_1$. Since $\mathcal{P}_1$ is spanned by matrices of the form $\begin{pmatrix} B & 0 \\ 0 & -B \end{pmatrix}$, with $B \in u(2^{N-1})$, a maximal Abelian subalgebra has dimension 2^{N-1}. It is spanned by matrices of the form $\sigma_z \otimes F$ where F is in a maximal commuting set in $u(2^{N-1})$, for example diagonal matrices. The key observation, at this point, is that $\mathcal{K}_1$ is the same as $su(2^{N-1})$, and therefore the procedure can be applied recursively starting from the decomposition of $K_{1,1}$, $K_{1,2}$. A scheme of the procedure for this decomposition is given in Figure 9.7.

In [162] it is shown how to choose the Cartan subalgebras so that, in the final decomposition, all the nonlocal transformations can be obtained by local transformations and two qubit interactions, which are more easily implemented than multi-body interactions. This fact makes this procedure appealing for the direct applications of decompositions in the generation of given evolutions. Of course, one could combine the techniques of [72] and [162] and

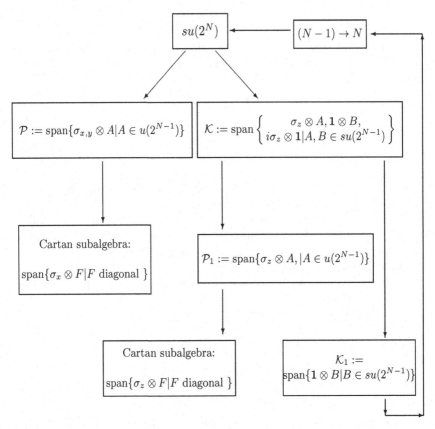

FIGURE 9.7: Procedure for the decomposition of $su(2^N)$ of Khaneja and Glaser [162]. The last iteration is for $N = 1$.

decompose some of the factors obtained with [72] with the technique of [162], or vice versa. More, in general these types of recursive decompositions are routinely used in the design of quantum circuits when one wants to obtain arbitrary unitary operations on quantum bits starting from local operations on the single qubits and a fixed set of operations on two qubits (see, e.g., [200]).

The recursive decompositions in [72], [162] only deal with multipartite systems of qubits. For the case of systems of arbitrary dimensions, a recursive decomposition was given by R. Romano and the author in [86] for a bipartite system. Such a decomposition uses an odd-even decomposition at the first step and recursive decompositions of $so(n)$ in the following recursive procedure.

9.3 Local Equivalence of States

A problem of great interest in entanglement theory is to find methods to determine whether or not two states are **locally equivalent**. This means that it is possible to transfer between the two states by only local transformations. Any measure of entanglement will give then the same value on the two states. This problem can be seen as a special case of a general controllability problem (cf. Chapter 3). Given two density matrices ρ_1 and ρ_2 and a (dynamical) Lie algebra $\mathcal{L}$, how can we check whether ρ_1 is in the orbit (cf. (3.35))

$$\mathcal{O}_\mathcal{L}(\rho_2) := \{X\rho_2 X^\dagger | X \in e^\mathcal{L}\} \ ? \tag{9.39}$$

In the case where $\mathcal{L}$ is the direct sum of Lie algebras $su(n_j)$, where n_j is the dimension of the j-th subsystem in a multipartite system, we have the problem of local equivalence of states. This problem has a vast literature (e.g., [15], [17], [121], [183], [190], [226]), and to a large extent is still an open problem.

9.3.1 General considerations on dimensions

The space of density matrices for a general multipartite system with dimensions $n_1,...,n_N$ is a manifold with boundary with dimension $(n_1 n_2 \cdots n_N)^2 - 1$, i.e., the dimension of the space of Hermitian matrices modulo the constraint $\text{Tr}(\rho) = 1$. Any orbit (9.39) (with $\mathcal{L} = \bar{\oplus}_{j=1}^N su(n_j)$) is a sub-manifold. The structure of the orbits in the set of density matrices is, in general, quite complicated although some low-dimensional cases have been studied in detail [173]. The dimension of an orbit $\mathcal{O}_\mathcal{L}(\rho_2)$ in (9.39) can be calculated explicitly by noticing (cf. section 3.3.4) that $\mathcal{O}_\mathcal{L}(\rho_2)$ is diffeomorphic to $e^\mathcal{L}/e^{\mathcal{C}_{\rho_2} \cap \mathcal{L}}$ where $\mathcal{C}_{\rho_2}$ is the subspace of matrices in $u(n_1 n_2 \cdots n_N)$ which commute with ρ_2. The tangent space of $\mathcal{O}_\mathcal{L}(\rho_2)$ at ρ_2 can be identified with $[i\rho_2, \mathcal{L}]$ and therefore

$$\dim \mathcal{O}_\mathcal{L}(\rho_2) = \dim[i\rho_2, \mathcal{L}] = \dim \mathcal{L} - \dim \mathcal{C}_{\rho_2} \cap \mathcal{L}.$$

If we consider the linear operator $ad_{i\rho_2}$ acting on $\mathcal{L}$ as $ad_{i\rho_2}(A) := [i\rho_2, A]$, $\mathcal{C}_{\rho_2} \cap \mathcal{L} = \ker(ad_{i\rho_2})$. For example, if $\rho_2 = \frac{1}{n_1 n_2 \cdots n_N} \mathbf{1}$, $\ker(ad_{i\rho_2}) = \mathcal{L}$, and the number on the right-hand side is zero. This is the generic case. In fact, we have the following proposition [183].

Proposition 9.3.1 Assume the number of subsystems N is ≥ 2. For almost every density matrix ρ_2, $\ker(ad_{i\rho_2}) = \{0\}$. Therefore, since $\mathcal{L}$ is equal to the direct sum of $su(n_j)$, $j = 1, ..., N$, for almost every ρ_2 the dimension of the orbit (9.39) is $\sum_{j=1}^N n_j^2 - N$.

We shall not completely prove this result here and refer to [183] which uses a somewhat different language. However, we shall indicate an alternative proof and illustrate this proof for the 2×2 case.[22]

We have to prove that, except for a set of measure zero of density matrices, the rank of the linear transformation $ad_{i\rho_2}$ is equal to $\dim \mathcal{L}$. Since the determinant is an analytic function, the set of its zeros is a set of measure zero. Therefore, the rank of the matrix corresponding to $ad_{i\rho_2}$ is maximum except at a set of measure zero. The proof is obtained if we find a density matrix ρ_2 for which $rank(ad_{i\rho_2}) = \sum_{j=1}^{N} n_j^2 - N$, which is the maximum possible, given the dimension of the Lie algebra $\mathcal{L}$. Consider for example the case of two qubits. Recall the definition of the Pauli matrices $\sigma_{x,y,z}$ in (1.20). For α sufficiently small the matrix

$$\rho_2 := \frac{1}{4}\mathbf{1} + \alpha\sigma_z \otimes \mathbf{1} + \alpha\mathbf{1} \otimes \sigma_z + \alpha\sigma_z \otimes \sigma_z$$

is a density matrix, i.e., it has trace 1, it is Hermitian, and positive semidefinite. If we choose as basis in $\mathcal{L}$, $\{i\sigma_{x,y,z} \otimes \mathbf{1}, i\mathbf{1} \otimes \sigma_{x,y,z}\}$, in that order and as basis in $su(4)$ $\{i\sigma_{x,y,z} \otimes \mathbf{1}, i\mathbf{1} \otimes \sigma_{x,y,z}, i\sigma_x \otimes \sigma_{x,y,z}, i\sigma_y \otimes \sigma_{x,y,z}, i\sigma_z \otimes \sigma_{x,y,z}\}$, in that order, the matrix $ad_{i\rho_2}$ is given by

$$ad_{i\rho_2} := \begin{pmatrix}
0 & -\alpha & 0 & 0 & 0 & 0 \\
\alpha & 0 & 0 & 0 & 0 & 0 \\
0 & 0 & 0 & 0 & 0 & 0 \\
0 & 0 & 0 & 0 & -\alpha & 0 \\
0 & 0 & 0 & \alpha & 0 & 0 \\
0 & 0 & 0 & 0 & 0 & 0 \\
0 & 0 & 0 & 0 & 0 & 0 \\
0 & 0 & 0 & 0 & 0 & -\alpha \\
0 & 0 & 0 & 0 & \alpha & 0 \\
0 & 0 & -\alpha & 0 & 0 & 0 \\
0 & 0 & 0 & 0 & 0 & 0 \\
0 & 0 & 0 & 0 & 0 & 0 \\
0 & \alpha & 0 & 0 & 0 & 0 \\
0 & 0 & 0 & 0 & 0 & 0 \\
0 & 0 & 0 & 0 & 0 & 0
\end{pmatrix},$$

which has rank 6 for $\alpha \neq 0$.

We notice that the above considerations already suggest a test for local equivalence of states. In fact, if two states ρ_1 and ρ_2 are equivalent the corresponding orbits $\mathcal{O}_\mathcal{L}(\rho_1)$ and $\mathcal{O}_\mathcal{L}(\rho_2)$ coincide by definition and therefore they have the same dimensions. If we add this to the obvious requirement that ρ_1 and ρ_2 must have the same spectrum, we have

---

[22]The dimension of the generic orbit under local transformations for the case of *pure* states is calculated in [58].

Proposition 9.3.2 If two density matrices ρ_1 and ρ_2 are locally equivalent, then

1. They have the same spectrum

2. $\dim \ker(ad_{i\rho_1}) = \dim \ker(ad_{i\rho_2})$.

Proposition 9.3.2 then reduces the problem to checking local equivalence for iso-spectral density matrices whose orbits have the same dimension. Given the dimension of a *generic* orbit as above, there will be $\sum_{j=1}^{N} n_j^2 - N$ parameters to identify the position of a density matrix on that orbit and

$$D := (n_1^2 n_2^2 \cdots n_N^2 - 1) - (\sum_{j=1}^{N} n_j^2 - N) \tag{9.40}$$

parameters to identify the orbit itself. Therefore, we will need to check at least D parameters to verify that two density matrices are on the same orbit, assuming that they are on orbits with maximal dimension. If they are on orbits with smaller dimensions we shall need to verify a larger number of parameters.

9.3.2 Invariants and polynomial invariants

Much of the study of the problem of local equivalence of states has focused on the search for local invariants. Let $\mathcal{B}$ denote the set of density matrices for a general multipartite system and $\mathcal{L}$, as above, be the Lie algebra corresponding to the Lie group of local transformations. A function $f : \mathcal{B} \to \mathbf{R}$ (or $\mathbb{C}$) is called a **local invariant** if for every $\rho \in \mathcal{B}$ and $X \in e^{\mathcal{L}}$

$$f(\rho) = f(X\rho X^\dagger).$$

If f is a local invariant, for every two local equivalent matrices ρ_1 and ρ_2, $f(\rho_1) = f(\rho_2)$, in particular every measure of entanglement must be a local invariant. An example of a local invariant for a system of N qubits is the N-qubits concurrence $C_{\Theta_{o-e}}$ of Remark 9.2.7. To distinguish two density matrices which are on generic (maximal dimension) orbits one needs (at least) a number D in (9.40) of local invariants. A set of local invariants $\mathcal{S} := \{f_1, ..., f_m\}$ is *redundant* if there are two different subsets of $\mathcal{S}$, $\mathcal{S}_1$ and $\mathcal{S}_2$, such that the fact that $\mathcal{S}_1$ is a set of invariants implies that $\mathcal{S}_2$ is also a set of invariants. If it is not redundant it is called *independent*. A set of invariants $\mathcal{S} := \{f_1, ..., f_m\}$ is called **complete** when ρ_1 and ρ_2 are locally equivalent *if and only if* $f_j(\rho_1) = f_j(\rho_2)$, $j = 1, ..., m$. One would like to be able to determine the local equivalence of density matrices using the minimum possible number of invariants and therefore a set of independent invariants. The crucial problem is, however, to provide a *complete* set of invariants.

Among functions, polynomials have remarkable algebraic properties. In particular, invariant polynomials with values in $\mathbf{R}$ or $\mathbb{C}$ form an algebra over

R or $\mathbb{C}$ since the linear combination or product of invariant polynomials is still an invariant polynomial. It follows from results in invariant theory [258] that the algebra of invariant polynomials is generated by a *finite* set of homogeneous invariant polynomials.[23] Therefore, the task is to find a set of homogeneous invariant polynomials. Examples are given by the coefficients of the characteristic polynomial of the matrix ρ.

In [121], [226] a method is described to construct systematically all the homogeneous invariant polynomials of degree $1, 2, \ldots$ and so on. Also methods are described to eliminate, in the process, some polynomials which are not independent. Notice in this regard that if p_1 and p_2 are homogeneous invariant polynomials of degree d_1 and d_2, $p_1 p_2$ is a homogeneous invariant polynomial of degree $d_1 + d_2$. Let $\bar{n}$ be the dimension of the total system $\bar{n} := n_1 n_2 \cdots n_r$. The starting point of the procedure of [121], [226] for the construction of polynomial invariants is the observation that, to every homogeneous polynomial $p = p(\rho)$ of degree k, there corresponds an $\bar{n}^k \times \bar{n}^k$ matrix F, such that

$$p(\rho) = \text{Tr}(F\rho^{\otimes k}), \qquad (9.41)$$

where $\rho^{\otimes k}$ denotes the tensor product of k copies of ρ. Moreover, p is an invariant polynomial if and only if (9.41) is satisfied for an F which commutes with $(e^{\mathcal{L}})^{\otimes k}$. In order to see this, assume (9.41) holds for some F which commutes with $(e^{\mathcal{L}})^{\otimes k}$. We have, with $X \in e^{\mathcal{L}}$,

$$p(X\rho X^\dagger) = \text{Tr}(F(X\rho X^\dagger)^{\otimes k}) = \text{Tr}(FX^{\otimes k}\rho^{\otimes k}(X^\dagger)^{\otimes k}) =$$

$$\text{Tr}((X^\dagger)^{\otimes k}FX^{\otimes k}\rho^{\otimes k}) = \text{Tr}(F\rho^{\otimes k}) = p(\rho),$$

where we used the commuting property of F. This shows that p is an invariant polynomial. Conversely, assume that p is an invariant polynomial and write p as in (9.41). From the invariance property of p, we have

$$p(\rho) = \text{Tr}(U^\dagger F U \rho^{\otimes k}), \qquad (9.42)$$

for every $U \in (e^{\mathcal{L}})^{\otimes k}$. If $d\mu$ denotes the invariant measures on the compact Lie group $(e^{\mathcal{L}})^{\otimes k}$ (see, e.g., [201], [293]) then we can integrate (9.42) with this measure over the compact Lie group $(e^{\mathcal{L}})^{\otimes k}$ and write

$$p(\rho) = \text{Tr}(\tilde{F}\rho^{\otimes k}),$$

with

$$\tilde{F} := \int_{(e^{\mathcal{L}})^{\otimes k}} U^\dagger F U d\mu.$$

Because of the invariance of the measure $d\mu$, $\tilde{F}$ commutes with all of $(e^{\mathcal{L}})^{\otimes k}$.

---

[23]That is, every invariant polynomial is obtained by linear combination and-or products of a finite set of homogeneous invariant polynomials.

The next observation is that there is no loss of generality in looking for invariant homogeneous polynomial p such that the corresponding matrix F is Hermitian. In fact, given that F commutes with every $U \in (e^{\mathcal{L}})^{\otimes k}$, we also have

$$U(F + F^\dagger)U^\dagger = F + F^\dagger,$$

and

$$U(iF - iF^\dagger)U^\dagger = iF - iF^\dagger.$$

Therefore, there are two Hermitian (linearly independent) matrices $M_1 := F + F^\dagger$ and $M_2 := iF - iF^\dagger$ which correspond to homogeneous invariant polynomials of degree k. Notice, moreover, that these polynomials are real.

The last observation is that if (and only if) the $\bar{n}^k \times \bar{n}^k$ Hermitian matrix M commutes with the elements of the Lie group $e^{\mathcal{L}^{\otimes k}}$, then it commutes with all the elements of the Lie algebra corresponding to $e^{\mathcal{L}^{\otimes k}}$. Such a Lie algebra is a $\bar{n}^k$ representation of $\mathcal{L}$. It is spanned by matrices of the form

$$G := A \otimes 1 \otimes \cdots \otimes 1 + 1 \otimes A \otimes 1 \otimes \cdots \otimes 1 + \cdots + 1 \otimes \cdots \otimes 1 \otimes A, \quad (9.43)$$

with A in $\mathcal{L}$, where the tensor products are taken k times.

In conclusion, one has the following procedure to calculate all the homogeneous polynomial invariants of degree k: First, one calculates a basis of $\mathcal{L}$ and therefore a maximal set of linearly independent matrices G in (9.43). Then one finds all the Hermitian, $\bar{n}^k \times \bar{n}^k$, matrices which commute with all the elements G. This involves the solution of a linear system of equations.

Although the above procedure only involves eventually linear algebra operations and allows us to calculate all the homogeneous invariant polynomials of a degree k, the dimension of the system of equations grows exponentially with k. Moreover, the procedure does not take into account the fact some homogeneous polynomials of degree k can be obtained from polynomials of degree smaller than k. Therefore, their computation is a waste of computational resources. We refer to [121] and [226] for a treatment of these problems.

Finally, there is always the problem, once a certain number of invariants has been found, to show that they form a complete set of invariants.

9.3.3 Some solved cases

For some particular situations, a complete set of local invariants have been found. These invariants allow us to decide whether two density matrices are locally equivalent. They are not necessarily polynomials. We mention here some cases presented in the literature without going into details of the specific results. In [190] the case of two qubits is solved and a complete set of 18 invariants is found. The case of $N \geq 2$ qubits is considered in [183] and generic (i.e., maximal dimensional) orbits are considered. In this case the invariants are the parameters of a canonical point on the orbit. These invariants allow us to distinguish density matrices belonging to two different generic orbits. The case of three qubits is considered in detail in [183] and generalizations are indicated for a larger number of qubits. A complete set of invariants is found for bipartite generic states in [17] while a complete set of invariants for a class of tripartite pure states is given in [15]. A special class of bipartite quantum systems is also considered in [16] while nondegenerate tripartite mixed states are considered in [114] using a method developed in [108].

9.4   Notes and References

In this chapter, we have given some of the main ideas concerning the quantum theory of entanglement from the point of view of the properties of states and their dynamics. There are several review papers (see, e.g., [137], [140], [216], [271], [300]) on quantum entanglement and textbooks (see, e.g., [167], [191], [207], [219]). We refer to these references for more details. These references also present a more operational point of view in the sense that they consider quantum entanglement as a resource in quantum communication, teleportation and computation. The paper [145] gives a treatment of the problem of entanglement detection from a computational perspective. A neighboring field is the study of quantum states and their classification [32]. It must be said that the theory is in rapid progress and several mathematical problems have received only a partial solution. In our treatment we have illustrated three interrelated mathematical problems:

- The problem of finding tests to check whether the state of a multipartite system is entangled or not.

- The problem of defining meaningful and useful measures of entanglement.

- The problem of testing whether it is possible to transfer a quantum state from one value to another by only local transformations.

In the study of entanglement several new concepts are being introduced and relations with other notions are constantly discovered. As for controllability, a concept of *indirect controllability* was introduced in [236], [237]. This concerns the possibility of driving the state of a system by manipulating the state of an ancillary system which then interacts with the target system. In this scheme of control, as opposed to the schemes treated in this book, the evolution is fixed and determined by a (total) Hamiltonian H, while the control degree of freedom is the initial state of the ancillary system. If H only gives local evolutions, clearly there is no possibility of controlling the target system with this method. In general, it is reasonable to expect that the controllability properties of this scheme will depend on the entangling properties of the Hamiltonian H. In fact, in [236] an explicit relation between controllability and entanglement was discovered in the simplest case where both target system and ancillary system are two-level systems.

The three entanglement criteria described in subsection 9.1.2 were among the strongest available at the moment of writing of the first edition of this book (November 2006). For example, the CCN criterion was proved in [240] to be not weaker than the reduction criterion in [138]. However, the list is not complete. More recent references on this topic are [4], [128], [182], [262].

Our main references have been [216] for measures of entanglement and [300], in particular for entanglement of formation and concurrences, to which we refer for further references. The main reference for the OED Decomposition was [81] and for the treatment of local invariant polynomial [121].

We believe that a treatment of entanglement based on dynamics, i.e., on the study of the unitary group action on the set of density matrices, along with application of the results on the structure of the unitary group is very promising. We have seen some evidence of this in the treatment of concurrences and generalized concurrences in subsections 9.2.1 and 9.2.2.

9.5 Exercises

Exercise 9.1 Prove Proposition 9.1.3. Use the last part of this proposition to give an algorithm in terms of matrix elements for the calculation of the partial trace in the case of two spin $\frac{1}{2}$ systems, where, for each subsystem, we consider the natural basis $\left\{ \begin{pmatrix} 1 \\ 0 \end{pmatrix}, \begin{pmatrix} 0 \\ 1 \end{pmatrix} \right\}$.

Exercise 9.2 Prove Proposition 9.1.4.

Exercise 9.3 Analyze the entanglement of the two qubits system in Example 9.1.12 using the PPT criterion of Theorem 9.1.7.

Exercise 9.4 Prove formula (9.17). Use this formula to prove that the entropy of the partial trace is a monotonically increasing function of the concurrence which has its minimum (maximum) at zero (one) and that the minimum (maximum) is equal to zero (one).

Exercise 9.5 Prove that $\tilde{\theta}_{II} \otimes \tilde{\theta}_{II}(\rho) = J \otimes J \bar{\rho} J \otimes J$ and that all the eigenvalues of $\rho \tilde{\theta}_{II} \otimes \tilde{\theta}_{II}(\rho)$ are real and nonnegative.

Exercise 9.6 Prove that (9.24) follows from (9.23) (Hint: can use contradiction by assuming that one quantity is strictly larger than the other in (9.24). Use (9.20)).

Exercise 9.7 Notice that the procedure described in subsection 9.3.2 to generate homogeneous invariant polynomials does not depend on the nature of the Lie algebra $\mathcal{L}$ (which in the case treated was the Lie algebra corresponding to local transformation on a multipartite system). Consider a two-level system and calculate all the homogeneous polynomials of degree 1 and 2 which are invariant with respect to the Lie sub-algebra $so(2)$ of $su(2)$.

Exercise 9.8 Reconsider the linear map (9.11) but without the $\tilde{\theta}_{II}(A)$ terms, showing that it is a positive map. The reduction criterion [138] [271] (cf. Theorem 9.1.10) says that if ρ is separable $(\Phi \otimes \mathbf{1})(\rho) \geq 0$ and $(\mathbf{1} \otimes \Phi)(\rho) \geq 0$. From these deduce the two necessary conditions of separability of the reduction criterion

$$\mathbf{1} \otimes Tr_S(\rho) - \rho \geq 0, \qquad Tr_P(\rho) \otimes \mathbf{1} - \rho \geq 0.$$

Chapter 10

Applications of Quantum Control and Dynamics

In this chapter, we shall discuss some practical applications of quantum control and dynamics. Our goal is not to give an exhaustive introduction to areas of application, but to point out the role of control and dynamics so as to further motivate the field as well as the techniques described in the previous chapters. The basics of mathematical modeling of quantum mechanical control systems were described in Chapter 2. Here we focus on more practical aspects in the areas of nuclear magnetic resonance and control of molecular and atomic systems. We also discuss a possible implementation of quantum information processing, namely the one with trapped ions. The dynamics here is used to implement quantum logic gates. We shall show how, changing the physical set-up, it is possible to create different Hamiltonians. The control is obtained by switching among these different Hamiltonians.

10.1   Nuclear Magnetic Resonance Experiments

10.1.1   Basics of NMR

The basic set-up of an experiment in nuclear magnetic resonance (NMR) consists of an ensemble of spin-$\frac{1}{2}$ particles in a magnetic field $\vec{B}$, which is constant and pointing in a fixed direction, say the z-direction (cf. Figure 10.1). If all the other degrees of freedom are neglected, the energy of one of the spin-$\frac{1}{2}$ particles is proportional to the dot product between the spin angular momentum and the magnetic field (cf. subsection 2.4.1).

Liouville's equation describing the dynamics of this system is given in (2.62) of Chapter 2. In this case, ($B_x = 0$, $B_y = 0$), this equation takes the form[1]

$$i\frac{d}{dt}\rho = \gamma\frac{1}{2}[\sigma_z B_z, \rho].$$

---

[1] For simplicity, we are using units such that $\hbar = 1$. We also drop the index e in $\vec{B}_e$ used in Chapter 2 to denote an 'external' field.

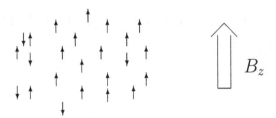

FIGURE 10.1: The initial equilibrium set-up of an NMR experiment consists of an ensemble of spin $\frac{1}{2}$ particles in a constant magnetic field B_z along the z direction. Most of the spins are aligned in the lowest energy configuration in the same direction as the field. Some of them are aligned in the opposite direction.

This equation has, as equilibrium points, the density matrices commuting with σ_z. For a single spin, $\rho_+ := \begin{pmatrix} 1 & 0 \\ 0 & 0 \end{pmatrix}$, $\rho_- := \begin{pmatrix} 0 & 0 \\ 0 & 1 \end{pmatrix}$ correspond to spin angular momentum aligned in the same direction as the field, the z direction, or in the opposite direction, respectively. These correspond to two values of the energy which are given by $\pm \frac{\gamma}{2} B_z$. In the absence of any other external field, the ensemble of spin particles will be in one of the two equilibrium states, with a majority in the state of lower energy, which corresponds to spin in the same direction as the magnetic field (while the state with higher energy corresponds to the spin in the opposite direction). Denoting by N_L the number of spins in the lower energy state and by N_H the number of spins in the higher energy value, the density matrix describing the ensemble will be

$$\rho_{EQ} := \begin{pmatrix} \frac{N_L}{N_H+N_L} & 0 \\ 0 & \frac{N_H}{N_H+N_L} \end{pmatrix}. \tag{10.1}$$

The ratio between the number N_L of spins in the lower energy state and the number N_H of spins in the higher energy state can be calculated using *Maxwell-Boltzmann statistics*. It is given by

$$\frac{N_H}{N_L} = e^{-\frac{\Delta}{kT}},$$

where $\Delta := |\gamma B_z|$ is the difference between the two values of the energy, k is the Boltzmann constant and T the absolute temperature. This ratio is always less than one although at room temperature it is almost one. As the temperature decreases, there are few spins with the high energy value.

In NMR experiments, the spin system is perturbed from the equilibrium state using the x and y components of the magnetic field (B_x and B_y in equation (2.62)) as controls. Once the state ρ is in a nonequilibrium position ρ_0 and the x and y components of the magnetic field have been switched back to zero, ρ will evolve according to

$$\rho(t) = e^{-i\frac{\omega_0}{2}\sigma_z t}\rho_0 e^{i\frac{\omega_0}{2}\sigma_z t}, \tag{10.2}$$

where the *Larmor frequency* ω_0 is given by $\omega_0 := \gamma B_z$. Define a *magnetization vector*, $\vec{M} := M_x\vec{i} + M_y\vec{j} + M_z\vec{k}$, with $M_{x,y,z} := \mathrm{Tr}(\sigma_{x,y,z}\rho)$. According to (10.2), the vector $\vec{M}$ will revolve around the z axis in a periodic motion with frequency given by the Larmor frequency ω_0. Such a motion of the spin system is called **precession**.

The system of spins has to be regarded as the quantum mechanical version of a magnet whose associated magnetic field is the magnetization vector $\vec{M}$. According to Faraday's induction law (2.6) the precession motion will induce a current in a nearby coil.[2] The signal detected is a sinusoid with magnitude proportional to the component of $\vec{M}(0)$ in the plane determined by the normal to the coil inner surface; see Figure 10.2. The frequency of the sinusoid is given by the Larmor frequency.

In reality, the signal detected by the coil is a damped sinusoid (rather than a perfect sinusoid) of the type in Figure 10.3. This signal is called *Free Induction Decay* (FID). The damping is due to the *relaxation* of the spin, that is, the interaction of the spin with its surroundings. The relaxation process eventually takes the spin system back to its original state ρ_{EQ} in (10.1).[3]

The FID signal is Fourier transformed and shows a peak at the Larmor frequency ω_0. This Larmor frequency depends on the type of spin particle

---

[2]In order to see this consider the integral version of Faraday's law, i.e., the one obtained from (2.6) by taking the surface integral on the inner surface of the coil S_c. This gives

$$\int_{S_c} (\nabla \times \vec{E}) \cdot \vec{n}dS = -\frac{\partial}{\partial t}\int_{S_c} \vec{B} \cdot \vec{n}dS,$$

and applying Stokes' theorem of vector calculus we transform the first integral into a line integral of the electric field along the border of S_c, ∂S_c. We have

$$\int_{\partial S_c} \vec{E} \cdot d\vec{l} = -\frac{\partial}{\partial t}\int_{S_c} \vec{B} \cdot \vec{n}dS.$$

The electric field along the coil produces the current.

[3]A phenomenological analysis of the relaxation process [37], [55] shows that it is mainly due to two processes: The *relaxation spin-lattice* is due to the fluctuations of the magnetic field in the lattice and takes the value of M_z back to its equilibrium value. The *spin-spin relaxation* is due to interactions with other spins as well as inhomogeneity of the field B_z, which causes the particles to rotate at slightly different Larmor frequencies. It is the process responsible for making the transversal components of the magnetization M_x and M_y go to zero. The characteristic times of the relaxation process depend on the external environment of the spin particle.

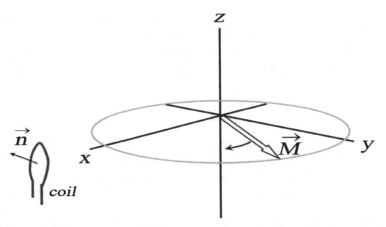

FIGURE 10.2: In NMR experiments, a precessing magnetization vector induces a sinusoidal current in a coil. This is a consequence of Faraday's induction law (2.6).

which has produced the FID signal. For example, it is in ratio 10 : 1 : 2.5 for the nuclei of the Hydrogen isotope ^{1}H, the Nitrogen isotope ^{15}N and the Carbon ^{13}C, respectively. Therefore, the evaluation of the Larmor frequency allows one to determine the type of nucleus that has generated the signal and therefore it is a tool for the determination of the structure of molecules.[4]

The motion of the electrons around the nucleus creates a magnetic field which (in most cases) opposes the constant external magnetic field. This magnetic field will be different according to the environment of the nucleus under consideration and the chemical bonds with the neighboring nuclei. It will result in a change in the Larmor frequency which is called *chemical shift*. The Larmor frequency is always proportional to the external field in the z direction, but the proportionality constant changes because of the chemical shift. Thanks to the chemical shift, it is possible to distinguish nuclei of the

---

[4]This mode of operation is sometimes called *Fourier spectroscopy* NMR. Another method of operation is *continuous wave NMR*. There are two ways to perform the NMR experiments in continuous wave NMR. In one of them, in addition to the magnetic field in the z-direction which is kept constant, the sample is subject to a rotating electromagnetic field in the x-y plane for a certain time. The frequency of this field is varied. At a frequency equal to the Larmor frequency the magnetization in the z-direction will reverse direction. In terms of energy, the magnetic field will induce population transfer from the lower to the higher energy level (cf. section (8.1)). The frequency at which this state transfer occurs gives information on the Larmor frequency of the spin particle. In another scheme, the magnetic field in the z-direction is varied while the frequency of the transversal field is kept constant. Analogously to the previous case a transition will occur when the frequency of the incoming field is equal to the Larmor frequency. Recall from Chapter 7 and in particular section 7.3 that a rotating field in the x-y plane, (7.46), (7.47), at the resonance frequency, optimizes the energy necessary for the given state transfer.

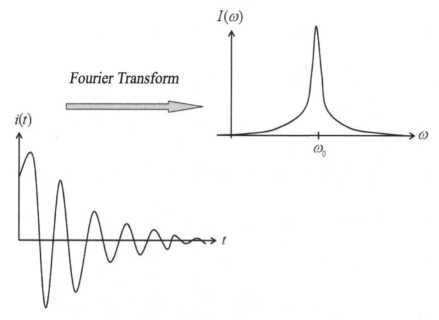

FIGURE 10.3: Free induction decay signal $i = i(t)$ in the time domain and its Fourier transform $I(\omega)$.

same type but in different positions in a given molecule and therefore to assign various peaks in a spectrum to the various nuclei in a molecule. This process is called **frequency labeling**.

A different way of looking at an NMR experiment is in terms of energy levels and transitions among different levels. A spin-$\frac{1}{2}$ particle can be in two distinguished states with energy $E = \pm\frac{\omega_0}{2}$. When stimulated with an external time-varying field the system undergoes some population transfer from the lower to the higher energy level. The energy so accumulated is then emitted during the precession-relaxation process and therefore the frequency detected in the Fourier analysis corresponds to the difference in energy levels. This picture extends to the case of several noninteracting spin particles in that every peak in the spectrum, corresponding to a precession-relaxation process of a single spin, is associated to a transition frequency between two different energy levels in the energy diagram. This picture is particularly useful in the case where the spin particles cannot be considered isolated and the coupling among them is not negligible. The principle that the lines in the spectrum correspond to transitions among different energy levels still holds and from these transitions it is possible to study the structure of a molecule. A typical energy diagram and spectrum for a two spin systems is presented in Figure 10.4.

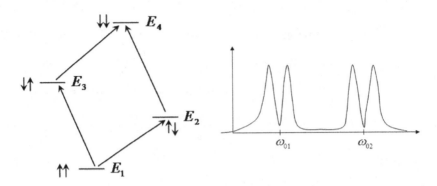

FIGURE 10.4: Energy diagram for a system of two spins with different Larmor frequencies $\omega_{0,1}$ and $\omega_{0,2}$ in the presence of a weak coupling. The transition from the level E_1 to the level E_2 as well as the transition from E_3 to E_4 correspond to a 'flip' of the second spin. In absence of coupling we would have $E_4 - E_3 = E_2 - E_1 = \omega_{0,2}$. This would correspond to a peak in the energy spectrum at $\omega_{0,2}$. In the presence of a weak coupling between the two spins this peak slightly splits to form a doublet centered at $\omega_{0,2}$ whose width depends on the size of the coupling (which is (in the weak coupling case) assumed much smaller in magnitude than the Larmor frequency). A similar doublet is formed around the Larmor frequency $\omega_{0,1}$ corresponding to transitions involving the first spin.

10.1.2   Two-dimensional NMR

In principle, in NMR experiments, one can analyze the geometry of a molecule by exciting one nucleus and studying the effect on neighboring nuclei. This assumes that the frequency labeling is performed accurately. For NMR of large proteins, [59], [302], the spectra are typically poorly resolved and the frequency labeling process is quite difficult if not impossible. This is due both to the larger number of nuclei, which make the spectrum overly crowded, and to the fact that larger relaxation rates make the peaks wider in the Fourier transform of the FID signal. A way to overcome this problem is to use two-dimensional NMR spectroscopy in which two nuclei are distinguished by studying their interaction with a third spin. A typical 2-*D* NMR experiment follows the scheme of Figure 10.5.

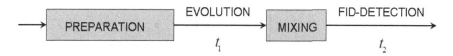

FIGURE 10.5: Various stages of a two-dimensional NMR experiment.

In the period called *preparation*, a sample is perturbed from equilibrium with one or more control pulses which may cause the spin of one of two nuclei, say spin 1, to transfer from being aligned along the z-axis to a position in the x-y plane. During the *evolution* time t_1 the system evolves for example by allowing the magnetization of spin 1 $M_{x,y,z} = \text{Tr}(\sigma_{x,y,z}\rho)$ to precess in the x-y plane according to (10.2). During the *mixing* period more control pulses are applied in order to modify the state of spin 2 by interaction with spin 1. This process is sometimes called *polarization transfer* or *magnetization transfer*. After the mixing period, the FID signal $f(t_2)$ is detected as a function of the time t_2 elapsed from the end of the mixing period. This signal will, however, depend on the time t_1, i.e., it will be $f(t_2) := f_{t_1}(t_2)$. The experiment is repeated for different values of t_1, starting with $t_1 = 0$ and with small increments of t_1. Typically between 50 and 500 repetitions are performed so as to obtain a function $s(t_1, t_2) := f_{t_1}(t_2)$. This function is then Fourier transformed with respect to t_1 and t_2, with a two-dimensional Fourier transform

$$S(\omega_1, \omega_2) := \int_{-\infty}^{\infty} \int_{-\infty}^{\infty} s(t_1, t_2) e^{-i2\pi(t_1\omega_1 + t_2\omega_2)} \, dt_1 \, dt_2.$$

The function $S(\omega_1, \omega_2)$ is represented as a two-dimensional function, in particular as a contour plot, and it will present various (two-dimensional) peaks. A peak at a frequency pair $(\omega_{01}, \omega_{02})$ is interpreted as due to a portion of the magnetization for spin 1 which was precessing with Larmor frequency ω_{01} which is transferred to spin 2 and therefore precesses at Larmor frequency ω_{02}. Therefore, the mixing time has the role of transferring signal from one spin to another. Part A of Figure 10.6 shows two peaks in a two-dimensional spectrum, one diagonal peak at $(\omega_{01}, \omega_{01})$ corresponds to signal which is not transferred from one spin to the other while the cross peak $(\omega_{01}, \omega_{02})$ corresponds to a signal transferred to a different frequency.

Using 2-D NMR it is possible to distinguish spins that are indistinguishable in conventional NMR because their Larmor frequencies are too similar, say $\approx \omega_{01}$. If they interact with two different nuclei having different Larmor frequencies equal to ω_{02}^A and ω_{02}^B, respectively, the two-dimensional Fourier

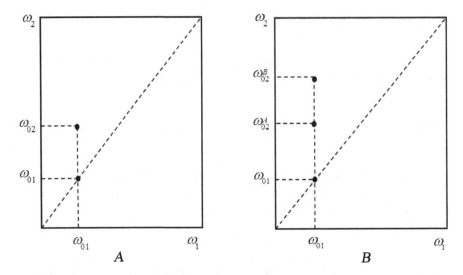

FIGURE 10.6:   Plots in 2-D NMR spectroscopy. In part A two peaks reveal that part of the signal has been transferred during the mixing time to a second spin. In part B two off diagonal peaks allow to distinguish two different spins with approximately equal Larmor frequency.

transform will have plots as in Part B of Figure 10.6, which allow to distinguish the two spins.

10.1.3   Control problems in NMR

Already in basic NMR spectroscopy several manipulation tasks of nuclear spins can be studied from the point of view of control theory. For example, assume we want to maximize the signal to noise ratio in the detection of the FID signal. Given a certain orientation for the coil detecting the *FID* signal, we require that the component of the magnetization vector in the plane orthogonal to the coil be maximum at the beginning of the precession. In fact, we have seen that the amplitude of the sinusoidal signal detected is proportional to such a component. This is a control problem of choosing the B_x and B_y functions in Liouville's equation (2.62) in the period preceding the precession.

There are many important control problems in 2-*D* NMR concerning the construction of control pulses to transfer magnetization from one spin to the other. Methods to obtain this by minimizing the effects of relaxation, maximizing the efficiency of transfer, and minimizing the time have been only recently rigorously tackled using tools of control theory. We refer in particular the work of N. Khaneja and co-workers ([164], [165], and references therein) which also treat specific applications to structure determination for large proteins. The mathematical models in the case of closed systems are bilinear control systems.

10.2   Molecular Systems Control

The mathematical modeling of quantum mechanical molecular and atomic control systems was discussed in Chapter 2. Our goal here is to indicate applications as well as to describe practical issues related to the laboratory implementation.

10.2.1   Pulse shaping

Molecular and atomic quantum control is achieved through appropriately shaped laser pulses. A laser emits electromagnetic radiation which may have different shapes. These shapes can be modified using *pulse shaping* techniques [294]. Modern pulse shapers use optical materials whose refractive indexes depend (in a nonlinear fashion) on the frequency and/or amplitude of the light which crosses them. By placing in parallel and-or in sequence several optical materials with different refractive indexes and-or refocusing the light in a given point, one can create various shapes, which match the controls designed according to the techniques described in the previous chapters. An example of a pulse shaper is given in [292]. This shaper is made of 128 pixels whose refractive index can be changed by modifying the voltages applied to each of them. The pixels form a panel and they are connected to a computer which has encoded the control algorithm to be applied. The scheme of [292] is particularly useful in cases where the control law is changed in different experiments because it allows one to obtain different control laws with the same set-up.

In principle, there is no restriction to the light shape which can be obtained with pulse shaping except that the electromagnetic field, once treated itself as a quantum mechanical system, has to obey the Heisenberg uncertainty relation for any two noncommuting observables (cf. 1.2.2.3). For a single sinusoidal pulse with energy content ΔH and duration Δt, the Heisenberg uncertainty principle gives

$$\Delta H \Delta t \geq \frac{\hbar}{2}.$$

This, using the relation between energy ΔH and frequency ω, i.e., $\Delta H = \hbar \omega$, gives that it is not possible to engineer very short pulses with small frequency.

In general, a pulse will be the superposition of several Fourier components. The Heisenberg uncertainty principle says that very short pulses must have a very high frequency content, which agrees with our intuition. This places a restriction in cases where the target of control evolves (and decays) with characteristic times which are very short.

10.2.2   Objectives and techniques of molecular control

The general objective of molecular control is to selectively break or make chemical bonds to achieve desired products in a reaction. In traditional methods, control is often obtained by varying the external conditions of a chemical reaction, by varying for example temperature or pressure, or by introducing a catalyst in the reaction. However, control by electromagnetic pulses appropriately shaped may be the best way to proceed when traditional methods fail. This is the case for example in the synthesis of molecules whose existence has been predicted theoretically but that cannot be obtained with conventional chemistry. One example might be a ring configuration of Ozone O_3 (cyclic Ozone) which is considered a high energy compound of great interest. It has been identified on a solid support [215] but it is of interest to obtain it in the gas phase in large quantity [269].

An example of a finite-dimensional molecular control system was given in [272]. The goal of that paper was to demonstrate an optimal control algorithm. The system under consideration is a five-level system whose Hamiltonian is given by

$$H = H_0 + H_1 u,$$

with control u, where the internal Hamiltonian H_0 is given by

$$H_0 := \begin{pmatrix} 1 & 0 & 0 & 0 & 0 \\ 0 & 1.2 & 0 & 0 & 0 \\ 0 & 0 & 1.3 & 0 & 0 \\ 0 & 0 & 0 & 2.0 & 0 \\ 0 & 0 & 0 & 0 & 2.15 \end{pmatrix},$$

and an interaction Hamiltonian H_1,

$$H_1 := \begin{pmatrix} 0 & 0 & 0 & 1 & 1 \\ 0 & 0 & 0 & 1 & 1 \\ 0 & 0 & 0 & 1 & 1 \\ 1 & 1 & 1 & 0 & 0 \\ 1 & 1 & 1 & 0 & 0 \end{pmatrix}. \tag{10.3}$$

There are two groups of eigenstates of the internal Hamiltonian. The first three levels correspond to vibrational levels of a ground state. These are levels with slightly different energies as compared to the lowest energy level, the ground state. The remaining two energy levels correspond to excited levels. In absence of a driving field there is no coupling between these levels. Moreover, it follows from (10.3) that the external field only couples eigenstates in the first group with eigenstates in the second group. For this model, Lie algebraic calculations as in Chapter 3 show that one has complete controllability over the possible state transfers [229]. One can then apply a method among the ones described in the previous chapters to obtain constructive control. In particular the method described in [272] is a numerical method based on optimal control theory.

The models arising in molecular control are often more complicated than the one described above and in several cases they are infinite-dimensional. For this reason, many of the proposed schemes in the literature use physical intuition and take into account laboratory constraints along with the mathematical model. Many examples of schemes in molecular control are presented in the books [233], [254]. Several schemes in the infinite-dimensional case use the *combined action* of two or more laser fields. One of the first ideas in this direction was put forward by Brumer and Shapiro [48] and it is developed in depth in the book [254], with a comprehensive review given in [253]. The main idea is as follows. Among the eigenstates in the continuum of energies there are some that evolve into states corresponding to the desired outcome. For example, the photo-dissociation of a three atomic molecule ABC can proceed according to two different 'channels'

$$ABC \to AB + C, \tag{10.4}$$

or

$$ABC \to AC + B. \tag{10.5}$$

There will be, at a given energy E, states $|E, n, 1^-\rangle$ which evolve according to the channel (10.4) and states $|E, n, 2^-\rangle$ which evolve according to the channel (10.5). Here n represents additional degrees of freedom and E is the energy of both types of states $|E, n, 1^-\rangle$ and $|E, n, 2^-\rangle$. The technique of [48] consists of exciting the system to energy E with *two* (or more) lasers. In first order perturbation theory, the probability of being in a state of the type $|E, n, 1^-\rangle$ will be the sum of two terms due to the two laser fields and an interference term depending on the relative parameters of the two fields. The control consists of adjusting these parameters so as to maximize (or minimize) the ratio between the probability of having a state of the type $|E, n, 1^-\rangle$ and the one of having a state of the type $|E, n, 2^-\rangle$.

One of the major obstacles in applications of control algorithms to molecular systems is that the models are not only very complex but also have unknown parameters. For this reason, a *learning approach* to the control of molecular processes has been recently introduced. This approach has shown great promise [222]. Several samples of the same reactants are used and in each experiment the result is recorded. On the basis of this result the control law is modified for the following experiment, and the process is repeated several times. Algorithms of genetic type [117] are often used for this purpose. Figure 10.7 (cf. [26]) presents a learning scheme for molecular control.

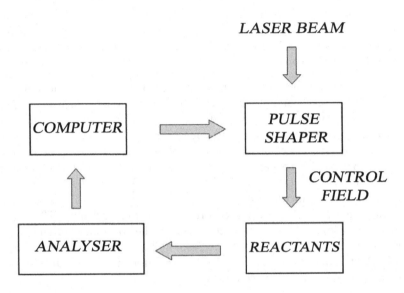

FIGURE 10.7: Scheme for the laboratory implementation of learning molecular control with pulse shaping.

10.3   Atomic Systems Control; Implementations of Quantum Information Processing with Ion Traps

Atomic systems such as *atoms in gas phase, atoms in optical cavities, trapped ions* provide a wealth of models for the application of techniques of control and dynamical analysis. These systems are relatively simple and typically have large coherence times (i.e., large intervals of time where they behave quantum mechanically). There are several applications of these systems which motivate the use of control techniques. These include *quantum metrology, quantum communication,* and *quantum information processing.* Using control methods, it is possible to achieve particular states of interest which are not available with other means. Moreover, control of atomic systems can be used to test physical theories and models. We refer to [186] for a discussion of the applications of control of atomic systems and for a list of references.

In the sequel of this section, we shall discuss a specific proposal of implementation of quantum information processing [65], [148], [259]. In particular we shall be concerned with the proposal using *trapped ions*. We use this as an example of the modeling techniques of Chapter 2 as well as of practical implementation of different Hamiltonians for the same system. It should be emphasized that this is only one of many proposals for the implementation of quantum computation which include, among others, the use of NMR experiments.

10.3.1 Physical set-up of the trapped ions quantum information processor

Trapped ions can be controlled by laser fields and used to perform quantum information processing. The **ion trap quantum information processor** consists of a number N of controlled ions which are confined in a region by an appropriate potential. Typically, the ions are placed along a line (linear trap) which we take as the x axis. Each ion is used to implement a quantum bit. In particular, two of the energy levels of each ion are used as representative of 0 and 1 in quantum information. Each ion has internal degrees of freedom (such as electron spin and nuclear spin) which contribute to the internal energy levels. Moreover, there exists an external *vibrational* degree of freedom given by the motion of the ions along the x axis. By various techniques (see, e.g., [96], [197]), it is possible to *cool* the ions to the point where only collective vibrations of the ions are significant and they are small. With a laser field it is possible to control the single ions and therefore perform one-quantum-bit operations. Two-quantum-bit operations can be performed by using the extra degree of freedom given by the position of the center of mass of the ions as an auxiliary qubit. One first performs a quantum logic operation by shining a laser on the j-th ion. This mathematically corresponds to an operation on the space spanned by the eigenvectors corresponding to the internal degrees of freedom of the j-th ion and the eigenvectors of the position of center of mass. There is a coupling between these two 'subsystems' [5] because the electromagnetic field seen by the j-th ion depends on its position. Similarly, a quantum evolution can be performed between the k-th ion and the subsystem corresponding to the position of center of mass. The net result is a two qubit operation between the j-th and the k-th ions and it is possible to show that

---

[5]They do not correspond physically to two quantum systems but mathematically are treated as such.

the operations so obtained are universal for quantum computation. A scheme of the ion trap quantum information processor is described in Figure 10.8 (cf. [65]).

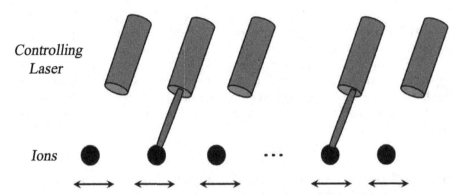

FIGURE 10.8: Ion Trap quantum information processor consisting of a string of collectively vibrating ions in a linear trap controlled through laser sources. In the figure two laser sources control two ions. These can then be switched off and other laser sources can be activated to control other ions.

10.3.2 Classical Hamiltonian

The classical Hamiltonian describing the motion of the ions is given by (2.34) where the external energy term V_e in (2.33) is replaced by a harmonic potential term, that is,

$$V_e := \sum_{j=1}^{N} \frac{m_j}{2} (\omega_x^2 x_j^2 + \omega_y^2 y_j^2 + \omega_z^2 z_j^2),$$

imposed by the external set up. This is the energy term describing the *collective motion* of the ions. It does not take into account the internal dynamics of each ion and the interaction of the ions with the laser field. These will be considered in the next two subsections.

Assuming that the ions all have the same mass M and the same charge e and neglecting the terms in (2.40) corresponding to the energy of the field only, we rewrite the Hamiltonian describing the motion of the N ions as

$$H = \sum_{j=1}^{N} \frac{1}{2} M \left(\left(\frac{d\vec{r}_j}{dt} \right)^2 + \omega_x^2 x_j^2 + \omega_y^2 y_j^2 + \omega_z^2 z_j^2 \right) + \frac{1}{8\pi\epsilon_0} \sum_{k \neq j} \frac{e^2}{|\vec{r}_k - \vec{r}_j|}. \quad (10.6)$$

The design is such that $\omega_y, \omega_z \gg \omega_x$. Moreover, the ions are cooled to very low temperature so that the term $M(\frac{d\vec{r}_j}{dt})^2$ representing the kinetic energy in

(10.6) is small. As the energy H is constant, we can assume that $z_j \approx y_j \approx 0$, so that the ions are aligned along the x axis and therefore $|\vec{r}_k - \vec{r}_j| \approx |x_k - x_j|$ in (10.6). Under these circumstances, the motion in the y-z plane and the motion along the x axis of the ions can be separated,[6] and we can study the motion of the ions along the x axis using the Hamiltonian

$$H_x := \sum_{j=1}^{N} \left(\frac{1}{2} M \dot{x}_j^2 + \frac{1}{2} M \omega_x^2 x_j^2 \right) + \frac{1}{8 \pi \epsilon_0} \sum_{k \neq j} \frac{e^2}{|\vec{x}_k - \vec{x}_j|}. \qquad (10.7)$$

An analysis based on classical Hamiltonian mechanics [259] shows that the motion of the ions is a superposition of oscillations about an equilibrium position. The frequencies of the various modes of oscillation can be calculated explicitly and they are nearly independent of N. The lowest frequency is ω_x. It corresponds to the frequency of the oscillation of the center of mass of the system of ions. In this mode, the ions oscillate together back and forth. This frequency is also clearly separated from other frequencies so that the effect of an external control tuned at this frequency could be decoupled from the other oscillation modes. These considerations suggest treating the whole system as a unique quantum harmonic oscillator with mass NM and frequency ω_x. In this setting the analysis is not different from the case of a *single* trapped ion except that the mass is NM rather than M. A rigorous analysis and justification of this assumption is given in ([148] (section 5)).

10.3.3 Quantum mechanical Hamiltonian

With the above approximations, the quantum mechanical system for the trapped ions consists of a harmonic oscillator subsystem corresponding to the motion of the center of mass of the ions and N subsystems each describing the internal degree of freedom of each ion. The associated Hilbert space will be therefore the tensor product of the Hilbert space for a harmonic oscillator and N Hilbert spaces for the internal degrees of freedom of the ions. The quantum mechanical Hamiltonian of the system of N ions under the control of N electromagnetic fields (lasers) has the form

$$H_{TOT} := \sum_{j=1}^{N} H_{0,j} + H_{ho} + \sum_{j=1}^{N} H_{I,j}.$$

In this expression, $H_{0,j}$ is the internal Hamiltonian of the j-th ion, H_{ho} is the Hamiltonian corresponding to the center of mass motion modeled as a

---

[6]Taking as canonical variables $q := \{x_j, y_j, z_j\}$, $j = 1, ..., N$ and canonical momentum $p := M\dot{q}$, and applying the Hamilton-Jacobi equations of Hamiltonian dynamics (B.13), (B.14) one easily sees that, to study the motion in the x-direction, one can take as Hamiltonian H_x in (10.7).

harmonic oscillator and H_{Ij} represents the interaction Hamiltonian of the j-th ion with the external control laser field.

Following the scheme of [65], each ion has three distinguished levels. The levels $|g\rangle$ and $|e_0\rangle$ (ground state $|g\rangle$ and excited state $|e_0\rangle$) are chosen to represent the 0 and 1 value of the qubit, respectively, while an auxiliary level $|e_1\rangle$ is used in the implementation of quantum logic gates.[7] The energy-level diagram for the internal degrees of freedom of a single ion is of the type reported in Figure 10.9.

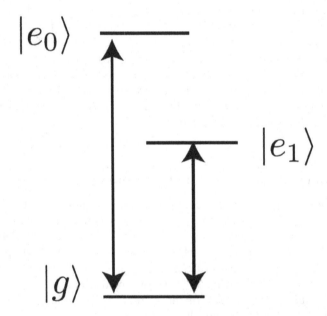

FIGURE 10.9: Energy levels for the internal degrees of freedom of a single ion. In absence of control and coupling it is a three-level system with Hamiltonian given in (10.8).

---

[7]Methods to perform quantum logic operations without the auxiliary level exist [196].

Transitions between two different levels are induced by the controlling laser field as we shall see in the following. The internal Hamiltonian $H_{0,j}$ for a single ion, using $\hbar = 1$, and omitting energy levels which are of no interest, has therefore the form

$$H_{0,j} = \frac{\omega_0}{2}(|e_0\rangle\langle e_0| - |g\rangle\langle g|) + (-\frac{\omega_0}{2} + \omega_{aux})|e_1\rangle\langle e_1|, \qquad (10.8)$$

for some given transition energies ω_0 and ω_{aux}.

The harmonic oscillator Hamiltonian H_{ho}, which corresponds to the lowest mode of oscillation where all the ions oscillate together, can be written in terms of the destruction and creation operators $\hat{a}$ and $\hat{a}^\dagger$ defined in subsection 2.6.1. Eliminating the shift in energy $\hbar\frac{\omega_x}{2}$ in (2.77), we can write the quantum harmonic oscillator Hamiltonian as (again, $\hbar = 1$)

$$H_{ho} := \omega_x \hat{a}^\dagger \hat{a}.$$

10.3.4 Practical implementation of different interaction Hamiltonians

The interaction Hamiltonian for the j-th ion H_{Ij} has the form of the electric dipole interaction discussed in 2.3.1.2. Consider only operators concerning the j-th ion and omit the index j for simplicity of notation. A control laser field (acting on the j-th ion) in *standing wave configuration* has the form

$$\vec{E} = \vec{\epsilon} E_0 \sin(kd(t)) \cos(\omega_L t + \phi).$$

In this expression, $\vec{\epsilon}$ is the polarization vector which gives the direction of the field, E_0 the maximum magnitude of the field and k is the wave-number given by $k := \frac{2\pi}{\lambda}$, with λ the (spatial) wave-length. $d = d(t)$ is the distance of the ion from the laser source. It depends on time since the ion oscillates about its equilibrium position. ω_L is the temporal frequency.

There are two schemes that are used, the one where the equilibrium position of the ion corresponds to the *antinode* (i.e., a point of maximum) of the standing wave and the one where it corresponds to a *node* of the standing wave (i.e., a point where the field is always zero). It is possible to go from a node to an antinode configuration, for example, by changing the inclination of a mirror. The antinode and node configurations are described in Figures 10.10 and 10.11, respectively.

In both cases, the distance d from the laser source is written as

$$d = d_{eq} + \cos(\theta)\delta x,$$

where d_{eq} is the distance of the equilibrium position from the laser source, while θ is the inclination angle of the laser field with respect to the x axis and δx is the displacement of the ion from its equilibrium position.

LASER SOURCE

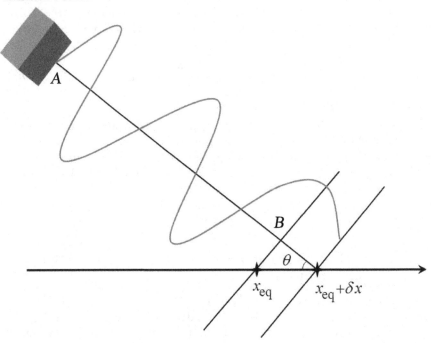

FIGURE 10.10: Scheme of laser control of a single ion in the **antinode** configuration. The distance between the two points A and B is d_{eq}.

In the *antinode configuration* $d_{eq} = \frac{4l+1}{4}\lambda$, for some integer l, so that

$$\sin(kd) = \sin\left(k\frac{4l+1}{4}\lambda + k\cos(\theta)\delta x\right)$$

$$-\sin\left(\frac{\pi}{2} + k\cos(\theta)\delta x\right) = \cos(k\cos(\theta)\delta x).$$

In the so-called *Lamb-Dicke limit* the displacement is much smaller than the wavelength λ and therefore $|k\delta x| << 1$ and omitting terms quadratic and higher in $k\cos(\theta)\delta x$, we can write $\cos(k\cos(\theta)\delta x) \approx 1$. Therefore, in the antinode configuration, the interaction Hamiltonian is obtained by quantization of the following classical Hamiltonian where $\vec{d}$ represents the electric dipole (cf. the last term in (2.47))

$$\tilde{H}_{Ia} := -\vec{d} \cdot \vec{\epsilon} E_0 \cos(\omega_L t + \phi).$$

In the *node configuration*, $d_{eq} = l\lambda$ for some integer l, and therefore

$$\sin(kd) = \sin(kl\lambda + k\cos(\theta)\delta x) = \sin(k\cos(\theta)\delta x).$$

LASER SOURCE

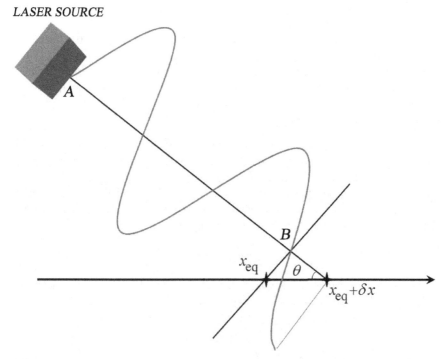

FIGURE 10.11: Scheme of laser control of a single ion in the **node** configuration. The distance between the two points A and B is d_{eq}.

Again, in the Lamb-Dicke limit, $\sin(k\cos(\theta)\delta x) \approx k\cos(\theta)\delta x$ so that the interaction Hamiltonian, is obtained by quantization of the classical Hamiltonian

$$\tilde{H}_{In} := -\vec{d} \cdot \vec{\epsilon} E_0 k \cos(\theta)\delta x \cos(\omega_L t + \phi). \tag{10.9}$$

Denote by $\hat{d}_\epsilon$ the operator (acting on the Hilbert space of the internal degrees of freedom of the ion) corresponding to the component of the dipole vector $\vec{d}$ along the polarization vector $\vec{\epsilon}$.[8] Therefore, in the antinode configuration and with polarization $\vec{\epsilon}$, the *total* quantum mechanical Hamiltonian (including the internal, the harmonic oscillator and interaction parts) can be written (for one ion) as

$$H_{a,\epsilon} = \frac{\omega_0}{2}(|e_0\rangle\langle e_0| - |g\rangle\langle g|) + (\omega_{aux} - \frac{\omega_0}{2})|e_1\rangle\langle e_1| +$$

$$\omega_x \hat{a}^\dagger \hat{a} - \frac{1}{2}\hat{d}_\epsilon E_0 \cos(\omega_L t + \phi).$$

---

[8]If $\vec{\epsilon} := \epsilon_x \vec{i} + \epsilon_y \vec{j} + \epsilon_z \vec{k}$, $\vec{d} = d_x \vec{i} + d_y \vec{j} + d_z \vec{k}$, and if $\hat{d}_{x,y,z}$ denote the Hermitian operators corresponding to d_x, d_y and d_z, then $\hat{d}_\epsilon := \hat{d}_x \epsilon_x + \hat{d}_y \epsilon_y + \hat{d}_z \epsilon_z$.

This Hamiltonian determines the evolution of one ion when a laser field acts on it.

It is possible to choose the polarization $\vec{\epsilon}$ so that $\langle e_1|\hat{d}_\epsilon|g\rangle = 0$. In this case, the Hamiltonian $H_{a,\epsilon}$ is effectively a two-level ($|g\rangle$ and $|e_0\rangle$) Hamiltonian. The third level $|e_1\rangle$ and the harmonic oscillator levels are decoupled in this case. By appropriately choosing the phase ϕ, the laser frequency ω_L and possibly the amplitude E_0, it is possible to perform any rotation in the space of the qubit spanned by $|g\rangle$ and $|e_0\rangle$.

The quantization of the node configuration Hamiltonian (10.9) is slightly more complicated because one not only introduces the operator $\hat{d}_\epsilon$ but also replaces the displacement δx with the corresponding quantum mechanical operator $\hat{\delta x}$. The operator $\hat{\delta x}$ can be written in terms of the creation and destruction operators $\hat{a}^\dagger$ and $\hat{a}$. In particular, using (2.74) and (2.75) with $\hat{\delta x}$ in place of $\hat{x}$, NM in place of m and ω_x in place of ω, with $\hbar = 1$, we have

$$\hat{\delta x} := \frac{1}{\sqrt{2NM\omega_x}}(\hat{a} + \hat{a}^\dagger).$$

In the node configuration, and with polarization $\vec{\epsilon}$, the *total* quantum mechanical Hamiltonian can be written (for one ion) as

$$H_{n,\epsilon} = \frac{\omega_0}{2}(|e_0\rangle\langle e_0| - |g\rangle\langle g|) + (\omega_{aux} - \frac{\omega_0}{2})|e_1\rangle\langle e_1| +$$

$$\omega_x\hat{a}^\dagger\hat{a} - \frac{k\cos(\theta)E_0}{\sqrt{2NM\omega_x}}\hat{d}_\epsilon(\hat{a} + \hat{a}^\dagger)\cos(\omega_L t + \phi).$$

This Hamiltonian determines the evolution of one ion when a laser field acts on it. With respect to the corresponding Hamiltonian in the antinode configuration, the evolution of the ion is now coupled (through the external field) to the harmonic oscillator.

It is possible to choose the polarization $\vec{\epsilon}$ so that $\langle e_1|\hat{d}_\epsilon|g\rangle = \langle e_1|\hat{d}_\epsilon|e_0\rangle = 0$. In this case, the level $|e_1\rangle$ is virtually decoupled from the rest of the system. Therefore, we write $H_{n,\epsilon}$ in a simplified form as

$$H_{n,\epsilon} = \frac{\omega_0}{2}(|e_0\rangle\langle e_0| - |g\rangle\langle g|) +$$

$$\omega_x\hat{a}^\dagger\hat{a} - \frac{1}{2}\frac{k\cos(\theta)E_0}{\sqrt{2NM\omega_x}}\hat{d}_\epsilon(\hat{a} + \hat{a}^\dagger)\left(e^{i(\omega_L t + \phi)} + e^{-i(\omega_L t + \phi)}\right).$$

To analyze the dynamics under this Hamiltonian, we perform a change of coordinates $|\psi\rangle \to |\tilde{\psi}\rangle := e^{iH_0 t}|\psi\rangle$, with

$$H_0 := \frac{\omega_0}{2}(|e_0\rangle\langle e_0| - |g\rangle\langle g|) + \omega_x\hat{a}^\dagger\hat{a}$$

and go to the interaction picture (cf. subsection 1.3.2). The state $|\tilde{\psi}\rangle$ evolves according to the Hamiltonian $H_I'(t) := e^{iH_0 t}H_I e^{-iH_0 t}$, with H_I given by

$$H_I := -\frac{1}{2}\frac{k\cos(\theta)E_0}{\sqrt{2NM\omega_x}}\hat{d}_\epsilon(\hat{a} + \hat{a}^\dagger)(e^{i(\omega_L t + \phi)} + e^{-i(\omega_L t + \phi)}).$$

We define $S_0 := |e_0\rangle\langle e_0| - |g\rangle\langle g|$ and calculate

$$H_I'(t) = e^{i\frac{\omega_0}{2}S_0 t}e^{i\omega_x\hat{a}^\dagger\hat{a}t}H_I e^{-i\frac{\omega_0}{2}S_0 t}e^{-i\omega_x\hat{a}^\dagger\hat{a}t}.$$

This gives

$$H_I'(t) = -\frac{1}{2}\frac{k\cos(\theta)E_0}{\sqrt{2NM\omega_x}}e^{i\frac{\omega_0}{2}S_0 t}\hat{d}_\epsilon e^{-i\frac{\omega_0}{2}S_0 t}\times$$

$$(e^{i\omega_x\hat{a}^\dagger\hat{a}t}\hat{a}e^{-i\omega_x\hat{a}^\dagger\hat{a}t} + e^{i\omega_x\hat{a}^\dagger\hat{a}t}\hat{a}^\dagger e^{-i\omega_x\hat{a}^\dagger\hat{a}t})(e^{i(\omega_L t+\phi)} + e^{-i(\omega_L t+\phi)}).$$

Using the fact (see Exercise 10.1) that

$$e^{i\omega_x\hat{a}^\dagger\hat{a}t}\hat{a}e^{-i\omega_x\hat{a}^\dagger\hat{a}t} = \hat{a}e^{-i\omega_x t}, \qquad (10.10)$$

we have

$$H_I'(t) = -\frac{1}{2}\frac{k\cos(\theta)E_0}{\sqrt{2NM\omega_x}}e^{i\frac{\omega_0}{2}S_0 t}\hat{d}_\epsilon e^{-i\frac{\omega_0}{2}S_0 t}\times$$

$$(\hat{a}e^{-i\omega_x t} + \hat{a}^\dagger e^{i\omega_x t})(e^{i\omega_L t}e^{i\phi} + e^{-i\omega_L t}e^{-i\phi}).$$

Given that $\omega_L > 0$ and $\omega_x > 0$, we do a first *rotating wave approximation* (cf. section 8.1) by neglecting the fast oscillating terms. Moreover, by defining

$$\alpha := \langle e_0|\hat{d}_\epsilon|g\rangle,$$

we obtain

$$H_I'(t) := -\frac{1}{2}\frac{k\cos(\theta)E_0}{\sqrt{2NM\omega_x}}\left(\alpha|e_0\rangle\langle g|e^{i\omega_0 t} + \alpha^*|g\rangle\langle e_0|e^{-i\omega_0 t}\right)\times$$

$$(\hat{a}e^{i(\omega_L-\omega_x)t}e^{i\phi} + \hat{a}^\dagger e^{-i(\omega_L-\omega_x)t}e^{-i\phi}).$$

At this point, by choosing

$$\omega_L = \omega_x - \omega_0,$$

and doing another rotating wave approximation, we finally obtain the Hamiltonian[9]

$$H_I'(t) = -\frac{1}{2}\frac{k\cos(\theta)E_0}{\sqrt{2NM\omega_x}}\left(\alpha|e_0\rangle\langle g|\hat{a}e^{-i\phi} + \alpha^*|g\rangle\langle e_0|\hat{a}^\dagger e^{i\phi}\right). \qquad (10.11)$$

Although this is an infinite-dimensional Hamiltonian, the two-dimensional subspace spanned by $|g\rangle|1\rangle$ and $|e_0\rangle|0\rangle$ is invariant (Exercise 10.3). Here $|0\rangle$ and $|1\rangle$ are the first two eigenstates of the harmonic oscillator corresponding to no phonon and one phonon respectively. Therefore, the dynamics given by this Hamiltonian can be studied as finite-dimensional dynamics. In an analogous manner, a similar Hamiltonian (Exercise 10.2) can be obtained for a coupling between $|e_1\rangle|0\rangle$ and $|g\rangle|1\rangle$.

---

[9]Note that the product between $|e_0\rangle\langle g|$ and $\hat{a}$ is a tensor product as these two operators act on different spaces.

10.3.5 The control problem: Switching between Hamiltonians

Summarizing the above discussion we are able to implement three different types of Hamiltonians:

1. A Hamiltonian in the antinode configuration which provides rotations for a single qubit, i.e., rotations in the space spanned by the eigenvectors $|e_0\rangle$ and $|g\rangle$ associated to the internal degree of a ion.

2. A Hamiltonian in the node configuration which couples the levels $|e_0\rangle$ and $|g\rangle$ to the harmonic oscillator motion of the center of mass of the ions.

3. A Hamiltonian in the node configuration which couples the levels $|e_1\rangle$ and $|g\rangle$, where $|e_1\rangle$ is the auxiliary level of the internal degree of one ion, to the harmonic oscillator motion of the center of mass of the ions.

With the above Hamiltonians in the node and antinode configuration and with the various polarizations of the field, it is possible to perform local operations on the qubits (in the antinode case) or entangling operations between the qubits system whose Hilbert space is spanned by $|e_0\rangle$ and $|g\rangle$ and the harmonic oscillator system with Hilbert space spanned by $|0\rangle$ and $|1\rangle$. For a two qubit operation between two ions j and m, one uses entangling operations first for the j-th ion and the harmonic oscillator and then between the harmonic oscillator and the m-th ion. This way, it is possible to perform evolutions corresponding to quantum logic operations between the two quantum bits which, along with single rotations of the qubits, are universal for quantum computation [65].

10.4   Notes and References

In this chapter, we have discussed several applications of quantum control and dynamics. These physical examples, along with the modeling considerations in Chapter 2, should provide physical motivation for the analysis presented in the remaining chapters. A survey on physical applications of quantum control along with a long list of references can be found in [186]. This paper also contains a discussion of NMR experiments from the control perspective. An introductory book on spin dynamics and NMR is [180]. Another popular text on NMR is [257], which uses the density matrix approach followed in our description.

Molecular control is treated in the books [233], [254] as well as the review papers [222], [303].

The original proposal for quantum information processing using ion traps was put forward in [65]. After that, several papers have analyzed and improved that scheme. In our treatment, we have greatly benefited from [148] and [179]. The paper [230] presents results on the controllability of the scheme of ion trap quantum information processor, in particular in reference to earlier results on controllability of infinite-dimensional quantum mechanical systems [274], [275]. We refer to [207] for a survey of methods to implement quantum information and comparison of the trapped ion scheme with other schemes.

In 2012 D. Wineland was awarded the Nobel Prize in Physics for his studies on the experimental implementation of quantum control and measurement of ion traps [312]. Such implementation uses many of the theoretical ideas we have described in section 10.3, in particular the coupling between vibrational degrees of freedom and internal (electronic) degrees of freedom of the ions. Although, the main objective of the first proposals using this scheme was implementation of quantum information processing [65], the research of D. Wineland and co-authors considered a larger array of application s which goes from optical clocks [64] to the experimental investigation of the quantum to classical transition [178] [311]. The interaction with the external environment, which is neglected in the models considered in this book, becomes crucial in these experiments both as a tool for control (for example, in cooling) and as the main phenomenon to investigate responsible for the quantum to classical transition. S. Haroche shared the Nobel Prize with D. Wineland in 2012 for quantum control and measurements experiments having the same objectives but using a different physical apparatus. In these experiments the main object of interest were optical cavities [255] [256], that is, systems of (quantized) electro-magnetic field trapped between mirrors. The state of the electromagnetic field is manipulated-probed through the interaction with an atom. Several important experiments have been carried out with this scheme including a 'recording' of quantum to classical transition [93] and the implementation of quantum feedback via quantum nondemolition measurement [247]. More explanations and introduction to the 2012 Nobel Prize in Physics can be found in [312].

10.5 Exercises

Exercise 10.1 Prove formula (10.10) using the commutation relation (2.76).

Exercise 10.2 Consider the interaction Hamiltonian in the node configuration in subsection 10.3.4 and assume the polarization vector $\vec{\epsilon}$ is chosen so that $\langle e_0|\hat{d}_\epsilon|g\rangle = \langle e_0|\hat{d}_\epsilon|e_1\rangle = 0$. Then obtain the Hamiltonian corresponding to (10.11), coupling the levels $|e_1\rangle$ and $|g\rangle$ to the harmonic oscillator.

Exercise 10.3 Verify the invariance of the vector space spanned by $|e_0\rangle|0\rangle$ and $|g\rangle|1\rangle$ under the Hamiltonian (10.11).

Appendices

Appendix A

Positive and Completely Positive Maps, Quantum Operations and Generalized Measurement Theory

A.1   Positive and Completely Positive Maps

Consider a finite-dimensional[1] Hilbert space $\mathcal{H}$ and the set of linear operators $\mathcal{H} \to \mathcal{H}$, $\mathcal{B}(\mathcal{H})$, which is a Hilbert space with the inner product $\langle A, B \rangle := \text{Tr}(B^\dagger A)$. A linear operator $\mathcal{B}(\mathcal{H}) \to \mathcal{B}(\mathcal{H})$ is called a **superoperator** although we shall more often use the word **map**.

Definition A.1.1 A map $\Gamma : \mathcal{B}(\mathcal{H}) \to \mathcal{B}(\mathcal{H})$ is called **positive** (or **positivity preserving**) if and only if $A \geq 0$ implies $\Gamma(A) \geq 0$.[2] Γ is called n-**positive** if and only if the map $\Gamma \otimes 1_n : \mathcal{B}(\mathcal{H} \otimes \mathcal{H}_n) \to \mathcal{B}(\mathcal{H} \otimes \mathcal{H}_n)$, where 1_n is the identity map over the n-dimensional Hilbert space $\mathcal{H}_n$, is positive. A map Γ is **completely positive** if it is n-positive for any positive integer n.

In particular notice that 1-positivity is equivalent to positivity and therefore clearly a completely positive map is also positive. A simple example of a completely positive map is the unitary evolution $\Gamma(\rho_S) := U\rho_S U^\dagger$. In fact we have for any $\rho \geq 0$, $\rho \in \mathcal{B}(\mathcal{H} \otimes \mathcal{H}_n)$ and any n, $(\Gamma \otimes 1_n)(\rho) = (U \otimes 1_n)\rho(U^\dagger \otimes 1_n) \geq 0$.

Historically positive and complete positive maps were introduced in quantum mechanics in the study of the dynamics of open systems. When a system evolves interacting with the environment, a 'reduced' dynamics is specified for the system [20],[47]. This dynamics is different from the unitary evolution of closed systems studied in this book. It is a one parameter ($t \in \mathbf{R}$) family of maps $\{\Gamma_t\}$ acting on the set of density matrices ρ. Each map necessarily has to be positive since, if ρ is a density matrix, $\Gamma_t(\rho)$ has to be another density

---

[1]These definitions can be extended to the infinite-dimensional case (see, e.g., [235]).
[2]$A \geq 0$ means that $\langle \psi | A | \psi \rangle \geq 0$, for all $|\psi\rangle$ in $\mathcal{H}$, or, equivalently, that all the eigenvalues of A are nonnegative.

matrix, in particular it must have nonnegative eigenvalues. However, sometimes the stronger requirement of complete positivity is imposed so that the
system coupled with an auxiliary system (e.g., an environment), which has no
dynamics, still gives a density matrix as a result of the evolution. This topic
is discussed in depth in the thesis [235] both in theoretical terms and with
physical examples.

A.2   Quantum Operations and Operator Sum Represen- tation

The state of quantum systems is modified not only by evolutions but also
by measurements. All the possible physical transformations on a state ρ of
a quantum system can be described introducing, in an abstract manner, the
concept of quantum operations.

Definition A.2.1 A **quantum operation** Γ is a linear map $\Gamma : \mathcal{B}(\mathcal{H}) \to$
$\mathcal{B}(\mathcal{H})$, which satisfies the following axioms:

1. Γ is *trace-nonincreasing*, i.e.,

$$\mathrm{Tr}(\Gamma(\rho)) \leq \mathrm{Tr}(\rho).$$

2. Γ is *completely positive*.

Examples of quantum operations are the unitary evolution $\rho \to X\rho X^\dagger$ for
$X \in U(n)$, or the transformation (cf. (1.38)) due to a Von Neumann-Lüders
measurement with result j,

$$\rho \to P_j \rho P_j,$$

where P_j is the associated projection. Another operation is obtained by letting
the system (of dimension n_S) in state ρ first couple with an auxiliary system
(of dimension n_P) in state σ and then evolve according to a unitary evolution
$X \in U(n_S n_P)$. Then the state of the main system is extracted via the partial
trace operation with respect to the second system P. Therefore, the operation
Γ is

$$\Gamma : \rho \to \mathrm{Tr}_P(X\rho \otimes \sigma X^\dagger).$$

Remark A.2.2 It is possible to extend Definition A.2.1 to maps Γ between
two different spaces, that is $\Gamma : \mathcal{B}(\mathcal{H}_1) \to \mathcal{B}(\mathcal{H}_2)$ with $\mathcal{H}_1 \neq \mathcal{H}_2$. In this case,
one more example of operation Γ is the addition of an uncorrelated system,
i.e.,

$$\Gamma : \rho \to \rho \otimes \sigma.$$

A general and very useful *representation of operations* is given by the Kraus **operator sum representation theorem**.

Theorem A.2.3 *([170], [207](Theorem 8.1)) Every quantum operation* Γ *as defined in Definition A.2.1 can be written as*

$$\Gamma(\rho) := \sum_k \Omega_k \rho \Omega_k^\dagger, \qquad (A.1)$$

for a countable set of operators Ω_k *on* $\mathcal{H}$ *satisfying*[3]

$$\sum_k \Omega_k \Omega_k^\dagger \leq 1.$$

A.3 Generalized Measurement Theory

As it is mentioned at the beginning of section 1.2, there exists several types of measurement and the Von Neumann-Lüders measurement is just one special case which, however, describes correctly most experimental situations. The formalism of operations allows us a unified treatment of measurements which goes under the name of **generalized measurement theory**. Central to this theory are the concepts of operations and effects.

Given a measurement scheme and a measurable set of possible outcomes $\mathcal{M}$, with every result $m \in \mathcal{M}$ is associated a positive operator F_m, called an **effect**. If ρ is the current state of the system, the probability of obtaining the result m (or of an event m to occur) is

$$\Pr(m) = \mathrm{Tr}(F_m \rho).$$

After a result m (or, more generally an event m) has occurred, the state is modified according to

$$\rho \to \Pr(m)^{-1} \Phi_m(\rho).$$

The positive maps Φ_m are an example of *operations* and $\mathrm{Tr}(\Phi_m(\rho)) = \Pr(m) = \mathrm{Tr}(F_m \rho)$. They can be expressed in operator sum representation according to Theorem A.2.3. The Von Neumann-Lüders measurement with a discrete set of outcomes is a special case where the effects are given by the projections P_m and the associated operations Φ_m are given by $\Phi_m(\rho) := P_m \rho P_m$.

Another example is the *indirect Von Neumann-Lüders measurement*, where a system S with state ρ is first coupled with an auxiliary system P in state σ. Then the whole system evolves with evolution X and an observable is

---

[3]Recall that, for two operators A and B, $A \leq B$ means that $B - A$ is a positive operator.

measured on S. Assume this measurement gives a result m whose associated projection is P_m. The operation Γ is given by

$$\Gamma(\rho) := \mathrm{Tr}_P \left(P_m \otimes 1 X \rho \otimes \sigma X^\dagger P_m \otimes 1 \right). \tag{A.2}$$

To obtain the operation sum representation of this operation, we use the property (9.5) of the partial trace along with its Definition 9.1.2. Given an orthonormal basis $\{|p_k\rangle\}$, $k = 1, \ldots, n_P$ of the Hilbert space associated to system P, and writing σ as

$$\sigma := \sum_{l=1}^{n_P} \alpha_l |p_l\rangle\langle p_l|, \qquad \alpha_l \geq 0,$$

we can write

$$\Gamma(\rho) := P_m \left(\sum_{k=1}^{n_P} \sum_{l=1}^{n_P} \alpha_l 1_{n_S} \otimes \langle p_k | \left(X\rho \otimes |p_l\rangle\langle p_l| X^\dagger \right) 1_{n_S} \otimes |p_k\rangle \right) P_m,$$

which we can write as

$$\Gamma(\rho)$$

$$= P_m \left(\sum_{k=1}^{n_P} \sum_{l=1}^{n_P} \alpha_l 1_{n_S} \otimes \langle p_k | X 1_{n_S} \otimes |p_l\rangle (\rho \otimes 1_1) 1_{n_S} \otimes \langle p_l | X^\dagger 1_{n_S} \otimes |p_k\rangle \right) P_m.$$

Since $\rho \otimes 1_1 = \rho$, the operator sum representation (A.2.1) holds for Γ with the Ω_k's given by the Ω_{kl}

$$\Omega_{kl} := P_m \sqrt{\alpha_l} 1_{n_S} \otimes \langle p_k | X 1_{n_S} \otimes |p_l\rangle.$$

For more details on the general measurement theory in terms of effects and operations we refer to [47].

Appendix B

Lagrangian and Hamiltonian Formalism in Classical Electrodynamics

B.1 Lagrangian Mechanics

Euler-Lagrange equations

Consider a system with n degrees of freedom $x_1, ..., x_n$. The state of the system at every instant and its future evolution is determined by $x_1, ..., x_n$ and the time derivatives $\dot{x}_1, ..., \dot{x}_n$. The *equations of motion* are differential equations involving $x_1, ..., x_n$. From their integration it is possible to obtain the state of the system at every time.

In **Lagrangian mechanics** (see, e.g., [175]) one derives the equations of motion from a function $L := L(x_1, ..., x_n, \dot{x}_1, ..., \dot{x}_n, t)$ called the **Lagrangian**. The *principle of least action* states that the physical path connecting initial $(x_j(t_0))$ and final $(x_j(t_f))$, $j = 1, ..., n$, configuration is an extremum for the integral

$$S := S(x_j) := \int_{t_0}^{t_f} L(x_j(t), \dot{x}_j(t), t)\, dt. \tag{B.1}$$

The integral S is called the *action*. The fact that x_j is an extremum means that

$$\frac{d}{d\epsilon} S(x_j + \epsilon \delta x_j)\bigg|_{\epsilon=0} = 0,$$

for every admissible perturbation of the trajectory δx_j with $\delta x_j(t_0) = \delta x_j(t_f) = 0$. From this variational principle, one obtains[1] that the true trajectory x_j,

---

[1]Consider

$$\frac{d}{d\epsilon} S(x_j + \epsilon \delta x_j)\bigg|_{\epsilon=0} = \int_{t_0}^{t_f} \frac{d}{d\epsilon} L(x_j + \epsilon \delta x_j, \dot{x}_j + \epsilon \delta \dot{x}_j, t)\, dt \bigg|_{\epsilon=0}.$$

Applying the chain rule we obtain

$$\frac{d}{d\epsilon} S(x_j + \epsilon \delta x_j)\bigg|_{\epsilon=0} = \int_{t_0}^{t_f} L_x \delta x + L_{\dot{x}} \delta \dot{x}\, dt \bigg|_{\epsilon=0}.$$

DOI: 10.1201/9781003051268-B

$j = 1, ..., n$, has to satisfy the **Euler-Lagrange equations**

$$\frac{d}{dt}\frac{\partial L}{\partial \dot{x}_j} = \frac{\partial L}{\partial x_j}, \qquad j = 1, ..., n, \tag{B.2}$$

which give the *equations of motion*.

The choice of the Lagrangian is crucial for the correct determination of the equations of motion. For a system of N particles, let the positions $\vec{r}_j$, $j = 1, ..., N$, be functions of the *independent* degrees of freedom $x_1, ..., x_n$ and time t. The l-th component of the *generalized force*, Q_l, $l = 1, ..., n$, is defined as

$$Q_l := \sum_{j=1}^{N} \vec{F}_j \cdot \frac{\partial \vec{r}_j}{\partial x_l},$$

where $\vec{F}_j$ is the force applied to the j-th particle. If there exists a function $V = V(x_j, \dot{x}_j)$ such that

$$Q_l := -\frac{\partial V}{\partial x_l} + \frac{d}{dt}\frac{\partial V}{\partial \dot{x}_l}, \tag{B.3}$$

then the Lagrangian L can be chosen as

$$L = L(x_j, \dot{x}_j, t) := T - V, \tag{B.4}$$

where T is the kinetic energy of the system of particles and V is the same as in (B.3) and it is called *generalized potential*. A special but important case is when V does not depend on $\dot{x}_j$. In this case, it is called *potential* and the generalized forces Q_l, $l = 1, ..., n$, are given by

$$Q_l := -\frac{\partial V}{\partial x_l}.$$

The general situation of equation (B.3) is, however, relevant in several cases as in the following example in electrodynamics [118], [245].

Example B.1.1 Consider a particle with charge q moving in an electromagnetic field. As there are no constraints to the motion of the particle, the degrees of freedom x_1, x_2, x_3 are taken equal to the components of the position vector $\vec{r}$. The generalized forces Q_1, Q_2, Q_3 are taken to be the components

---

Using integration by part on the second term in the integral we have

$$\left.\frac{d}{d\epsilon}S(x_j + \epsilon\delta x_j)\right|_{\epsilon=0} = \left[L_{\dot{x}}\delta x\right]_{t_0}^{t_f} + \left.\int_{t_0}^{t_f} L_x \delta x - \left(\frac{d}{dt}L_{\dot{x}}\right)\delta x\, dt\right|_{\epsilon=0}.$$

Since $\delta x(t_0) = \delta x(t_f) = 0$ the term outside the integral is zero. Moreover, since the integral has to be zero for every δx, one obtains (B.2).

of the applied force $\vec{F} := [F_1, F_2, F_3]^T$ which is equal to the Lorentz force in (2.10). The Lorentz force (2.10) can be derived as in (B.3), i.e.,

$$F_j = \frac{\partial V}{\partial x_j} + \frac{d}{dt}\frac{\partial V}{\partial \dot{x}_j}, \qquad j = 1, 2, 3, \tag{B.5}$$

if we use $(\vec{v} := \dot{\vec{r}})$

$$V := q\phi - q\vec{A}\cdot\vec{v}, \tag{B.6}$$

where ϕ and $\vec{A}$ are the scalar and vector potential, respectively.[2]

Therefore, the Lagrangian for a particle in an electromagnetic field can be written as

$$L := T - V = \frac{1}{2}m\vec{v}\cdot\vec{v} - q\phi + q\vec{A}\cdot\vec{v}.$$

Remark B.1.2 The Lagrangian L giving the correct equations of motion is *not* unique. Given a Lagrangian L, consider a function L' obtained by adding to L the total derivative of a function f of x_j and t, i.e.,

$$L' := L + \frac{d}{dt}f(x_j(t), t) = L + \sum_j \frac{\partial f}{\partial x_j}\dot{x}_j + \frac{\partial f}{\partial t}. \tag{B.9}$$

The values of the action S in (B.1), calculated for L' and L, differ only by a constant term independent of the trajectory. Therefore, the trajectories corresponding to the extrema in the two cases are the same. Equivalently one can verify that if a trajectory x_j is such that Euler-Lagrange equations (B.2) are verified with L then these equations are verified with L' given in (B.9).

---

[2]This can be easily verified for each component of $\vec{F}$ (cf. [118] Chpt. 1). For example, for the first component F_1, we have from (2.10)

$$F_1 = qE_1 + q(\vec{v}\times\vec{B})_1, \tag{B.7}$$

while calculating F_1 using (B.5) and (B.6), we obtain

$$F_1 = -q\frac{\partial\phi}{\partial x_1} - q\frac{dA_1}{dt} - q\frac{\partial}{\partial x_1}(\vec{A}\cdot\vec{v}). \tag{B.8}$$

Expanding the second term on the right-hand side, we obtain

$$F_1 = -q\frac{\partial\phi}{\partial x_1} - q\frac{\partial A_1}{\partial t} - q\sum_{j=1,2,3}\frac{\partial A_1}{\partial x_j}\dot{x}_j - q\frac{\partial}{\partial x_1}(\vec{A}\cdot\vec{v}).$$

Using (2.15), the first two terms are equal to qE_1 while a straightforward calculation shows that

$$q\sum_{j=1,2,3}\frac{\partial A_1}{\partial x_j}\dot{x}_j - q\frac{\partial}{\partial x_1}(\vec{A}\cdot\vec{v}) = \left(\vec{v}\times(\nabla\times\vec{A})\right)_1$$

which along with (2.12) shows that F_1 in (B.8) is equal to F_1 in (B.7).

Conjugate variables, Hamiltonian and Hamilton-Jacobi equations

Given the Lagrangian for a particular problem with degrees of freedom $x_1, ..., x_n$, one defines a *conjugate canonical momentum* $\vec{p} := [p_1, ..., p_n]^T$ by

$$p_j := \frac{\partial L}{\partial \dot{x}_j}, \qquad j = 1, ..., n. \tag{B.10}$$

With this definition, the Euler-Lagrange equations (B.2) can be written as

$$\frac{d}{dt} p_j = \frac{\partial L}{\partial x_j}, \qquad j = 1, ..., n. \tag{B.11}$$

With the introduction of the conjugate momentum $\vec{p}$, the Euler Lagrange equations, which are n second order differential equations, can be transformed into a set of $2n$ first order differential equations. One introduces the **Hamiltonian function**

$$H := \sum_{j=1}^{n} \dot{x}_j p_j - L(x_j, \dot{x}_j, t). \tag{B.12}$$

Notice that because of (B.10), H in (B.12) does not depend explicitly on $\dot{x}_j$, that is, $H = H(x_j, p_j, t)$.

By taking the partial derivative with respect to x_j of both sides of equation (B.12) and using (B.11), we obtain

$$\dot{p}_j = -\frac{\partial H}{\partial x_j}, \qquad j = 1, ..., n. \tag{B.13}$$

Taking the partial derivative with respect to p_j in both sides of (B.12), we obtain

$$\dot{x}_j = \frac{\partial H}{\partial p_j}, \qquad j = 1, ..., n. \tag{B.14}$$

Equations (B.13),(B.14) are the **Hamilton-Jacobi equations.**

Remark B.1.3 Refer to the situation of Remark B.1.2. If $\vec{p}$ is the conjugate momentum for the Lagrangian L, the conjugate momentum for the Lagrangian L' is, using (B.9),

$$p'_j = \frac{\partial L'}{\partial \dot{x}_j} = \frac{\partial L}{\partial \dot{x}_j} + \frac{\partial f}{\partial x_j} = p_j + \frac{\partial f}{\partial x_j}, \qquad j = 1, ..., n. \tag{B.15}$$

The corresponding Hamiltonian H' is by definition (B.12)

$$H' := \sum_{j=1}^{n} \dot{x}_j p'_j - L' = \sum_{j=1}^{N} \dot{x}_j (p_j + \frac{\partial f}{\partial x_j}) - L - \sum_{j=1}^{N} \frac{\partial f}{\partial x_j} \dot{x}_j - \frac{\partial f}{\partial t}.$$

This formula shows that, if f does not depend explicitly on time, we have

$$H'(x_j, p_j + \frac{\partial f}{\partial x_j}) = H(x_j, p_j),$$

i.e., by replacing the definition (B.15) in the Hamiltonian H' we obtain the Hamiltonian H.

In the case where the Hamiltonian $H =$ does not depend explicitly on time, i.e., $H = H(x_j, p_j)$, we have, using (B.13) and (B.14),

$$\frac{d}{dt}H(x_j, p_j) = \sum_j \frac{\partial H}{\partial x_j}\dot{x}_j + \sum_j \frac{\partial H}{\partial p_j}\dot{p}_j = \sum_j \frac{\partial H}{\partial x_j}\frac{\partial H}{\partial p_j} - \sum_j \frac{\partial H}{\partial p_j}\frac{\partial H}{\partial x_j} = 0.$$

Therefore, H is a constant of motion which has usually the physical meaning of the *energy* of the system.

Example B.1.4 Consider a set of particles with coordinates $x_1, ..., x_n$ in a potential $V = V(x_1, ..., x_n)$. The Lagrangian can be written as

$$L = T - V = \frac{1}{2}\sum_{j=1}^{n} m_j \dot{x}_j^2 - V,$$

where m_j is the mass of the particle with coordinate x_j. The conjugate momentum from equation (B.10) coincides with the *kinematical momentum*, i.e., $p_j = m_j\dot{x}_j$. Using the definition of Hamiltonian (B.12) we have $H = \frac{1}{2}\sum_j m_j\dot{x}_j^2 + V$ namely the Hamiltonian is the sum of the kinetic and potential energy.

There are, however, examples of Lagrangian functions which give the correct equations of motions but whose corresponding Hamiltonian is not equal to the energy (see Exercise 1 in Complement D_{II} in [67]).

Remark B.1.5 The optimal control problem of Lagrange (cf. subsection 7.1.1)), with no bound on the control, can be seen as the problem to find vector functions x and u which minimize

$$J(u) = \int_0^T L(x, u, t)\, dt,$$

subject to the differential constraint

$$\dot{x} = f(x, u).$$

If $\{x, u\}$ is optimal, then there exists a vector function $\lambda = \lambda(t)$ of *Lagrange multipliers* such that $\{\lambda, x, u\}$ is optimal for the unconstrained problem of minimizing

$$J'(u) = \int_0^T L + \lambda^T (\dot{x} - f)\, dt.$$

The augmented problem has Lagrangian

$$L' := L + \lambda^T(\dot{x} - f).$$

The Euler-Lagrange equations give

$$L'_\lambda = \dot{x}^T - f^T = 0,$$

$$\frac{d}{dt}L'_{\dot{x}} = \dot{\lambda}^T = L_x - \lambda^T f_x,$$

and

$$L'_u = L_u - \lambda^T f_u = 0,$$

where the last two correspond to (7.31) (with $\mu = -1$) and minimization with respect to u of h in (7.32). In particular, only the canonical moment conjugate to x is not identically zero and it is equal to $L'^T_{\dot{x}} = \lambda$. Using the definition (B.12), we obtain

$$H = \lambda^T \dot{x} - L' = \lambda^T f - L,$$

which is equal to the optimal control Hamiltonian h in (7.32). More on a unified study of Lagrangian mechanics, optimal control and calculus of variations can be found in textbooks such as [110], [153], [181].

B.2   Extension of Lagrangian Mechanics to Systems with Infinite Degrees of Freedom

An electromagnetic field is specified by the value of the field at any point in space. The degrees of freedom of this system are uncountably infinite being given by the values of the field at every point in space. The coupled system of particles and field is what interests us. However, in this section, we outline the general formalism. Our treatment in this and the following section follows at times [284] and is largely based on [67] to which we refer for a complete exposition.

Euler-Lagrange equations

In the case of an infinite set of variables, the generalized coordinates (replacing $x_1, ..., x_n$ above) will be denoted by $X(j, \vec{r}, t)$. Here j varies on a countable set, which we shall assume finite, and $\vec{r}$ varies in an uncountable set. In view of the application to electrodynamics we shall assume this set to be the space $\mathbf{R}^3$ so that $\vec{r}$ denotes a point in space. In $X(j, \vec{r}, t)$, j and $\vec{r}$ have to be seen as a couple of indexes one varying in a discrete set and one (uncountable) varying in a continuous set $\mathbf{R}^3$.

In analogy with what was done for the case of finite degrees of freedom, we consider a *Lagrangian L* depending on all of the $X(j, \vec{r}, t)$'s, their (partial) derivatives with respect to time $\frac{\partial}{\partial t} X(j, \vec{r}, t)$ and possibly explicitly on time, i.e.,

$$L := L\left(X(j, \vec{r}, t), \frac{\partial}{\partial t} X(j, \vec{r}, t), t \right).$$

Given a path, i.e., specified for every $X(j, \vec{r}, \cdot)$ the functional dependence on the third argument t, L will be a function of t only which we can integrate between initial and final time to obtain the *action* (cf. (B.1))

$$S := \int_{t_0}^{t_f} L\left(X(j, \vec{r}, t), \frac{\partial}{\partial t} X(j, \vec{r}, t), t \right) dt.$$

The principle of least action still holds true for a system with an infinite number of degrees of freedom. It states that the true path will be an extremum for the action.

In applications to electrodynamics, it is sufficient to restrict ourselves to Lagrangian functions which satisfy the following conditions.

1. The Lagrangian L has the partial derivatives with respect to the spatial degrees of freedom explicitly appearing as arguments so that

$$L := L\left(X(j, \vec{r}, t), \frac{\partial X(j, \vec{r}, t)}{\partial t}, \frac{\partial X(j, \vec{r}, t)}{\partial (x, y, z)}, t \right).$$

   These partial derivatives are not new independent variables as they can be derived from $X(j, \vec{r}, t)$ but we only assume that they appear explicitly in the expression of the Lagrangian.

2. The Lagrangian can be written as the triple integral of a *Lagrangian density function*

$$\mathcal{L} := \mathcal{L}\left(X(j, \vec{r}, t), \frac{\partial X(j, \vec{r}, t)}{\partial t}, \frac{\partial X(j, \vec{r}, t)}{\partial (x, y, z)}, \vec{r}, t \right), \text{ i.e.,}$$

$$L = \int_{\mathbf{R}^3} \mathcal{L} \, d\vec{r}.$$

Moreover, it is assumed that the variables $X(j, \vec{r}, t)$, which in electrodynamics represent potentials, vanish as $|\vec{r}| \to \infty$.

Under these assumptions, from the principle of least action, one can derive the equations corresponding to the Euler-Lagrange equations (B.2) [67].[3]

---

[3]The derivation follows the same steps as that of Euler-Lagrange equation (B.2) except that one needs to introduce the functional derivative, in order to deal with functions of an uncountable number of variables.

Define $X_j := X_j(\vec{r}, t) := X(j, \vec{r}, t)$, $\dot{X}_j := \dot{X}_j(\vec{r}, t) := \frac{\partial X(j, \vec{r}, t)}{\partial t}$, $X_{j(x,y,z)} :=$ $X_{j(x,y,z)}(\vec{r}, t) := \frac{\partial}{\partial(x,y,z)} X(j, \vec{r}, t)$, so that

$$\mathcal{L} := \mathcal{L}\left(X_j, \dot{X}_j, X_{jx}, X_{jy}, X_{jz}, \vec{r}, t\right),$$

with $j = 1, ..., n$, the Euler-Lagrange equations are

$$\frac{d}{dt} \frac{\partial \mathcal{L}}{\partial \dot{X}_j} = \frac{\partial \mathcal{L}}{\partial X_j} - \sum_{l=x,y,z} \frac{\partial}{\partial l} \frac{\partial \mathcal{L}}{\partial(X_{jl})}. \tag{B.16}$$

Remark B.2.1 A property analogous to the one discussed in Remark B.1.2 holds in the case of infinite degrees of freedom. We can modify the Lagrangian density by adding the total derivative of a function $f = f(X_j, \vec{r}, t)$ without modifying the resulting equations of motion (B.16) and the corresponding trajectory. A new Lagrangian density $\mathcal{L}'$ is obtained from $\mathcal{L}$ as

$$\mathcal{L}' = \mathcal{L} + \frac{d}{dt} f(X_j, \vec{r}, t).$$

Conjugate variables, Hamiltonian and Hamilton-Jacobi equations

As in the discrete case, one associates to every degree of freedom X_j a conjugate momentum $P_j := P_j(\vec{r}, t) := P(j, \vec{r}, t)$ which is defined using the Lagrangian density function as

$$P_j := \frac{\partial \mathcal{L}}{\partial \dot{X}_j}, \qquad j = 1, ..., n. \tag{B.17}$$

With this definition, the Euler-Lagrange equations (B.16) are written as

$$\frac{d}{dt} P_j = \frac{\partial \mathcal{L}}{\partial X_j} - \sum_{l=x,y,z} \frac{\partial}{\partial l} \frac{\partial \mathcal{L}}{\partial(X_{jl})}, \qquad j = 1, ..., n \tag{B.18}$$

The derivative $\frac{d}{dt}$ is written as a total derivative to emphasize that in the equations (B.18) j and $\vec{r}$ need to be treated as fixed indexes.

To extend the definition of Hamiltonian function (B.12) we need to 'sum' over all the indexes j and $\vec{r}$, the products of $P_j = P(j, \vec{r}, t)$ and $\dot{X}_j := \frac{\partial}{\partial t} X(j, \vec{r}, t)$. Since $\vec{r}$ is a continuous index, we replace the sum with an integral and define

$$H := \int_{\mathbf{R}^3} \sum_j P_j \dot{X}_j \, d\vec{r} - L = \int_{\mathbf{R}^3} \sum_j P_j \dot{X}_j - \mathcal{L} \, d\vec{r}. \tag{B.19}$$

This suggests to define an *Hamiltonian density*

$$\mathcal{H} := \sum_j P_j \dot{X}_j - \mathcal{L}.$$

By taking the partial derivative of both sides with respect to X_j and using (B.18), we obtain

$$\frac{d}{dt}P_j = -\frac{\partial \mathcal{H}}{\partial X_j} - \sum_{l=x,y,z}\frac{\partial}{\partial l}\frac{\partial \mathcal{H}}{\partial (X_{jl})}, \qquad j = 1,...,n. \qquad (B.20)$$

This equation corresponds to the first Hamilton-Jacobi equation (B.13). By differentiating the definition of the Hamiltonian (B.19) with respect to P_j, one obtains the Hamilton-Jacobi equation corresponding to (B.14), namely

$$\frac{d}{dt}X_j = \frac{\partial \mathcal{H}}{\partial P_j}, \qquad j = 1,...,n. \qquad (B.21)$$

Equations (B.20) and (B.21) are the Hamilton-Jacobi equations for a system with an infinite uncountable number of degrees of freedom. They are not formally identical to (B.13) and (B.14). This can be obtained by introducing the notion of functional derivative (see [67]).

Remark B.2.2 In this and the previous section we have assumed the dynamical variables (indexed by a discrete or a continuous set) used to describe the system form an *independent* set of variables. The case of dependent and redundant variables appears naturally in electrodynamics and will be further discussed in the following Remarks B.3.1, B.3.2.

B.3   Lagrangian and Hamiltonian Mechanics for a System of Interacting Particles and Field

Consider a system of N particles in an electromagnetic field in free space. This system is a combination of the two types of systems in the previous two sections, i.e., a system with a discrete set of degrees of freedom (the positions of the particles) and a system with a continuous set of degrees of freedom (the value of the field at every point in space). We choose a set of $3N$ variables given by the positions $\vec{r}_j$, of the j-th particle $j = 1,...,N$ for the degrees of freedom of the particles and the potentials $A_{x,y,z}(\vec{r})$, and $\phi(\vec{r})$ as the dynamical variables of the field. The Lagrangian associated to the system has to be such that the Euler-Lagrange equations (B.2) and (B.16) give the correct equations of motion which are the Lorentz equation (2.10) (with $\vec{B}$ and $\vec{E}$ given by (2.12) and (2.15), respectively) and the equations (2.16), (2.17).

Lagrangian and equations of motion

In terms of the variables $A_{x,y,z}$ and ϕ, the correct Lagrangian is given by

$$L = \sum_{j=1}^{N} \frac{1}{2} m_j \dot{\vec{r}}_j^{\,2} + \int_{\mathbf{R}^3} \mathcal{L} \, d\vec{r}, \qquad (B.22)$$

where the Lagrangian density $\mathcal{L}$ is given by (refer to the notations in Chapter 2)

$$\mathcal{L} := \frac{\epsilon_0}{2} \left(\left(\nabla \phi + \frac{\partial \vec{A}}{\partial t} \right)^2 - c^2 \left(\nabla \times \vec{A} \right)^2 \right) + \vec{J}(\vec{r}) \cdot \vec{A}(\vec{r}) - \rho(\vec{r}) \phi(\vec{r}),$$

where $\vec{J}(\vec{r})$ and $\rho(\vec{r})$ depend on the coordinates $\vec{r}_j$ because of (2.8) and (2.9). Notice that this Lagrangian density satisfies the conditions set at the beginning of the previous section. In particular, it is an explicit function of the spatial partial derivatives of the vector potential $\vec{A}$ and scalar potential ϕ.

To show that this Lagrangian gives the correct equations of motion, we can derive them using the Lagrangian in the Euler Lagrange equations. Let us detail the calculations for the Euler Lagrange equations for the discrete variables $\vec{r}_j$, which are of the form (B.2). To simplify notations, fix j and consider the equation corresponding to the x coordinate of the j-th particle, x_j. Using (2.8) and (2.9) we can write $\int_{\mathbf{R}^3} \mathcal{L} \, d\vec{r}$ as

$$\int_{\mathbf{R}^3} \mathcal{L} \, d\vec{r} = \sum_{j=1}^{N} (q_j \dot{\vec{r}}_j \cdot \vec{A}(\vec{r}_j) - q_j \phi(\vec{r}_j)) \qquad (B.23)$$

$$+ \frac{\epsilon_0}{2} \int_{\mathbf{R}^3} \left(\left(\nabla \phi + \frac{\partial \vec{A}}{\partial t} \right)^2 - c^2 \left(\nabla \times \vec{A} \right)^2 \right) d\vec{r}.$$

Noticing that the integral in the expression on the right-hand side does not depend on $\vec{r}_j$, nor on $\dot{\vec{r}}_j$, and using (B.22), the right-hand side of (B.2) is

$$\frac{\partial L}{\partial x_j} = q_j \dot{\vec{r}}_j \cdot \frac{\partial \vec{A}(\vec{r}_j)}{\partial x_j} - q_j \frac{\partial \phi(\vec{r}_j)}{\partial x_j},$$

while the left-hand side of B.2) gives

$$\frac{d}{dt} \frac{\partial L}{\partial \dot{x}_j} = \frac{d}{dt} \left(m_j \dot{x}_j + q_j A_x(\vec{r}_j) \right) = m_j \ddot{x}_j + q_j \nabla A_x(\vec{r}_j) \cdot \dot{\vec{r}}_j + q_j \frac{\partial A_x}{\partial t}(\vec{r}_j).$$

Equating these two expressions, we obtain

$$m_j \ddot{x}_j = -q_j \left(\frac{\partial}{\partial x_j} \phi(\vec{r}_j) + \frac{\partial}{\partial t} A_x(\vec{r}_j) \right) + q_j \dot{\vec{r}}_j \cdot \left(\frac{\partial \vec{A}(\vec{r}_j)}{\partial x_j} - \nabla A_x(\vec{r}_j) \right). \quad (B.24)$$

It can be verified directly that for any vector $\vec{v}$ and vector field $\vec{A}$,

$$\left(\vec{v} \times \left(\nabla \times \vec{A}\right)\right)_1 = \vec{v} \cdot \left(\frac{\partial \vec{A}}{\partial x} - \nabla A_x\right),$$

where $(\cdot)_1$ denotes the first component. Using this and (2.12) and (2.15), we see that (B.24) is the same as the first component of Lorentz equation (2.10).

Using the Euler-Lagrange equations (B.16) one can obtain analogously (2.16), (2.17) (cf. [67] II.B.2).

Conjugate variables and Hamiltonian

Consider the variables $\vec{r}_j$. The conjugate momentum $\vec{p}_j$ associated with $\vec{r}_j$ can be calculated from the definition (B.10) using (B.22), (B.23). We obtain

$$\vec{p}_j = m_j \dot{\vec{r}}_j + q_j \vec{A}(\vec{r}_j). \qquad (B.25)$$

The Lagrangian (B.22) does not depend on the derivative of ϕ with respect to time. Therefore, the conjugate momentum associated with ϕ is zero. This shows that the variable ϕ is not an independent variable but it can be expressed in terms of the other variables and their time derivatives. This dependence is given by equation (2.16) which is obtained from the Euler Lagrange equations (B.16) with the Lagrangian (B.22), (B.23). In principle therefore, using equation (2.16) we can eliminate the variable ϕ from the Lagrangian and reduce ourselves to only the variables $A_{x,y,z}$. This is not straightforward, however, as (2.16) is a differential constraint. One way to deal with this problem is to go to *reciprocal space* where all the variables are replaced by their spatial Fourier transforms and differential operators are replaced by algebraic operations (see [67]).

The variables $A_{x,y,z}$ still form, however, a redundant set of variables as the choice of the gauge is not specified. This further redundancy can be solved by imposing the Coulomb gauge constraint

$$\nabla \cdot \vec{A} = 0. \qquad (B.26)$$

$A_{x,y,z}$ with the constraint $\nabla \cdot \vec{A} = 0$ form a nonredundant number of independent variables describing the field.

Remark B.3.1 Recall that $\vec{r}$ in $\vec{A}(\vec{r})$ has to be considered as an *index* for the dynamical variables. A constraint such as (B.26) relates the value of $\vec{A}(\vec{r}_0)$ to the value of $\vec{A}(\vec{r})$ for values of $\vec{r}$ (infinitesimally) near $\vec{r}_0$. Lagrangian and Hamiltonian mechanics have to be modified in order to deal with problems with constraints. There are several approaches to do that. One of them is to use Lagrange multipliers (see, e.g., [174]). The Euler-Lagrange equations are not valid in the presence of constraints because in deriving them one assumes that the variation to the trajectory is unconstrained. Therefore, a different approach followed in [129], Chapter 3, consists of incorporating the constraints in the derivation of the Euler-Lagrange equations by imposing that the variation from the actual path also satisfies the constraint. This gives modified equations. A third approach is to reduce the number of variables to the truly independent ones by expressing the redundant variables in terms of the independent ones or their time derivatives. The Lagrangian is then rewritten in terms of the minimum number of independent variables. If the constraints are of differential type this can be done only in an implicit form. However, it can be done explicitly transforming the variables into their spatial Fourier transforms. The spatial Fourier transform $\vec{\mathcal{A}}(\vec{k})$ of a vector field $\vec{A}(\vec{r})$ defined in Exercise 2.4 can be seen as a *re-parametrization* of the field variables, where the parameter $\vec{r}$ is replaced by the parameter $\vec{k}$. For Fourier transforms *differential* constraints transform into *algebraic* constraints. For instance, the constraint $\nabla \cdot \vec{A} = 0$ transforms into the algebraic constraints $\vec{k} \cdot \vec{\mathcal{A}}(\vec{k}) = 0$.[4] Therefore, adopting the description in terms of Fourier transform the only two degrees of freedom for every $\vec{k}$ are the two components of the vector $\vec{\mathcal{A}}(\vec{k})$ perpendicular to $\vec{k}$ while the component parallel to $\vec{k}$ is zero (cf. Figures 2.1 and 2.2).

Remark B.3.2 We have given the form of the Lagrangian (B.22), without considering the dependence among the dynamical variables, only based on the fact that the Euler-Lagrange equations give the correct equations of motion. Then we have found that ϕ is not an independent variable and that we could make a change in the gauge without changing the equations of motion. A more rigorous approach would have been to first identify the truly independent variables. By adopting the Coulomb gauge, these can be taken in reciprocal space as the transversal components of the spatial Fourier transform of $\vec{A}$, while the longitudinal component is set equal to zero. The Fourier transform of ϕ is obtained from the Fourier transform of equation (2.16). The Lagrangian in (B.22) can be written in reciprocal space in terms of the Fourier transform of the various variables and then the number of variables reduced by eliminating the redundant variables. The resulting Lagrangian, called the *reduced Lagrangian*, is written in terms of only the truly independent variables. It gives the correct equation of motion, which correspond to (2.17). The conjugate momentum of the Fourier transform $\mathcal{A}(\vec{k})$, denoted by $\vec{\mathcal{P}}(\vec{k})$, can then be calculated using the definition (B.17) (opportunely adapted to accommodate complex variables). It is given by

$$\vec{\mathcal{P}}(\vec{k}) = \epsilon_0 \dot{\vec{\mathcal{A}}}(\vec{k}).$$

---

[4]In general, ∇ is replaced by $i\vec{k}$. The Fourier transform $\vec{\mathcal{A}}(\vec{k})$ is a complex vector. However, this does not lead to an increase in the number of variables since, from the fact that $\vec{A}(\vec{r})$ is real, it follows that

$$\vec{\mathcal{A}}(\vec{k}) = \vec{\mathcal{A}}^*(-\vec{k}).$$

If $\vec{P}(\vec{r})$ is the inverse transform of $\vec{\mathcal{P}}$, we have

$$\vec{P}(\vec{r}) = \epsilon_0 \dot{\vec{A}}(\vec{r}). \tag{B.27}$$

From Parseval-Plancherel theorem[5] it follows that

$$\int_{\mathbf{R}^3} \vec{\mathcal{P}}^*(\vec{k}) \cdot \dot{\vec{\mathcal{A}}}(\vec{k}) \, d\vec{k} = \int_{\mathbf{R}^3} \vec{P}^*(\vec{r}) \cdot \dot{\vec{A}}(\vec{r}) \, d\vec{r}. \tag{B.28}$$

In the Coulomb gauge, the definition of the momentum conjugate to $\vec{A}$ can be given rigorously in reciprocal space according to the discussion in the previous remark. The Hamiltonian is then written in reciprocal space according to the definition (B.19) and then can be written in real space $(\vec{r})$ using (B.28), where $\vec{P}(\vec{r})$ is given in (B.27). We have in real space,

$$H = \sum_{j=1}^{N} \vec{p}_j \cdot \dot{\vec{r}}_j + \int_{\mathbf{R}^3} P(\vec{r}) \cdot \dot{\vec{A}}(\vec{r}) d\vec{r} - L, \tag{B.29}$$

where L is given in (B.22), (B.23). By replacing the expressions for the conjugate momenta (B.25) and (B.27), we obtain[6]

$$H = \frac{1}{2} \sum_{j=1}^{N} m_j \dot{\vec{r}}_j + \frac{\epsilon_0}{2} \int_{\mathbf{R}^3} \dot{\vec{A}}^2 + c^2 (\nabla \times \vec{A})^2 d\vec{r} + \int_{\mathbf{R}^3} \rho \phi - \frac{\epsilon_0}{2} (\nabla \phi)^2 d\vec{r}. \tag{B.30}$$

The last term has been calculated in Remark 2.1.1 in Chapter 2. It is equal to V_{coul} in (2.27). Using (2.12) and (2.15) and the fact that we have adopted the Coulomb gauge, we obtain

$$H = \frac{1}{2} \sum_{j=1}^{n} m_j \dot{\vec{r}}_j^{\,2} + \frac{\epsilon_0}{2} \int_{\mathbf{R}^3} \vec{E}_T^2 + c^2 \vec{B}^2 d\vec{r} + V_{coul}. \tag{B.31}$$

This is the energy of the system of particles and electromagnetic field. The first term represents the kinetic energy of the particles while the remaining two terms represent the energy needed to establish the field.

$$\int_{\mathbf{R}^3} \vec{F}^*(\vec{r}) \vec{G}(\vec{r}) d\vec{r} = \int_{\mathbf{R}^3} \vec{\mathcal{F}}^*(\vec{k}) \cdot \vec{\mathcal{G}}(\vec{k}) d\vec{k},$$

where $\vec{\mathcal{F}}$ and $\vec{\mathcal{G}}$ are the spatial Fourier transforms of the fields $\vec{F}$ and $\vec{G}$, respectively.
[6] In applying the expression for the Lagrangian (B.23), we have used the fact that

$$\int_{\mathbf{R}^3} \dot{\vec{A}} \cdot \nabla \phi d\vec{r} = 0,$$

since $\nabla \phi$ is purely longitudinal and $\dot{\vec{A}}$ is purely transversal because of the Coulomb gauge assumption (cf. (2.26)).

Lagrangian and Hamiltonian formulation in the presence of an external field

In the presence of an external field the Lagrangian L in (B.22), (B.23) has to be modified as follows

$$L_e := L + \int_{\mathbf{R}^3} \mathcal{L}_e \, d\vec{r},$$

with $\mathcal{L}_e$ defined by

$$\mathcal{L}_e = J(\vec{r}) \cdot \vec{A}_e(\vec{r}, t) - \rho(\vec{r}) \phi_e(\vec{r}, t), \tag{B.32}$$

where $\vec{A}_e$ and ϕ_e are the vector and scalar external potentials. This Lagrangian is justified by the fact that it gives the correect equations of motion. In particular (cf. the diagram in 2.1.1.5) these are 1) the Lorentz equation (2.10) with $\vec{E}$ replaced by $\vec{E} + \vec{E}_e$ and $\vec{B}$ replaced by $\vec{B} + \vec{B}_e$, with $\vec{E}_e$ and $\vec{B}_e$ the external electric and magnetic field and 2) equations (2.16) and (2.17) (where the external field does not appear). Starting with this Lagrangian, the discussion in the previous subsection can be repeated with minor modifications. In particular the moment conjugate to $\vec{r}_j$, $j = 1, ..., N$, is given by (2.37) and the Hamiltonian is given by (2.34).

Appendix C

Complements to the Theory of Lie Algebras and Lie Groups

For concreteness, in the following discussion, we refer to a Lie algebra over the field $\mathbf{R}$, although much of what we shall say is valid for Lie algebras over general fields.

C.1 The Adjoint Representation and Killing Form

Consider a Lie algebra $\mathcal{L}$ and an element X of $\mathcal{L}$. X acts on $\mathcal{L}$ according to the linear map ad_X, defined by

$$ad_X Y := [X, Y].$$

Given a basis of $\mathcal{L}$, ad_X can be represented by a matrix. The map $X \to ad_X$ is called the **adjoint representation** of $\mathcal{L}$. It is a *representation* as it is a homomorphism from $\mathcal{L}$ to the Lie algebra of real matrices of dimension equal to $\dim(\mathcal{L})$. In fact we have

$$ad_{[X,Y]} = ad_X ad_Y - ad_Y ad_X.$$

The *Killing form* associated with a Lie algebra $\mathcal{L}$ is the bilinear form

$$\langle X, Y \rangle_K = \text{Tr}(ad_X ad_Y).$$

A bilinear form $\langle X, Y \rangle$ over a vector space $\mathcal{L}$ is said to be *nondegenerate* if $\langle X, Y \rangle = 0$ for every $Y \in \mathcal{L}$ implies $X = 0$.

C.2 Cartan Semisimplicity Criterion

The following fundamental theorem due to Cartan allows us to check whether a Lie algebra is semisimple or not.

Theorem C.2.1 *A Lie algebra $\mathcal{L}$ is semisimple if and only if its Killing form $\langle \cdot, \cdot \rangle_K$ is nondegenerate.*

An alternative test for semisimplicity says that $\mathcal{L}$ is semisimple if and only if

$$[\mathcal{L}, \mathcal{L}] = \mathcal{L}. \tag{C.1}$$

C.3 Correspondence between Ideals and Normal Subgroups

A **normal subgroup** H of a group G is a subgroup $II \subseteq G$ such that

$$ghg^{-1} \in H,$$

for every $h \in H$ and every $g \in G$. This definition applies in particular to Lie groups. Given the definition of ideal in a Lie algebra 3.1.5, we have the following correspondence between ideals and normal subgroups (see, e.g., [244] Corollary 7.16).

Theorem C.3.1 *Let $\mathcal{L}$ and $\mathcal{G}$ be Lie algebras and $e^{\mathcal{L}}$ be a Lie subgroup of $e^{\mathcal{G}}$. Then $e^{\mathcal{L}}$ is a normal subgroup of $e^{\mathcal{G}}$ if and only if $\mathcal{L}$ is an ideal in $\mathcal{G}$.*

C.4 Quotient Lie Algebras

Consider a Lie algebra $\mathcal{L}$ and an ideal $\mathcal{R}$ of $\mathcal{L}$ and write $\mathcal{L}$ as

$$\mathcal{L} = \mathcal{S} \oplus \mathcal{R}.$$

Consider the quotient vector space $\mathcal{L}/\mathcal{R}$ defined as the set of equivalence classes $\{(V)\}$ of elements V in $\mathcal{L}$ which differ by an element in $\mathcal{R}$.[1] $\mathcal{L}/\mathcal{R}$ can be given the structure of a Lie algebra by defining

$$[(V_1), (V_2)] := ([V_1, V_2]).$$

It is easily seen that, since $\mathcal{R}$ is an ideal, this definition is well posed as $([V_1, V_2])$ does not depend on the choice of the representative V_1 and V_2 in the equivalence classes (V_1) and (V_2).

---

[1] The basic vector space operation is defined by

$$\alpha(V_1) + \beta(V_2) = (\alpha V_1 + \beta V_2).$$

C.5   Levi Decomposition

Levi's decomposition theorem states that solvable and semisimple Lie algebras are the basic types of Lie algebras in the following sense:

Theorem C.5.1 *(**Levi Decomposition**) Every Lie algebra $\mathcal{L}$ is the direct sum (as vector spaces) of a semisimple Lie algebra $\mathcal{S}$ and a solvable Lie algebra $\mathcal{Z}$, i.e.,*

$$\mathcal{L} = \mathcal{S} \oplus \mathcal{Z}, \tag{C.2}$$

with $\mathcal{Z}$ an ideal in $\mathcal{L}$. The semisimple Lie algebra $\mathcal{S}$ is isomorphic to the quotient $\mathcal{L}/\mathcal{Z}$.

Remark C.5.2 The fact that $\mathcal{Z}$ is an ideal in $\mathcal{L}$ along with (C.2) is sometimes expressed by saying that $\mathcal{L}$ is the *semidirect* sum of $\mathcal{S}$ and $\mathcal{Z}$.

Levi decomposition is not unique. However, the solvable part $\mathcal{Z}$ is unique and it is called the *solvable radical* of $\mathcal{L}$. The semisimple part of $\mathcal{L}$, $\mathcal{S}$, is called the *Levi subalgebra*. Moreover, for any pair of Levi decompositions,

$$\mathcal{L} = \mathcal{S}_1 \oplus \mathcal{Z}, \qquad \mathcal{L} = \mathcal{S}_2 \oplus \mathcal{Z},$$

the isomorphism of $\mathcal{S}_1$ and $\mathcal{S}_2$ to $\mathcal{L}/\mathcal{R}$ provides an isomorphism between $\mathcal{S}_1$ and $\mathcal{S}_2$. We refer to [92] for a treatment of Levi decomposition from a computational point of view.

We have seen in Theorem 3.1.8 that if $\mathcal{L}$ is a Lie subalgebra of $u(n)$, Levi decomposition specializes in that $\mathcal{Z}$ is Abelian and the semidirect sum in (C.2) becomes a direct sum, i.e., $[\mathcal{S}, \mathcal{Z}] = 0$.

Appendix D

Proof of the Controllability Test of Theorem 3.2.1

The statement of the theorem refers to the situation where the class of controls is that of piecewise continuous functions. This is a subset of the more general class of *locally bounded and measurable functions*[1] for which the solution of (3.1) exists unique in $e^{\mathcal{L}}$ (cf. Lemma 2.1 in [154]). This gives the inclusion $\mathcal{R} \subseteq e^{\mathcal{L}}$. Alternatively, the inclusion $\mathcal{R} \subseteq e^{\mathcal{L}}$ follows directly from the fact that for a piecewise constant control the solution of (3.1) is the product of exponentials $e^{-iH(u_j)t_j}$ and since $-iH(u_j) \in \mathcal{L}$ and the product belongs to $e^{\mathcal{L}}$, because of the way $e^{\mathcal{L}}$ was defined.

We now consider the proof that $\mathcal{R}$ is dense in $e^{\mathcal{L}}$ and it coincides with $e^{\mathcal{L}}$ if $e^{\mathcal{L}}$ is a Lie subgroup of $U(n)$. Density has to be intended in the subset topology of $e^{\mathcal{L}}$ induced by the topology of $U(n)$. Such topology coincides with the topology of $e^{\mathcal{L}}$ if $e^{\mathcal{L}}$ is a Lie subgroup of $U(n)$. The proof presented here uses arguments borrowed from [74], [154], [291]. Related papers are [142], [229], [264].

$\mathcal{R}$ is dense in $e^{\mathcal{L}}$

Let $\mathcal{S} := \{A_1, ..., A_s\}$ be a set of linearly independent generators of $\mathcal{L}$ obtained by setting the controls u equal to appropriate values in $-iH(u)$. Clearly, all the elements in the *semigroups*

$$\{X \in e^{\mathcal{L}} | X = e^{A_j t}, t \geq 0\}, \quad j = 1, ..., s,$$

as well as their products, are in $\mathcal{R}$. Below we are going to prove that every matrix in $e^{\mathcal{L}}$ can be expressed as a finite product of matrices of the form e^{At}, with $A \in \mathcal{S}$ and $t \in \mathbf{R}$ (finite generation of $e^{\mathcal{L}}$). Therefore, to show the denseness property, it is enough to show that, for every matrix e^{At} with $A \in \mathcal{S}$ and $t \in \mathbf{R}$, we can always choose a *positive* α such that $e^{A\alpha}$ is arbitrarily close to e^{At}. In fact, for each element $e^{-A|t|}$, there exists a sequence of positive values $t_k \geq 0$ such that

$$\lim_{k \to \infty} e^{At_k} = e^{-A|t|}. \tag{D.1}$$

---

[1]Other more restricted classes can also be considered. See [154].

DOI: 10.1201/9781003051268-D

To construct such a sequence, pick the sequence $e^{jA|t|}$, which by compactness of $U(n)$ has a converging subsequence $e^{j(k)A|t|}$. By setting $e^{At_k} := e^{(j(k+1)-j(k)-1)A|t|}$ we obtain (D.1).

Finite generation of $e^{\mathcal{L}}$

Lemma 1 Every matrix X_f in $e^{\mathcal{L}}$ can be expressed as a finite product of matrices of the form e^{At}, with $A \in \{A_1, ..., A_s\}$ and $t \in \mathbf{R}$. The number of factors depends on X_f.

This lemma is often expressed by saying that $e^{\mathcal{L}}$ is *generated* by $\{A_1, ..., A_s\}$. In other words, if the matrices $A_1, ..., A_s$ generate the Lie algebra $\mathcal{L}$, we say that they (or the one-dimensional subgroups associated with them) also generate the Lie group $e^{\mathcal{L}}$

Proof. First notice that, if a set $\{F_1, F_2, ..., F_m\}$ is a *basis* for $\mathcal{L}$, a neighborhood of the identity in $e^{\mathcal{L}}$ can be obtained by varying $t_1, ..., t_m$ in a neighborhood of the origin in $\mathbf{R}^m$ in the expression

$$K := e^{F_1 t_1} e^{F_2 t_2} \cdots e^{F_m t_m}. \tag{D.2}$$

This follows from the Inverse Function Theorem (see, e.g., [260], Lemma 3.1. Chp. 5, for a statement of this result).

We now show how to obtain a basis of $\mathcal{L}$ by using similarity transformations involving only elements in the set $\mathcal{S} := \{A_1, ..., A_s\}$.

Assume $\mathcal{S} = \{A_1, ..., A_s\}$ is *not* a basis of $\mathcal{L}$. There exist two elements, A_j and A_k, and a time $\tilde{t}$ (arbitrarily small) such that $e^{A_k \tilde{t}} A_j e^{-A_k \tilde{t}}$ is linearly independent from $\{A_1, ..., A_s\}$. If this was not the case, we would have

$$e^{A_k t} A_j e^{-A_k t} = \sum_{l=1}^{s} a_l(t) A_l, \tag{D.3}$$

for some functions $a_l(t)$ and for every t. Differentiating (D.3) at $t = 0$, we obtain

$$[A_k, A_j] = \sum_{l=1}^{s} \dot{a}_l(0) A_l,$$

for every $k, j = 1, 2, ...s$, which (if $\dim(\mathcal{L}) > s$) contradicts the fact that $A_1, ..., A_s$ are generators of $\mathcal{L}$. Set

$$A_{s+1} := e^{A_k \tilde{t}} A_j e^{-A_k \tilde{t}}. \tag{D.4}$$

Of course, $A_1, ..., A_{s+1}$ still forms a set of linearly independent generators and therefore, as above, there exist two elements A_j and A_k in the set $\{A_1, ..., A_s, A_{s+1}\}$ and a time $\tilde{t}$ (arbitrarily small) such that $e^{A_k \tilde{t}} A_j e^{-A_k \tilde{t}}$ is linearly independent from $A_1, ..., A_{s+1}$. As in (D.4), we obtain a new element of $\mathcal{L}$ that we denote by A_{s+2}, such that $\{A_1, A_2, ..., A_{s+2}\}$ is a basis of an $s+$ two-dimensional subspace of $\mathcal{L}$. Proceeding this way, we obtain a basis $A_1, ..., A_s, A_{s+1}, ..., A_m$ of the Lie algebra $\mathcal{L}$ where the first s elements are the generators we started with and the elements $A_{s+1}, ..., A_m$ are obtained via similarity transformations with the iterative procedure we have described. Every element A_j, $j = 1, ..., m$, can be written in the form

$$A_j = e^{\tilde{A}_r t_r} e^{\tilde{A}_{r-1} t_{r-1}} \cdots e^{\tilde{A}_1 t_1} \tilde{A}_k e^{-\tilde{A}_1 t_1} \cdots e^{-\tilde{A}_{r-1} t_{r-1}} e^{-\tilde{A}_r t_r},$$

where the indeterminates $\tilde{A}_1, ..., \tilde{A}_r, \tilde{A}_k$ belong to $\mathcal{S}$, and r is finite.

Replacing now in (D.2) the expressions of $e^{F_j t}$, $j = 1, ..., m$, in terms of the matrices in $\mathcal{S}$ we obtain that all the elements in a neighborhood N of the identity in $e^{\mathcal{L}}$ can be written as

$$K := e^{\tilde{A}_1 t_1} e^{\tilde{A}_2 t_2} \cdots e^{\tilde{A}_p t_p}. \tag{D.5}$$

The indeterminates $\tilde{A}_j$ are in the set $\mathcal{S}$, $t_j \in \mathbf{R}$, $j = 1, ..., p$. Now, every element X_f in $e^{\mathcal{L}}$ can be written as a product of elements of the type e^B, with $B \in \mathcal{L}$, but each one of these can be written as K in (D.5) because so can $e^{\frac{B}{M}}$ for a sufficiently large positive integer M. This concludes the proof of the lemma. $\qquad\square$

Now we assume that $e^{\mathcal{L}}$ is a Lie subgroup of $U(n)$, in particular the topology of $e^{\mathcal{L}}$ coincides with the subset topology induced by $U(n)$. We show that $e^{\mathcal{L}} = \mathcal{R}$.

$\mathcal{R}$ contains a neighborhood of the identity in $e^{\mathcal{L}}$

Recall that m denotes the dimension of $\mathcal{L}$. From the proof of Lemma 1, we can choose $m - s$ elements in $e^{\mathcal{L}}$, $U_1, ..., U_{m-s}$, and $m - s$ elements in the set $\{A_1, ..., A_s\}$, say $\bar{A}_1, ..., \bar{A}_{m-s}$, such that,

$$\{A_1, A_2, ..., A_s, U_1 \bar{A}_1 U_1^{-1}, ..., U_{m-s} \bar{A}_{m-s} U_{m-s}^{-1}\},$$

form a basis of $\mathcal{L}$. Given m elements $V_1, ..., V_m$ in $\mathcal{R}$, consider the function $\Phi := \Phi(j_1, ..., j_m)$ from a neighborhood of the point $(1, ..., 1)$ in $\mathbf{R}^m$ to $e^{\mathcal{L}}$,

$$\Phi(j_1, ..., j_m) := e^{A_1 j_1} V_1 e^{A_2 j_2} V_2 \cdots e^{A_s j_s} V_s e^{\bar{A}_1 j_{s+1}} V_{s+1} \cdots e^{\bar{A}_{m-s} j_m} V_m.$$

All the elements obtained by $\Phi(j_1, ..., j_n)$ with $j_1, ..., j_n \geq 0$ are in $\mathcal{R}$. We can show that, if we choose appropriately $V_1, ..., V_m$, the set of these elements contains a neighborhood of the point $U_0 := e^{A_1} V_1 e^{A_2} V_2 \cdots e^{A_s} V_s e^{\bar{A}_1} V_{s+1} \cdots e^{\bar{A}_{n-s}} V_n$ by showing that the Jacobian at $(1, ..., 1)$ is not zero and applying the Implicit Function Theorem (cf. [291]).

Differentiating at $(1, ..., 1)$ the function $\Phi(j_1, ..., j_m)$, we obtain

$$\frac{\partial \Phi(j_1, ..., j_m)}{\partial j_1}\bigg|_{j_1,...,j_m=1,...,1} = A_1 e^{A_1} V_1 e^{A_2} V_2 \cdots e^{A_s} V_s e^{\bar{A}_1} V_{s+1} \cdots e^{\bar{A}_{m-s}} V_m,$$

$$\frac{\partial \Phi(j_1, ..., j_m)}{\partial j_2}\bigg|_{j_1,...,j_m=1,...,1} = e^{A_1} V_1 A_2 e^{A_2} V_2 \cdots e^{A_s} V_s e^{\bar{A}_1} V_{s+1} \cdots e^{\bar{A}_{m-s}} V_m,$$

$$\vdots$$

$$\frac{\partial \Phi(j_1, ..., j_m)}{\partial j_{s+1}}\bigg|_{j_1,...,j_m=1,...,1} = e^{A_1} V_1 e^{A_2} V_2 \cdots e^{A_s} V_s \bar{A}_1 e^{\bar{A}_1} V_{s+1} \cdots e^{\bar{A}_{m-s}} V_m,$$

$$\vdots$$

$$\frac{\partial \Phi(j_1, ..., j_m)}{\partial j_m}\bigg|_{j_1,...,j_m=1,...,1} = e^{A_1} V_1 e^{A_2} V_2 \cdots e^{A_s} V_s e^{\bar{A}_1} V_{s+1} \cdots \bar{A}_{m-s} e^{\bar{A}_{m-s}} V_m.$$

Choosing (recall that $\mathcal{R}$ is dense in $e^{\mathcal{L}}$)

$$V_1 \approx e^{-A_1},$$

$$\vdots$$

$$V_{s-1} \approx e^{-A_{s-1}},$$
$$V_s \approx e^{-A_s} U_1,$$
$$V_{s+1} \approx e^{-\bar{A}_1} U_1^{-1} U_2,$$

$$\vdots$$

$$V_{m-1} \approx e^{-\bar{A}_{m-s-1}} U_{m-s-1}^{-1} U_{m-s},$$
$$V_m \approx e^{-\bar{A}_{m-s}} U_{m-s}^{-1},$$

we obtain

$$\frac{\partial \Phi(j_1, ..., j_m)}{\partial j_1}\bigg|_{j_1,...,j_m=1,...,1} \approx A_1,$$

$$\vdots$$

$$\frac{\partial \Phi(j_1, ..., j_m)}{\partial j_s}\bigg|_{j_1,...,j_m=1,...,1} \approx A_s,$$

$$\frac{\partial \Phi(j_1, ..., j_m)}{\partial j_{s+1}}\bigg|_{j_1,...,j_m=1,...,1} \approx U_1 \bar{A}_1 U_1^{-1},$$

$$\vdots$$

$$\frac{\partial \Phi(j_1, ..., j_m)}{\partial j_m}\bigg|_{j_1,...,j_m=1,...,1} \approx U_{m-s} \bar{A}_{m-s} U_{m-s}^{-1},$$

which are linearly independent and this proves our claim.

Now, for some sufficiently small ϵ, consider the open ball (in $U(n)$) centered at U_0, $B_\epsilon(U_0)$. We have that $B_\epsilon(U_0) \cap e^{\mathcal{L}}$ is a subset of $\mathcal{R}$. We show that this implies that $\mathcal{R}$ contains à neighborhood of the identity. Using the fact that $\mathcal{R}$ is dense in $e^{\mathcal{L}}$, choose $\bar{U} \in \mathcal{R}$, such that $||\bar{U} - U_0^{-1}|| < \frac{\epsilon}{2}$. This implies that

$$||U_0 - \bar{U}^{-1}|| < \frac{\epsilon}{2}.$$

Then, if $F \in B_{\frac{\epsilon}{2}}(I) \cap e^{\mathcal{L}}$, writing $F = \bar{U}X$, we have

$$\frac{\epsilon}{2} > ||F - I|| = ||\bar{U}X - I|| = ||X - \bar{U}^{-1}|| \geq ||X - U_0|| - ||U_0 - \bar{U}^{-1}||,$$

and therefore

$$||X - U_0|| < \epsilon,$$

which implies $X \in \mathcal{R}$ and therefore $F = \bar{U}X \in \mathcal{R}$, since $\mathcal{R}$ is a semigroup.

Conclusion

The semigroup $\mathcal{R}$ contains a neighborhood of the identity in $e^{\mathcal{L}}$. Since $e^{\mathcal{L}}$ is connected, it follows that $\mathcal{R} = e^{\mathcal{L}}$. Alternatively, for every $A \in \mathcal{L}$, $e^{\frac{A}{M}}$ belongs to an arbitrarily small neighborhood of the identity choosing M large and therefore it is in the reachable set $\mathcal{R}$ and so is $(e^{\frac{A}{M}})^M = e^A$.

Appendix E

The Baker-Campbell-Hausdorff Formula and Some Exponential Formulas

There are several versions of the **Baker-Campbell-Hausdorff** (BCH) formula. One version, particularly useful, is as follows. Consider two matrices A and B, we have

$$e^A B e^{-A} = \sum_{k=0}^{\infty} \frac{1}{k!} ad_A^k B, \tag{E.1}$$

where $ad_A^k B$ is defined recursively by $ad_A^0 B := B$, $ad_A^k B := [A, ad_A^{k-1} B]$, if $k \geq 1$. This expression can be obtained by calculating the McLaurin series of the matrix function $F(t) := e^{At} B e^{-At}$ at $t = 0$. Exponentiating both sides of (E.1), we obtain the formula

$$e^A e^B e^{-A} = e^{\sum_{k=0}^{\infty} \frac{1}{k!} ad_A^k B}. \tag{E.2}$$

If A and B commute, i.e., $[A, B] = 0$, we have

$$e^A e^B = e^B e^A = e^{A+B}. \tag{E.3}$$

This formula has a generalization to the case where both A and B commute with $[A, B]$, i.e.,

$$[B, [A, B]] = [A, [A, B]] = 0. \tag{E.4}$$

In this case, (E.2) reduces to

$$e^A e^B e^{-A} = e^{B + \frac{1}{2}[A,B]},$$

which using the commutativity properties (E.4) becomes

$$e^A e^B = e^B e^A e^{\frac{1}{2}[A,B]},$$

that is, property (E.3) is modified by adding the extra factor $e^{\frac{1}{2}[A,B]}$.

We now consider ways to estimate how e^{A+B} differs from $e^A e^B$, in particular when A and B are small. Let us introduce a parameter $t \in \mathbf{R}$. We have that there exists an $\epsilon > 0$ such that for $|t| < \epsilon$

$$e^{tA} e^{tB} = e^{tA + tB + \frac{t^2}{2}[A,B] + O(t^3)}. \tag{E.5}$$

DOI: 10.1201/9781003051268-E

Moreover we can approximately move in the direction of $[A, B]$ by moving back and forth along the directions of A and B since we have

$$e^{-tA}e^{-tB}e^{tA}e^{tB} = e^{t^2[A,B]+O(t^3)}. \tag{E.6}$$

In (E.5) and (E.6) $O(t^3)$ is a matrix function of t such that $\frac{O(t^3)}{t^3}$ is bounded for $|t| < \epsilon$. Another useful formula is

$$e^{A+B} = \lim_{n \to \infty} (e^{\frac{A}{n}} e^{\frac{B}{n}})^n. \tag{E.7}$$

This is known as *Lie product formula* or *Trotter formula*. The proofs of formulas (E.5), (E.6) and (E.7) can be found in [244] along with several other exponential formulas.

Appendix F

Proof of Theorem 7.2.1

We shall follow [110] (section II-10, II-11) to which we refer for details.

We rewrite the definition of strong variation u^ϵ. Fix $\tau \in (0, T]$. A strong variation of $u = u(t)$ is a function $u_\epsilon = u_\epsilon(t)$, such that $u^\epsilon(t) = u(t)$, for $t \in [0, \tau - \epsilon]$ and for $t \in (\tau, T]$, $u^\epsilon(t) \equiv v$, for $t \in (\tau - \epsilon, \tau]$, where v is a value in the set of admissible controls, $\mathcal{U} \subseteq \mathbf{R}^m$.

We denote by x^ϵ the trajectory corresponding to u^ϵ. Condition (7.16) gives

$$\frac{d}{d\epsilon} J(u^\epsilon)\bigg|_{\epsilon=0} \geq 0.$$

We have

$$\frac{d}{d\epsilon} J(u^\epsilon)\bigg|_{\epsilon=0} = \frac{d}{d\epsilon} \phi(x^\epsilon(T))\bigg|_{\epsilon=0} = \phi_x(x(T)) \frac{d}{d\epsilon} x^\epsilon(T)\bigg|_{\epsilon=0}, \qquad \text{(F.1)}$$

where ϕ_x is the gradient with respect to x of the function ϕ, i.e.,

$$\phi_x := \left[\frac{\partial \phi}{\partial x_1}, ..., \frac{\partial \phi}{\partial x_n} \right].$$

In the following, we calculate $\frac{d}{d\epsilon} x^\epsilon(T)|_{\epsilon=0}$. Defining $x_\epsilon := x^\epsilon(\tau)$, we can write

$$\frac{d}{d\epsilon} x^\epsilon(T)\bigg|_{\epsilon=0} = \frac{\partial x^\epsilon(T)}{\partial x_\epsilon} \frac{\partial x_\epsilon}{\partial \epsilon}\bigg|_{\epsilon=0}. \qquad \text{(F.2)}$$

On the right-hand side of the above expression, the term $\frac{\partial x_\epsilon}{\partial \epsilon}|_{\epsilon=0}$ describes how $x(\tau)$ varies because of the variation of the control, while the term $\frac{\partial x^\epsilon(T)}{\partial x_\epsilon}$ describes the propagation of this change of the state in the interval from time τ to time T.

Effect of a strong variation of the control

We write (cf. (7.1))

$$x^\epsilon(\tau) = x^\epsilon(\tau - \epsilon) + \int_{\tau-\epsilon}^{\tau} f(x^\epsilon(s), v) ds,$$

$$x(\tau) = x(\tau - \epsilon) + \int_{\tau-\epsilon}^{\tau} f(x(s), u(s)) ds.$$

DOI: 10.1201/9781003051268-F

Since $x^\epsilon(\tau - \epsilon) = x(\tau - \epsilon)$, we have

$$x^\epsilon(\tau) = x(\tau) + \int_{\tau-\epsilon}^{\tau} (f(x^\epsilon(s), v) - f(x(s), u(s)))ds.$$

Adding and subtracting $f(x(s), v)$ inside the integral, we obtain

$$x^\epsilon(\tau) = x(\tau) + \int_{\tau-\epsilon}^{\tau} (f(x(s), v) - f(x(s), u(s))) ds$$

$$+ \int_{\tau-\epsilon}^{\tau} (f(x^\epsilon(s), v) - f(x(s), v)) ds.$$

As $\epsilon \to 0$, the second integral goes to zero faster than ϵ since both the integrand and the interval of integration go to zero. Therefore, using a Taylor series for the first integral, with initial point $\epsilon = 0$, we can write

$$x_\epsilon := x^\epsilon(\tau) = x(\tau) + (f(x(\tau), v) - f(x(\tau), u(\tau))) \epsilon + o(\epsilon).$$

Differentiating this with respect to ϵ, at $\epsilon = 0$, we obtain

$$\left.\frac{\partial x_\epsilon}{\partial \epsilon}\right|_{\epsilon=0} = f(x(\tau), v) - f(x(\tau), u(\tau)). \tag{F.3}$$

This gives the second factor in the right-hand side of (F.2).

Variational equation

On the interval $(\tau, T]$, x^ϵ and x satisfy the same differential equation (7.1), that is,

$$\dot{x} = f(x(t), u(t)), \tag{F.4}$$

but with different initial conditions. For a general differential equation of the form (F.4), write x as $x := x(t, y)$ where the dependence on the initial condition y is made explicit. Differentiating both sides of (F.4) with respect to y at the actual value of the initial condition x_0, and changing the order of differentiation, we obtain

$$\left.\frac{d}{dt}\frac{\partial}{\partial y}x(t, y)\right|_{y=x_0} = f_x(x(t, x_0), u(t))\left.\frac{\partial}{\partial y}x(t, y)\right|_{y=x_0}.$$

If V is defined as $V(t) := \frac{\partial}{\partial y}x(t, y)|_{y=x_0}$, it measures the *sensitivity* of the solution $x(t, x_0)$ to variations of the initial condition x_0. It satisfies the **variational equation**

$$\dot{V} = f_x(x(t, x_0), u(t))V, \tag{F.5}$$

with initial condition equal to the identity matrix.

The Pontryagin maximum principle

In (F.2), $\frac{\partial x^\epsilon(T)}{\partial x_\epsilon}\big|_{\epsilon=0}$ is equal to $V(T, \tau)$, where V is the solution of the variational equation (cf. (F.5))

$$\frac{\partial V(t, \tau)}{\partial t} = f_x(x(t, x(\tau)), u(t))V(t, \tau), \tag{F.6}$$

with initial condition equal to the identity matrix. The expression for $\frac{\partial x_\epsilon}{\partial \epsilon}\big|_{\epsilon=0}$ is given in (F.3). Therefore, we obtain for (F.2)

$$\frac{d}{d\epsilon}x^\epsilon(T)\bigg|_{\epsilon=0} = V(T, \tau)\left(f\left(x(\tau), v\right) - f\left(x(\tau), u(\tau)\right)\right).$$

Plugging this into (F.1), we obtain

$$\frac{d}{d\epsilon}J(u^\epsilon)\bigg|_{\epsilon=0} = \frac{d}{d\epsilon}\phi(x^\epsilon(T))\bigg|_{\epsilon=0}$$

$$= \phi_x\left(x(T)\right)V(T, \tau)\left(f\left(x(\tau), v\right) - f\left(x(\tau), u(\tau)\right)\right) \geq 0.$$

We summarize this in the following theorem.

Theorem F.0.1 *Assume u is the optimal control and x is the corresponding trajectory, both of them defined in $[0, T]$. Then, for every $\tau \in (0, T]$ and v in the set of admissible values of the control, if $V(\cdot, \tau)$ is the solution of the variational equation (F.6), starting at τ, with $V(\tau, \tau)$ equal to the identity, we have*

$$\phi_x(x(T))V(T, \tau)\left(f\left(x(\tau), v\right) - f\left(x(\tau), u(\tau)\right)\right) \geq 0. \tag{F.7}$$

The necessary conditions for optimality are usually not expressed in the form stated in Theorem F.0.1. They are reformulated as follows. Given an optimal control u and corresponding trajectory x, define an n-dimensional time-varying vector function λ, called *the costate*, satisfying the linear time-varying differential equation

$$\dot{\lambda}^T = -\lambda^T f_x(x(t), u(t)). \tag{F.8}$$

These equations are called *adjoint equations*. If z is a solution of the linear time-varying equation

$$\dot{z} = f_x(x(t), u(t))z, \tag{F.9}$$

a straightforward calculation shows that $\frac{d}{dt}\lambda^T z = 0$ so that

$$\lambda^T(t)z(t) = constant.$$

Now, fix τ and consider the solution of (F.9) with initial condition

$$z(\tau) = f\left(x(\tau), v\right) - f\left(x(\tau), u(\tau)\right).$$

We have
$$z(t) = V(t, \tau) \left(f\left(x(\tau), v\right) - f\left(x(\tau), u(\tau)\right) \right)$$

and
$$\lambda^T(t) V(t, \tau) \left(f\left(x(\tau), v\right) - f\left(x(\tau), u(\tau)\right) \right) = constant. \tag{F.10}$$

In particular, if we choose
$$\lambda^T(T) = -\phi_x(x(T)), \tag{F.11}$$

we have
$$\lambda^T(t) V(t, \tau) \left(f\left(x(\tau), v\right) - f\left(x(\tau), u(\tau)\right) \right) \tag{F.12}$$
$$= -\phi_x\left(x(T)\right) V(T, \tau) \left(f\left(x(\tau), v\right) - f\left(x(\tau), u(\tau)\right) \right) \leq 0,$$

for every $t \in [\tau, T]$, where we have used (F.7). Writing (F.12) for $t = \tau$, using the fact that $V(\tau, \tau)$ is equal to the identity, we obtain

$$\lambda^T(\tau)(f(x(\tau), v) - f(x(\tau), u(\tau))) \leq 0. \tag{F.13}$$

Equation (F.13), along with (F.8) and (F.11), gives equations (7.19), (7.17) and (7.18) of Theorem 7.2.1, respectively. Equation (F.10) with $\tau = t$ gives equation (7.20).

List of Acronyms and Symbols

$[A, B] := AB - BA$, commutator of A and B.

$\{A, B\} := AB + BA$, anti-commutator of A and B.

$\langle \cdot, \cdot \rangle_K$ Killing form

$\bar{\oplus}$ Direct sum of Lie algebras

$\mathbf{1}$ Identity matrix or identity operator

$\mathbf{1}_n$ $n \times n$ Identity matrix or identity operator on a vector space of dimension n.

$\| \cdot \|$ Norm of an operator or a vector.

$\|A\|_1$ Trace norm of the matrix A (sum of the singular values)

$A^\dagger$ transpose conjugate (adjoint) of an operator A.

$a^\dagger$ conjugate of an element a in the group algebra $\mathbb{C}[G]$.

A^a anti-transpose of the matrix A.

$ad_X(Q) := [X, Q]$

$\vec{A}$ vector potential.

$\mathbb{C}[G]$ group algebra associated with the finite group G

δ_{jk} Kronecker delta symbol

$e^{\mathcal{L}}$ Lie group associated to the Lie algebra $\mathcal{L}$

$End(V)$ Algebra of Endomorphisms on the vector space V.

$End^G(V)$ Algebra of Endomorphisms on the vector space V commuting with the group G.

$\text{Im}(x)$ Imaginary part of x.

$\mathcal{L}^G(n)$ Lie subalgebra of $\mathcal{L}$ of matrices commuting with matrices in G.

$\text{Re}(x)$ Real part of x.

ρ density matrix.

$\rho = \rho(\vec{r})$ charge density as a function of the space variable $\vec{r}$. .

$su^G(n)$ Lie subalgebra of $su(n)$ of matrices commuting with matrices in G.

$u^G(n)$ Lie subalgebra of $u(n)$ of matrices commuting with matrices in G.

$\hat{S}$ Von Neumann Entropy

S_n group of permutations of n objects.

a.e. almost everywhere

h.o.t. higher order terms

BCH Baker-Campbell-Hausdorff

CCD Concurrence Canonical Decomposition

CSA Cartan subalgebra

DMC Density Matrix Controllability

ESC Equivalent State Controllability

FID Free Induction Decay

GYS Generalized Young Symmetrizer

LOCC Local Operation along with Classical Communication

LU Local Unitary

NMR Nuclear Magnetic Resonance

OC Operator Controllability

OED Odd-Even Decomposition

PMP Pontryagin Maximum Principle

PSC Pure State Controllability

$rref(A)$ reduced row echelon form of the matrix A.

References

[1] A. F. Abouraddy, B. E. A. Saleh, A. V. Sergienko and M. C. Teich, Degree of entanglement for two qubits, *Phys. Rev. A*, 64, 050101, (2001).

[2] A. Abragam, *Principles of Nuclear Magnetism*, International Series of Monographs on Physics, Vol. 32, Oxford Science Publications, New York, Reprint 2002.

[3] A. A. Agrachev and Y. Sachkov, *Control Theory from the Geometric Viewpoint*, Encyclopedia of Mathematical Sciences, Vol. 87, Springer, 2004.

[4] Y. Akbari-Kourbolagh and M. Azhdargalam, Entanglement criterion for multipartite systems based on quantum Fisher information, *Phys. Rev. A*, 99, 012304, (2019).

[5] N. I. Akhiezer and I. M. Glazman, *Theory of Linear Operators in Hilbert Spaces*, Pitman, Boston, 1981.

[6] F. Albertini and D. D'Alessandro, The Lie algebra structure and controllability of spin systems, *Linear Algebra Appl.*, 350, 213–235, (2002).

[7] F. Albertini and D. D'Alessandro, Notions of controllability for bilinear multilevel quantum systems, *IEEE T. Automat. Contr.*, 48, No. 8, 1399–1403, (2003).

[8] F. Albertini and D. D'Alessandro, Control of the evolution of Heisenberg spin systems, *Eur. J. Control*, Special issue on Lagrangian and Hamiltonian Methods for Nonlinear Control, 10/5, 497–504, (2004).

[9] F. Albertini and D. D'Alessandro, Model identification for spin networks, *Linear Algebra Appl.*, 394, 237–256, (2005).

[10] F. Albertini and D. D'Alessandro, Input-output model equivalence of spin systems: a characterization using Lie algebra homomorphisms, *SIAM J. Control Optim.*, 47, No. 4, 2016–2043, (2008).

[11] F. Albertini and D. D'Alessandro, On symmetries in time optimal control, sub-Riemannian geometries and the K-P problem, *J. Dyn. Control Syst.*, 23, No. 1, pp. 13-38, (2018).

[12] F. Albertini and D. D'Alessandro, Controllability of symmetric spin networks, *J. Math. Phys.*, 59, No. 5, 052102, (2018).

[13] F. Albertini and D. D'Alessandro, Subspace controllability of bipartite symmetric spin networks, *Linear Algebra Appl.*, 585 , 1–23, (2020).

[14] F. Albertini, D. D'Alessandro and B. Sheller, Sub-Riemannian geodesics on $SU(n)/S(U(n-1) \times U(1))$ and optimal control of three level quantum systems, *IEEE T. Automat. Contr.*, vol. 65, no. 3, 1176–191, (2020).

[15] S. Albeverio, L. Cattaneo, S. M. Fei and X. H. Wang, Equivalence of tripartite quantum states under local unitary transformations, *Int. J. Quantum Inf.*, 3, No. 4, 603–609, (2005).

[16] S. Albeverio, S. M Fei and D. Goswami, Local invariants for a class of mixed states, *Phys. Lett. A.*, 340, 37–42, (2005).

[17] S. Albeverio, S. M. Fei, P. Parashar and W. L. Yang, Nonlocal properties and local invariants for bipartite systems, *Phys. Rev. A,* 68, 010313 (R), (2003).

[18] J. Alcock-Zeilinger and H. Weigert, Compact Hermitian Young projection operators, *J. Math. Phys.*, 58, No. 5, (October 2016).

[19] S. T. Ali. and M. Engliš, Quantization methods: a guide for physicists and analysts, *Rev. Math. Phys.*, 17, 391–490, (2005).

[20] R. Alicki and K. Lendi, *Quantum Dynamical Semigroups and Applications*, Lecture Notes in Physics, Vol. 286, Springer-Verlag, 1987.

[21] C. Altafini, Controllability of quantum mechanical systems by root space decompositions of $su(n)$, *J. Math. Phys.*, 43, 2051–2062, (2002).

[22] C. Altafini and F. Ticozzi, Modeling and control of quantum systems; an introduction, *IEEE T. Automat. Contr.*, 57, No. 8, 1898–1917, (2012).

[23] C. Arenz, G. Gualdi, and D. Burgarth, Control of open quantum systems: case study of the central spin model, *New J. Phys.*, 16 065023 (2014).

[24] G. B. Arfken and H. J. Weber, *Mathematical Methods for Physicists*, 4th edition, Academic Press, San Diego, 1995.

[25] V. I. Arnold, *Mathematical Methods of Classical Mechanics*, Springer-Verlag, New York, 1978.

[26] A. Assion, T. Baumert, M. Bergt, T. Brixner, B. Kiefer, V. Seyfried, M. Strehle and G. Gerber, Control of chemical reactions by feedback-optimized phase-shaped femtosecond laser pulses, *Science*, 282, No. 30, 919–922, (1998).

[27] K. Audenaert, F. Verstraete and B. De Moor, Variational characterisation of separability and entanglement of formation, *Phys. Rev. A*, 64, 052304, (2001).

[28] J. E. Avron, R. Seiler and L. G. Yaffe, Adiabatic theorems and applications to the quantum Hall effect, *Comm. Math. Phys.*, 110, 33–49, (1987).

[29] V. Ayala and A. Da Silva, About the continuity of reachable sets of restricted affine control systems, *Chaos, Solitons Fractals*, 94, 37–43, (2017).

[30] K. Beauchard and P. Rouchon, Bilinear control of Schrödinger PDEs, in *Encyclopedia of Systems and Control*, J. Baillieul and T. Samad eds., Springer, London, 2015.

[31] G. Benenti, G. Casati and G. Strini, *Principles of Quantum Computation and Information*, Volume I: Basic Concepts, World Scientific Publishing Co., River Edge, NJ, 2004.

[32] I. Bengtsson and K. Zyczkowski, *Geometry of Quantum States: an Introduction to Quantum Entanglement*, Cambridge University Press, Cambridge, New York, 2006.

[33] C. H. Bennett, H. J. Bernstein, S. Popescu and B. Schumacher, Concentrating partial entanglement by local operations, *Phys. Rev. A*, 53, 2046, (1996).

[34] C. H. Bennett, D. P. DiVincenzo, C. A. Fuchs, T. Mor, E. Rains, P. W. Shor, J. Smolin and W. K. Wootters, *Phys. Rev. A*, 59, 1070, (1999).

[35] K. Bergmann, H. Theuer and B. W. Shore, Coherent population transfer among quantum states of atoms and molecules, *Rev. Mod. Phys.*, **70**, 3, 1003–1025, (1998).

[36] A. M. Bloch, R. W. Brockett and C. Rangan, Finite controllability of infinite-dimensional quantum systems, *IEEE T. Automat. Contr.*, 55, No. 8, 1797–1805, (Aug. 2010).

[37] F. Bloch, W. Hansen and M. Packard, Nuclear induction, *Phys. Rev.*, 69, 127 (1946).

[38] R. Blume-Kohout, Optimal, reliable estimation of quantum states, *New J. Phys.*, 12, No. 4, 043034, (2010).

[39] W. M. Boothby, *An Introduction to Differentiable Manifolds and Riemannian Geometry*, Academic Press, Orlando, FL, 1986.

[40] U. Boscain, Resonance of minimizers for *n*-level quantum systems, *Proceedings of the 42-nd Conference on Decision and Control*, Maui, Hawaii, Dec 9–12, 2003, Vol. 4, pp. 416–421.

[41] U. Boscain, T. Chambrion and J. P. Gauthier, On the K+P problem for a three-level quantum system: optimality implies resonance, *J. Dyn. Control Syst.*, No. 8, 547–572, (2002).

[42] U. Boscain, T. Chambrion and G. Charlot, Nonisotropic 3-level quantum systems: complete solutions for minimum time and minimum energy, *Discrete Cont. Dyn. Syst.-B*, 5, No. 4, 957–990, (2005).

[43] U. Boscain and B. Piccoli, *Optimal Syntheses for 2-D Manifolds*, Springer SMAI, Vol. 43, 2004.

[44] D. W. Boukhvalov, M. Al-Seqer, E. Z. Kurmaev, A. Moewes, V. R. Galakhov, L. D. Finkelstein, S. Chiuzbaian, M. Neumann, V. V. Dobrovitski, M. I. Katsnelson, A. I. Lichtenstein, B. N. Harmon, K. Endo, J. M. North and N. S. Dalal, Electronic structure of a Mn_{12} molecular magnet: theory and experiments, *Phys. Rev. B*, 75, 014419, (2007).

[45] H. P. Breucr, Optimal entanglement criterion for mixed quantum states, *Phys. Rev. Lett.*, 97, 080501, (2006).

[46] H. P. Breuer, Separability criteria and bounds for entanglement measures, *J. Phys. A, Math. Gen.*, 39, 11847, (2006).

[47] H. P. Breuer and F. Petruccione, *The Theory of Open Quantum Systems*, Oxford University Press, Oxford; New York, 2002.

[48] P. Brumer and M. Shapiro, Control of unimolecular reactions using coherent light, *Chem. Phys. Lett.*, 126, 541–546, (1986).

[49] R. K. Brylinski and G. Chen eds., *Mathematics of Quantum Computation*, Chapman and Hall, CRC, Boca Raton, FL, 2002.

[50] S. S. Bullock and G. K. Brennen, Canonical decompositions of n-qubits quantum computations and concurrence, *J. Math. Phys.*, 45, No. 6, 2447, (2004).

[51] S. S. Bullock, G. K. Brennen and D. P. O'Leary, Time reversal and n-qubit canonical decompositions, *J. Math. Phys.*, 46, 062104, (2005).

[52] A. Bunse-Gerstner, R. Byers and V. Mehrmann, A chart of numerical methods for structured eigenvalue problems, *SIAM J. Matrix Anal. Appl.*, 13, 419, (1992).

[53] A. G. Butkovskiy and Y. I. Samoilenko, *Control of Quantum Mechanical Processes and Systems*, Mathematics and its Applications, Kluwer Academic Publishers, Dordrecht, 1990.

[54] R. N. Cahn, *Semi-simple Lie Algebras and Their Representations*, Frontiers in Physics, Lecture notes series 59, The Benjamin/Cummings Publishing Company, Inc., Menlo Park, California, 1984.

[55] H. Y. Carr and E. M. Purcell, Effects of diffusion on free precession in nuclear magnetic resonance experiments. *Phys. Rev.*, 94, 630–638, (1954).

[56] E. Cartan, Sur une class remarkable d'espaces de Riemann, *Bull. Suc. Math. France*, 54, 214–264, (1926).

[57] E. Cartan, Sur une class remarkable d'espaces de Riemann, *Bull. Suc. Math. France*, 55, 114–134, (1927).

[58] H. A. Carteret, A. Higuchi and A. Sudbery, Multipartite generalisation of the Schmidt decomposition, *J. Math. Phys.*, 41, 7932, (2000).

[59] J. Cavanagh, W. J. Fairbrother, A. G. Palmer and N. J. Skelton, *Protein NMR Spectroscopy, Principles and Practice*, Academic Press, San Diego, 1996.

[60] R. Chakrabarti, H. l Rabitz, Quantum control landscapes, *Int. Rev. Phys. Chem.*, 26, (2007).

[61] J. Chen, H. Zhou, C. Duan, and X. Peng, Preparing GHZ and W states on a long-range Ising spin model by global control, *Physical Review A* 95, 032340 (2017)

[62] K. Chen, S. Albeverio and S. M. Fei, Concurrence of arbitrary dimensional bipartite quantum states, *Phys. Rev. Lett.,* 95, 040504, (2005).

[63] K. Chen and L. A. Wu, A matrix realignment method for recognizing entanglement, *Quantum Inf. Comput.*, 3, 193, (2003).

[64] C. W. Chou, D .B. Hume, T. Rosenband and D. J. Wineland, Optical clocks and relativity, *Science*, 329, 1630, (2010).

[65] J. I. Cirac and P. Zoller, Quantum computation with cold trapped ions, *Phys. Rev. Lett.*, 74, No. 20, 4091–4094, (1995).

[66] C. Cohen-Tannoudji, B. Diu and F. Laloe, *Quantum Mechanics*, Wiley, New York, 1977.

[67] C. Cohen-Tannoudji, J. Dupont-Roc and G. Grynberg, *Photons and Atoms; Introduction to Quantum Electrodynamics*, Wiley Professional Paperback Series, New York, 1997.

[68] J. B. Conway, *A Course in Functional Analysis*, Graduate Texts in Mathematics, 96, Springer-Verlag, New York, 1990.

[69] G. W. Coulston and K. Bergmann, Population transfer by stimulated Raman scattering with delayed pulses: analytical results for multilevel systems, *J. Chem. Phys.*, 96, 3467, (1992).

[70] M. L. Curtis, *Matrix Lie Groups*, Springer, New York, 1979.

[71] S. Daftuar and P. Hayden, Quantum state transformations and the Schubert calculus, *Ann. Phys.*, 315, 80-122, (2005).

[72] M. Dagli, D. D'Alessandro and J. D. H. Smith, A general framework for recursive decompositions of unitary quantum evolutions, *J. Phys. A, Math. Theor.*, **41**, 15, 155302, (2008).

[73] D. D'Alessandro, Algorithms for quantum control based on decompositions of Lie groups, in *Proceedings of the 39-th Conference on Decision and Control*, Sidney, Dec. 2000, pp. 967–968.

[74] D. D'Alessandro, Uniform finite generation of compact Lie groups, *Syst. Contr. Lett.*, 47, 87–90, (2002).

[75] D. D'Alessandro, On quantum state observability and measurement, *J. Phys. A: Math. Gen.*. 36, 9721–9735, (2003).

[76] D. D'Alessandro, Controllability of one spin and two interacting spins, *Math. Contr. Signals Syst.*, 16, 1–25, (2003).

[77] D. D'Alessandro, On the observability and state determination of quantum mechanical systems, *Proceedings of the 43-rd Conference on Decision and Control*, Paradise Island, Bahamas, Dec. 2004.

[78] D. D'Alessandro, Optimal evaluation of generalized Euler angles with application to control, *Automatica*, 40, 1997–2002, (2004).

[79] D. D'Alessandro, General methods to control right-invariant systems on compact Lie groups and multilevel quantum systems, *J. Phys. A: Math. Theor.*, **42**, 395301, (2009).

[80] D. D'Alessandro, Constructive decomposition of the controllability Lie algebra for Quantum systems, *IEEE T. Automat. Contr.*, 1416–1421, (June 2010).

[81] D. D'Alessandro and F. Albertini, Quantum symmetries and Cartan decompositions in arbitrary dimensions, *J. Phys. A: Math. Theor.*, 40, 2439–2453, (2007).

[82] D. D'Alessandro, F. Albertini and R. Romano, Exact algebraic conditions for indirect controllability in quantum coherent feedback schemes, *SIAM J. Control Optim.*, 53-3, 1509–1542, (2015).

[83] D. D'Alessandro and M. Dahleh, Otimal control of two-level quantum systems, *IEEE T. Automat. Cont.*, 45, No. 1, (2001).

[84] D. D'Alessandro and J. Hartwig, Dynamical decomposition of Bilinear control systems subject to symmetries, *J. Dyn. Control Syst.*, 27 (2021), no. 1, 1–30.

[85] D. D'Alessandro and R. Romano, Further results on the observability of quantum systems under general measurement, *Quantum Inf. Proc.*, 5, No. 3, 139–160, (2006).

[86] D. D'Alessandro and R. Romano, Decompositions of unitary evolutions and entanglement dynamics of bipartite quantum systems, *J. Math. Phys.*. 47, 082109, (2006).

[87] D. D'Alessandro, B. Sheller and Z. Zhu, Time-optimal control of quantum Lambda systems in the KP configuration, *J. Math. Phys.*, 61, 052107, (2020).

[88] R. W. Daniels, *An Introduction to Numerical Methods and Optimization Techniques*, North-Holland, New York, 1978.

[89] G. M. D'Ariano, M. G. A. Paris, and M. F. Sacchi, Quantum tomographic methods, in *Quantum State Estimation*, M. Paris and J. Řeháček eds., *Lect. Notes Phys.*, 649, 7–58, (2004).

[90] G. M. D'Ariano, L. Maccone and M. Paini, Spin tomography, *J. Opt. B: Quantum Semiclass. Opt.*, 5, 77–84, (2003).

[91] J. H. Davies, *The Physics of Low-Dimensional Semiconductors*, Cambridge University Press, Cambridge, U.K., 1998.

[92] W. A. de Graaf, *Lie Algebras; Theory and Algorithms*, North-Holland, 2000.

[93] S. Deleglise, I. Dotsenko, C. Sayrin, J. Bernu, M. Brune, J. M. Raimond and S. Haroche, Reconstruction of non-classical cavity field states with snapshots of their decoherence, *Nature*, 455, 510–514. (25 September 2008).

[94] P. Deligne, P. Etingof et al., *Quantum Fields and Strings: A Course for Mathematicians*, American Mathematical Society, IAS, 1999.

[95] M. Demiralp and H. Rabitz, Optimally controlled quantum molecular dynamics: a perturbation formulation and the existence of multiple solutions, *Phys. Rev. A*, 47, No. 2, 809, (1992).

[96] F. Diedrich, J. C. Bergqvist, W. M. Itano and D. J. Wineland, Laser cooling to the zero-point energy of motion, *Phys. Rev. Lett.* **62**, 403, (1989).

[97] J. D. Dixon, *Problems in Group Theory*, Dover Publications 2007, Mineola N. Y., reprinted from Blaidshell Publishing Company, Waltham, MA, 1967.

[98] A. C. Doherty, P. A. Parrilo and F. M. Spedalieri, Distinguishing separable and entangled states, *Phys. Rev. Lett.*, 88, 187904, (2002).

[99] M. J. Donald, M. Horodecki and O. Rudolph, The uniqueness theorem for entanglement measures, *J. Math. Phys.*, 43, 4252, (2002).

[100] J. Dongarra, J. Gabriel, D. Koelling and J. Wilkinson, The eigenvalue problem for Hermitian matrices with time reversal symmetry, *Linear Algebra App.*, 60, 27, (1984).

[101] E. B. Dynkin, Semi-simple subalgebras of semi-simple Lie algebras, *Am. Math. Soc. Trans. Ser. 2*, 6, 111–244, (1957).

[102] A. Einstein, B. Podolsky and N. Rosen, Can quantum mechanical description of physical reality be considered complete?, *Phys. Rev. A*, 47, 777, (1935).

[103] A. Elgart and J. E. Avron, Adiabatic theorem without a gap condition, *Commun. Math. Phys.*, 203, 445–463, (1999).

[104] K. Erdmann and M. J. Wildon, *Introduction to Lie Algebras*, Springer Undergraduate Mathematics Series, 2007.

[105] R. R. Ernst, G. Bodenhausen and A. Wokaun, *Principles of Nuclear Magnetic Resonance in One and Two Dimensions*, Oxford University Press, 1987.

[106] E. Farhi, J. Goldstone, S. Gutmann and M. Spiser, Quantum computation by adiabatic evolution, xxx.lanl.quant-ph/0001106.

[107] E. Farhi, J. Goldstone, G. Sam, J. Lapan, A. Lundgren and D. Preda, A quantum adiabatic evolution algorithm applied to random instances of an NP-complete problem, *Science*, 292, 472–474, (April 20-th 2001).

[108] Shao-Ming Fei and Naihuan Jing, Equivalence of quantum states under local unitary transformations, *Phys. Lett. A*, 342, 77–81, (2005).

[109] R. P. Feynman, R. B. Leighton and M. Sands, Volume III of *The Feynman Lectures on Physics*, Addison-Wesley, Reading, Mass, 1965.

[110] W. Fleming and R. Rishel, *Deterministic and Stochastic Optimal Control*, Applications of Mathematics, Springer-Verlag, New York, 1975.

[111] W. Fulton, *Young Tableaux: With Applications to Representation Theory and Geometry* (London Mathematical Society Student Texts, pp. I-Iv). Cambridge University Press, Cambridge, 1996.

[112] W. Fulton and J. Harris, *Representation Theory; A First Course*, Graduate Texts in Mathematics, No. 129, Springer, New York, 2004.

[113] A. Galindo and P. Pascual, *Quantum Mechanics I, Texts and Monographs in Physics*, Springer-Verlag, Heidelberg, 1990.

[114] X. H. Gao, S. Albeverio, S. M. Fei and Z. X. Wang, Matrix tensor product approach to the equivalence of multipartite states under local unitary transformations, *Commun. Theor. Phys.*, 45, 267–270, (2006).

[115] S. Gasiorowicz, *Quantum Physics*, 3rd edition, Wiley, Hoboken, N.J., 2003.

[116] S. J. Glaser, U. Boscain, T. Calarco, C. P. Koch, W. Köckenberger, R. Kosloff, I. Kuprov, B. Luy, S. Schirmer, T. Schulte-Herbrüggen, D.

Sugny and F. K. Wilhelm, Training Schrodinger cats: quantum optimal control, *Eur. Phys. J.* D 69, 279, (2015).

[117] D. E. Goldberg, *Genetic Algorithms in Search, Optimization and Machine Learning*, Addison-Wesley, Reading, UK, 1993.

[118] H. Goldstein, *Classical Mechanics*, 2nd edition, Addison-Wesley, Reading, MA, 1980.

[119] W. Gordy, *Theory and Applications of Electron Spin Resonance*, Wiley, New York, 1980.

[120] K. Gottfried, *Quantum Mechanics:Fundamentals*, Graduate Texts in Contemporary Physics, Springer, New York, 2003.

[121] M. Grassl, M. Rötteler and T. Beth, Computing local invariants of qubit systems, *Phys. Rev. A*, 58, 1833, (1998).

[122] S. Grivopoulos and B. Bamieh, Iterative algorithms for optimal control of quantum systems, in *Proceedings of the 41-st IEEE Conference on Decision and Control*, Las Vegas, Nevada, USA, December 2002, pp. 2687–2691.

[123] S. Grivopoulos and B. Bamieh, Lyapunov-based control of quantum systems, in *Proceedings of the 42-nd Conference on Decision and Control*, Maui, Hawaii, USA, December 2003, pp. 434–437.

[124] H. J. Groenewold, On the principles of elementary quantum mechanics, *physica*, 12, No. 7, 405–460, (1946).

[125] S. Gudder, Quantum computation, *Amer. Math. Monthly*, 110, No. 3, 181–201, (2003).

[126] G. A. Hagedorn and A. Joye, Elementary exponential error estimates for the adiabatic approximation, *J. Math. Anal. Appl.*, 267, 235–246, (2002).

[127] B. Hall, *Lie Groups, Lie Algebras, and Representations, An Elementary Introduction*, Graduate Texts in Mathematics, 222, Springer-Verlag, 2003.

[128] K. He, M. Huang and J. Hou, Entanglement criterion independent on observables for multipartite Gaussian states based on uncertainty principle, *Scientific Reports*, 9, 1314, (2019).

[129] W. P. Healy, *Non-Relativistic Quantum Electrodynamics*, Academic Press, New York, 1982.

[130] S. Helgason, *Differential Geometry, Lie Groups and Symmetric Spaces*, Academic Press, New York, 1978.

[131] R. Hermann, *Lie Groups for Physicists*, Benjamin, New York, 1966.

[132] S. Hill and W. Wootters, Entanglement of a pair of quantum bits, *Phys. Rev. Lett.*, 78, 5022, (1997).

[133] M. Hirose and P. Cappellaro, Time optimal control with finite bandwidth, *Quantum Inf. Proc.*, 17, No. 4, (2018).

[134] T. S. Ho and H. Rabitz, Why do effective quantum controls appear easy to find? *J. Photochemi. Photobiol. A: Chem.*, 180, 226-240. (2006).

[135] R. A. Horn and C. R. Johnson, *Matrix Analysis*, Cambridge University Press, Cambridge, New York, 1985.

[136] R. A. Horn and C. R. Johnson, *Topics in Matrix Analysis*, Cambridge University Press, Cambridge, New York, 1991.

[137] M. Horodecki, Entanglement measures, *Quantum Inf. Comput.*, 1, No. 1, 3–26, (2001).

[138] M. Horodecki and P. Horodecki, Reduction criterion of separability and limits for a class of protocols of entanglement distillation, *Phys. Rev. A*, 59, 4206–4216, (1999).

[139] M. Horodecki, P. Horodecki and R. Horodecki, Separability of mixed states: necessary and sufficient conditions, *Phys. Lett. A*, 223, 1–8, (1996).

[140] R. Horodecki, P. Horodecki, M. Horodecki, K. Horodecki, Quantum entanglement, *Rev. Mod. Phys.*, 865–942 (2009).

[141] S. Howison, *Practical Applied Mathematics. Modelling, Analysis, Approximation*, Cambridge Texts in Applied Mathematics, No. 38, Cambridge, U.K., New York, 2005.

[142] G. M. Huang, T. J. Tarn and J. W. Clark, On the controllability of quantum mechanical systems, *J. Math. Phys.*, 24, No. 11, 2608–2618, (1983).

[143] M. Hsieh, R. B. Wu and H. Rabitz, Topology of quantum control landscape for observables, *J. Chem. Phys.*, 130, 104109, (2009).

[144] J. E. Humphreys, *Introduction to Lie algebras and Representation Theory*, Springer-Verlag, New York, 1972.

[145] L. M. Ioannou, Computational complexity of the quantum separability problem, *Quantum Inf. Comput.*, 7, No. 4, 335–370, (2007).

[146] T. Ionescu, On the generators of semisimple Lie algebras, *Linear Algebra Appl.*, 15, No. 3, 271–292, (2007).

[147] J. D. Jackson, *Classical Electrodynamics*, 3rd edition, Wiley, New York, 1999.

[148] D. F. V. James, Quantum dynamics of cold trapped ions with application to quantum computation, *Appl. Phys. B*, 66, 181–190, (1998).

[149] A. Jamiołkowski, On observability of N-level quantum systems, *Rep. Math. Phys.*, 21, No. 1, 101–109, (1985).

[150] S. Jansen, M. B. Ruskai and R. Seiler, Bounds for the adiabatic approximation with applications to quantum computation, *J. Math. Phys.*, 48, 102111, (2007).

[151] P. Jorrand and M. Mhalla, Separability of pure N-qubit states: two characterizations, *Int. J. Foundations Comput. Sci. (IJFCS)*, 14, No. 5, 797–814, (2003).

[152] A. Joye and C. Pfister, Full asymptotic expansion of transition probabilities in the adiabatic limit, *J. Phys. A*, 24, 753–766, (1991).

[153] V. Jurdjević, *Geometric Control Theory*, Cambridge University Press, 1997.

[154] V. Jurdjević and H. Sussmann, Control systems on Lie groups, *J. Different. Equat.*, 12, 313–329, (1972).

[155] T. Kailath, *Linear Systems*, Prentice-Hall, Englewood Cliffs, NJ, 1980.

[156] T. Kato, On the adiabatic theorem of quantum mechanics, *J. Phys. Soc. Japan*, 5, 435–439, (1950).

[157] K. Kawakubo, *The Theory of Transformation Groups*, Oxford University Press, New York, 1991.

[158] S. Keppeler and M. Sjödal, Hermitian Young operators, *J. Math. Phys.*, 55, 021702, (2014).

[159] H. K. Khalil, *Nonlinear Systems*, 3rd edition, Prentice Hall, Upper Saddle River, NJ, 2002.

[160] N. Khaneja, *Geometric Control in Classical and Quantum Systems*, Ph. D. Thesis, Division of Engineering and Applied Sciences, Harvard University, Cambridge, MA, June 2000.

[161] N. Khaneja, R. Brockett and S. J. Glaser, Time optimal control of spin systems, *Phys. Rev. A*, 63, 032308, (2001).

[162] N. Khaneja and S. J. Glaser, Cartan decomposition of $SU(n)$; constructive controllability of spin systems and universal quantum computing, *Chem. Phys.*, 267, 11, (2001).

[163] N. Khaneja, S. J. Glaser and R. W. Brockett, Sub-Riemannian geometry and time optimal control of three spin systems: coherence transfer and quantum gates, *Phys. Rev. A*, 65, 032301, (2002).

[164] N. Khaneja and S. Glaser, Optimal control of coupled spin dynamics under cross-correlated relaxation, in *Proceedings of the 42-nd IEEE Conference on Decision and Control*, Maui, Hawaii, USA, December 2003, pp. 422–427.

[165] N. Khaneja, F. Kramer and S. Glaser, Optimal experiments for maximizing coherence transfer between coupled spins, *J. Magnet. Resonance*, 173, 116–124, (2005).

[166] D. E. Kirk, *Optimal Control Theory; An Introduction*, Prentice-Hall, Englewood Cliffs, NJ, 1970.

[167] A. Y. Kitaev, A. H. Shen and M. N. Vyalyi, *Classical and Quantum Computation*, Graduate Studies in Mathematics. Vol. 47, American Mathematical Society Translations, 2002.

[168] A. Knapp, *Lie Groups; Beyond an Introduction*, 2nd edition, Progress in Mathematics, Vol. 140, Birkäuser, Boston, 2002.

[169] R. M. Koch and F. Lowenthal, Uniform finite generation of three-dimensional linear Lie groups, *Canad. J. Math.*, 27, 396–417, (1975).

[170] K. Kraus, *States, Effects, and Operations*, Vol. 190 of Lecture Notes in Physics, Springer-Verlag, Berlin, 1983.

[171] H. Kunita, Supports of diffusion processes and controllability problems, *Proc. of Intern. Symp. Stochastic Differential Equations*, Kyoto 1976, pp. 163–185.

[172] M. Kuranishi, On everywhere dense imbeddings of free groups in Lie groups, *Nagoya Math. J.*, **2**, 63–71,(1951).

[173] M. Kus and K. Zyczkowski, Geometry of entangled states, *Phys. Rev. A*, 63, 032307–13, (2001).

[174] C. Lanczos, *The Variational Principles of Mechanics*, University of Toronto Press, 1970.

[175] L. D. Landau and E. M. Lifshitz, *Mechanics*, Pergamon, New York, 1976.

[176] D. C. Lay, S. R. Lay, J. McDonald and J. J. McDonald *Linear Algebra and its Applications*, 5th edition, Pearson, 2015.

[177] B. Lee and Markus, *Foundations of Optimal Control Theory*, Wiley, New York, 1967.

[178] Dynamics of the dissipative two state system, A. J. Leggett, S. Chakravarty, A. T. Dorsey, M. P. A. Fisher, A. Garg and W. Zwerger, *Rev. Mod. Phys.* 59, 1 (1987), Erratum *Rev. Mod. Phys.* 67, 725, (1995).

[179] D. Leibfried, R. Blatt, C. Monroe and D. Wineland, Quantum dynamics of single trapped ions, *Revi. Mod. Phys.*, 75, 281–324, (2003).

[180] M. H. Levitt, *Spin Dynamics: Basics of Nuclear Magnetic Resonance*, John Wiley and Sons, New York, 2001.

[181] F. L. Lewis and V. L. Syrmos, *Optimal Control*, 2nd edition, John Wiley and Sons, New York, 1995.

[182] J. L. Li and C. F. Quiao, A necessary and sufficient criterion for the separability of quantum states, *Sci. Rep.*, 8, 1442, (2018).

[183] N. Linden, S. Popescu and A. Sudbery, Non-local properties of multiparticle density matrices, *Phys. Rev. Lett.*, 83, 243, (1999).

[184] S. Lloyd, Almost any quantum logic gate is universal, *Phys. Rev. Lett.*, 75, No. 2, 346, (1995).

[185] R. Loudon, *The Quantum Theory of Light*, 3rd edition, Oxford University Press, New York, 1973.

[186] H. Mabuchi and N. Khaneja, Principles and applications of control in quantum systems, *Int. J. Robust Nonlinear Cont.*, 15, 647-667, (2005).

[187] R. MacKenzie, Path Integrals Methods and Applications, Lectures at Rencontres du Vietnam: VIth Vietnam School of Physics, Vung Tau, Vietnam, 27 December 1999 – 8 January 2000, arXiv:quant-ph/0004090.

[188] Y. Maday and G. Turinici, New formulations of monotonically convergent quantum control algorithms, *J. Chem. Phys.*, 118, No. 18, 8191–8196, (2003).

[189] W. Magnus, On the exponential solution of differential equations for a linear operator, *Commun. Pure Appl. Math.*, 7, 649–673, (1954).

[190] Y. Makhlin, Nonlocal properties of two-qubit gates and mixed states and optimization of quantum computation, *Quant. Inf. Proc.*, 1, 243–252, (2002).

[191] H. Masahito, *Quantum Information. An Introduction*, Springer-Verlag, Berlin, 2006.

[192] E. Merzbacher, *Quantum Mechanics*, John Wiley and Sons, New York, 1961.

[193] A. Messiah, *Quantum Mechanics*, North-Holland Pub. Co., Interscience Publishers, Amsterdam, New York, 1961-2.

[194] D. Miličić, *Lectures on Lie Groups*, Lecture notes available at https://www.math.utah.edu/ milicic/Eprints/lie.pdf.

[195] M. Mirrahimi, P. Rouchon and G. Turinici, Lyapunov control of bilinear Schrödinger equations, *Automatica*, 41, 1987–1994, (2005).

[196] C. Monroe, D. Leibfried, B. E. King, D. M. Meekhof, W. M. Itano and D. J. Wineland, Simplified quantum logic with trapped ions, *Phys. Rev. A*, 55, No. 4, R2489–R2491, (1997).

[197] C. Monroe, D. M. Meekhof, B. E. King, S. R. Jefferts, W. M. Itano, D. J. Wineland and P. Gould, Resolved-sideband Raman cooling of a bound atom to the 3D zero-point energy, *Phys. Rev. Lett.*, 75, 4011, (1995a).

[198] D. Montgomery and H. Samelson, Transformation groups of spheres, *Ann. Math.*, 44, 454–470, (1943).

[199] D. Montgomery and L. Zippin, *Topological Transformation Groups*, Vol. 1, Interscience Tracts in Pure and Applied Mathematics, New York, 1955.

[200] M. Mottonen and J. J. Vartiainen, Decompositions of general quantum gates, Ch. 7 in *Trends in Quantum Computing Research*, NOVA Publishers, New York, 2006.

[201] F. D. Murnaghan, *The Theory of Group Representations*, Johns Hopkins Press, Baltimore, 1938.

[202] F. D. Murnaghan, *The Orthogonal and Symplectic Groups*, Communications of the Dublin Institute for Advanced Studies. Series A, No. 13, 1958.

[203] F. D. Murnaghan, *The Unitary and Rotation Groups*, Lecture Notes in Applied Mathematics, Vol. 3, Spartan Books, Washington, D.C., 1962.

[204] B. S. Mityagin, The zero set of a real analytic function, Preprint, https://arxiv.org/pdf/1512.07276.pdf

[205] G. Nenciu, On the adiabatic theorem of quantum mechanics, *J. Phys. A*, 13, L15–L18, (1980).

[206] D. Nguyen and T. Odagaki, Quantum and classical electrons in a potential well with uniform electric field, *Am. J. Phys.*, 55, 466–469, (1987).

[207] M. A. Nielsen and I. L. Chuang, *Quantum Computation and Quantum Information*, Cambridge University Press, Cambridge, U.K., New York, 2000.

[208] T. Norsen, *Foundations of Quantum Mechanics: An Exploration of the Physical Meaning of Quantum Theory*, Undergraduate Lecture Notes in Physics, Springer, 2017.

[209] M. Ohya and D. Petz, *Quantum Entropy and Its Use*, Springer-Verlag, Series in Theoretical and Mathematical Physics, Berlin, New York, 2004.

[210] D. Oriti (Editor), *Approaches to Quantum Gravity; Toward a New Understanding of Space, Time and Matter*, Cambridge University Press, 2009.

[211] J. A. Oteo and J. Ros, From time-ordered products to Magnus expansion, *J. of Math. Phys.*, 41, No. 5, 3268–3277, (2000).

[212] A. Peirce, M. Dahleh and H. Rabitz, Optimal control of quantum mechanical systems: existence, numerical approximations and applications, *Phys. Rev. A*, 37, No. 12, 4950, (1987).

[213] A. Peres, Separability criterion for density matrices, *Phys. Rev. Lett.*, 77, No. 8, 1413–1415, (1996).

[214] E. Pinch, *Optimal Control and the Calculus of Variations*, Oxford University Press, Oxford, U.K., New York, 1993.

[215] R. Plass, K. Egan, C. Collazo-Davila, D. Grozea, E. Landree, L. D. Marks, and M. Gajdardziska-Josifovska, Cyclic ozone identified in magnesium oxide (111) surface reconstructions, *Phys. Rev. Lett.*, 81, 4891, (1998).

[216] M. B. Plenio and S. Virmani, An introduction to entanglement measures, *Quant. Inf. Comp.*, 7, 1–51, (2007).

[217] T. Polack, H. Suchowski and D. Tannor, Uncontrollable quantum systems: a classification scheme based on Lie subalgebras, *Phys. Rev. A*, 79, 053403, (2009)

[218] L. S. Pontryagin, V. G. Boltyanski, R. S. Gamkrelidze and E. F. Mishchenko, *Mathematical Theory of Optimal Processes*, Interscience, (1962).

[219] J. Preskill, Lecture Notes on Quantum Computation, http://www.theory.caltech.edu/ preskill/ph229/

[220] J. F. Price, *Lie Groups and Compact Groups*, London Mathematical Society Lecture Notes Series, 25, Cambridge University Press, Cambridge, U.K., 1977.

[221] B. Qi, Z. Hou, L. Li, D. Dong, G.-Y. Xiang, and G.-C. Guo, Quantum state tomography via linear regression estimation, *Sci. Rep.*, 3, 3496, (2013).

[222] H. Rabitz, R. de Vivie-Riedle, M. Motzkus and K. Kompa, Whither the future of controlling quantum phenomena, *Science*, 288, 824–828, (2000).

[223] H. Rabitz, M. Hsieh and C. Rosenthal, Quantum optimally controlled transition landscapes, *Science*, 303, 998, (2004).

[224] H. Rabitz, M. Hsieh and C. Rosenthal, Landscape for optimal control of quantum-mechanical unitary transformations, *Phys. Rev. A*, 72, 052337, (2005).

[225] H. Rabitz, R. B. Wu, T. S. Ho, K. M. Tibbetts and X. Feng, Fundamental principles of control landscapes with applications to quantum mechanics, chemistry and evolution in *Recent Advances in the Theory and Application of Fitness Landscapes*, H. Richter and A. Engelbrecht eds., Emergence, Complexity and Computation, Vol. 6. Springer, Berlin, Heidelberg, 2014.

[226] E. M. Rains, Polynomial invariants of quantum codes, *IEEE T. Inform. Theory*, 46, 54, (2000).

[227] V. Ramakrishna, K. L. Flores, H. Rabitz and R. J. Ober, Quantum control by decompositions of $SU(2)$, *Phys. Rev. A*, 62, 053409, (2000).

[228] V. Ramakrishna, R. J. Ober, K. L. Flores and H. Rabitz, Control of a coupled two spin systems without hard pulses, *Phys. Rev. A*, 65, 063405, (2002).

[229] V. Ramakrishna, M. V. Salapaka, M. Dahleh, H. Rabitz and A. Peirce, Controllability of molecular systems, *Phys. Rev. A*, 51, No. 2, 960–966, (1995).

[230] C. Rangan, A. M. Bloch, C. Monroe and P. H. Bucksbaum, Control of trapped-ion quantum states with optical pulses, *Phys. Rev. Lett.*, 92, No. 11, 113004, (2004).

[231] K. N. S. Rao, *Linear Algebra and Group Theory for Physicists*, New Age International, New Delhi, 1996.

[232] B. Reichardt, The quantum adiabatic optimization algorithm and local minima, in *Proceedings of the 36-th Annual ACM Symposium on Theory of Computing (STOC)*, Chicago, IL, USA, June 13–16, pp. 502–510, 2004.

[233] S. A. Rice and M. Zhao, *Optical Control of Molecular Dynamics*, Wiley, New York, 2000.

[234] R. T. Rockafellar, *Convex Analysis*, Princeton Landmarks in Mathematics and Physics, Princeton University Press, Princeton, NJ, 1970.

[235] R. Romano, *Dissipative Dynamics in Particle Physics*, Ph. D. Thesis, Universita' degli Study di Trieste, Italy, 2002, arXiv:hep-ph/0306164.

[236] R. Romano and D. D'Alessandro, Incoherent control and entanglement for two dimensional coupled systems, *Phys. Rev. A*, 73, 022323, (2006).

[237] R. Romano and D. D'Alessandro, Environment mediated control of a quantum system, *Phys. Rev. Lett.*, 97, No. 8, 080401, (2006).

[238] W. Rossmann, *Lie Groups. An Introduction Through Linear Groups*, Oxford Graduate Texts in Methematics-5, Oxford University Press Inc., Oxford, 2002.

[239] W. Rudin, *Real and Complex Analysis*, 3rd edition, Mc Graw-Hill, New York, 1987.

[240] O. Rudolph, Further results on the cross norm criterion for separability, *Quantum Inf. Process.*, **4**, 219–239, (2005).

[241] O. Rudolph, A note on 'A matrix realignment method for recognizing entanglement', arXiv:quant-ph/0205017 v1.

[242] O. Rudolph, Some properties of the computable cross-norm criterion for separability, *Phys. Rev. A*, 67, 032312, (2003).

[243] P. Rungta, V. Buzek, C. M. Caves, M. Hillery and G. J. Milburn, Universal state inversion and concurrence in arbitrary dimensions, *Phys. Rev. A*, 64, 042315, (2001).

[244] A. A. Sagle and R. E. Walde, *Introduction to Lie Groups and Lie Algebras*, Academic Press, New York, 1973.

[245] J. J. Sakurai, *Modern Quantum Mechanics*, Addison-Wesley Pub. Co., Reading, MA, c1994.

[246] J. Salomon, Limit points of the monotonic schemes in quantum control, in *Proceedings of the 44-st IEEE Conference on Decision and Control*, Seville, December 2005.

[247] C. Sayrin, I. Dotsenko, X. Zhou, B. Peaudecerf, T. Rybarczyk, S. Gleyzes, P. Rouchon, M. Mirrahimi, H. Amini, M. Brune, J. M. Raimond and S. Haroche, Real-time quantum feedback prepares and stabilizes photon number states, *Nature*, 477, 73–77, (2011).

[248] G. Schaller, S. Mostame and R. Schützold, General error estimate for adiabatic quantum computing, *Phys. Rev. A*, 73, 062307, (2006).

[249] S. G. Schirmer, Quantum control using Lie group decompositions, in *Proceedings of the 40-th IEEE Conference on Decision and Control*, 4–7 Dec. 2001, pp. 298–303, Vol. 1.

[250] S. G. Schirmer, A. D. Greentree, V. Ramakrishna and H. Rabitz, Constructive control of quantum systems using factorization of unitary operators, *J. Phys. A,* 35, 8315–8339, (2002).

[251] S. G. Schirmer, J. V. Leahy and A. I. Solomon, Degrees of controllability for quantum systems and applications to atomic systems, *J. Phys. A*, 35, 4125–4141, (2002).

[252] A. N. Sengupta, *Representing Finite Groups; A Semisimple Introduction*, Springer Verlag, New York, 2012.

[253] M. Shapiro and P. Brumer, Quantum control of chemical reactions, *J. Chem. Soc., Faraday Trans.*, 93, No. 7, 1263–1277, (1997).

[254] M. Shapiro and P. Brumer, *Quantum Control of Molecular Processes*, 2nd edition, Wiley-VCH Verlag GmbH & Co. KGaA, Weinheim, Germany, 2012.

[255] A. E. Siegman, Laser beams and resonators: the 1960s, *IEEE J. Sel. Top. Quant.* 6, No. 6, 1380–1388, (2000).

[256] A. E. Siegman, Laser beams and resonators: Beyond the 1960s, *IEEE J. Sel. Top. Quant.* 6, No. 6, 1389–1399. (2000).

[257] C. P. Slichter, *Principles of Magnetic Resonance*, Springer-Verlag, New York, 1996.

[258] T. A. Springer, *Invariant Theory*, Lecture Notes in Mathematics, Vol. 585, Springer, Berlin, 1997.

[259] A. Steane, The ion trap quantum information processor, *Appl. Phys. B*, 64, 623–642, (1997).

[260] S. Sternberg, *Lectures on Differential Geometry*, Prentice Hall, Englewood Cliffs, NJ, 1964.

[261] W. Strauss, *Partial Differential Equations: An Introduction*, 2nd edition, John Wiley and Sons, 2008.

[262] AiniSyahida Sumairi, S. N. Hazmin and C. H. Raymond Ooi, Quantum entanglement criteria, *J. Mod. Optic.*, 60, No. 7, 589–597.

[263] H. J. Sussmann, Lie brackets, real analyticity and geometric control, in *Differential Geometric Control Theory*, R. W. Brockett, R. S. Millman and H. J. Sussmann eds., pp. 1–116, Birkhauser, Boston, 1983.

[264] H. J. Sussmann and V. Jurdjevic, Controllability of nonlinear systems, *J. Differ. Equations*, 12, 95–116, (1972).

[265] J. R. Sylvester, Determinants of block matrices, *Math. Gazette*, 84, No. 501, 460–467, (2000).

[266] A. Szabo and N. S. Ostlund, *Modern Quantum Chemistry: Introduction to Advanced Electronic Structure Theory*, McGraw-Hill Publishing Company, New York, 1989.

[267] P. I. Tamborenea and H. Metiu, Localization and entanglement of two interacting electrons in a quantum-dot molecule, *Europhys. Lett.*, 53, No. 6, 776–782, (2001).

[268] D. J. Tannor, V. Kazakov and V. Orlov, Control of photochemical branching. Novel procedures for finding optimal pulses and global upper bounds, in *Time Dependent Quantum Molecular Dynamics*, J. Broeckhove and L. Lathouwers eds., pp. 347–360, Plenum, 1992.

[269] Temple University. Temple researcher attempting to create cyclic ozone. *Science Daily*, 8 February 2005.

[270] Y. S. Teo, H. Zhu, B.-G. Englert, J. Řeháček and Z. Hradil, Quantum-state reconstruction by maximizing likelihood and entropy, *Phys. Rev. Lett.*, 107, No. 2, 020404, (2011).

[271] B. Terhal, Detecting quantum entanglement, *Theor. Comput. Sci.*, 287, No. 1, 313, (2002).

[272] S. H. Tersigni, P. Gaspard and S. Rice, On using shaped light pulses to control the selectivity of product formation in a chemical reaction: an application to a multiple level system, *J. Chem. Phys.*, 93, No. 3, 1670–1680, (1990).

[273] W. Tung, *Group Theory in Physics*, World Scientific, Singapore, 1985.

[274] G. Turinici, Controlabilite exacte de la population des etats propres dans les systemes quantiques bilineaires (French), *C. R. Acad. Sci. Paris Ser. I Math.*, 330, No. 4, 327–332, (2000).

[275] Gabriel Turinici and Herschel Rabitz, Quantum wave function control-lability, *Chem. Phys.*, 267, 1–9, (2001).

[276] G. Turinici and H. Rabitz, Wavefunction controllability for finite-dimensional bilinear quantum systems, *J. Phys. A: Math. Gen.*, 36, 2565–2576, (2003).

[277] A. Uhlmann, Entropy and optimal decompositions of states relative to a maximal commutative subalgebra, *Open Systems Inf. Dyn.*, 5, No. 3, 209–228, (1998).

[278] A. Uhlmann, Optimizing entropy relative to a channel or a subalgebra, *Open Systems Inf. Dyn.*, 5, 209–227, (1998).

[279] A. Uhlmann, Fidelity and concurrence of conjugated states, *Phys. Rev. A*, 62, 032307, (2000).

[280] U. Vaidya, D. D'Alessandro and I. Mezić, Control of Heisenberg spin systems; Lie algebraic decompositions and action-angle variables, in *Proceedings 42-nd Conference on Decision and Control*, Vol. 4, pp. 4174–4178, Maui HI, Dec. 2003.

[281] W. van Dam, M. Mosca and U. Vazirani, How powerful is adiabatic quantum computation?, in *Proceedings 42-nd IEEE Symposium on Foundations of Computer Science*, Las Vegas, NV, 2001, 279–287.

[282] R. Van Handel, J. K. Stockton and H. Mabuchi, Feedback control of quantum state reduction, *IEEE Trans. Automat. Control*, 50, No. 6, 768–780, (2005).

[283] P. Vettori, On the convergence of a feedback control strategy for multi-level quantum systems, in *Proceedings 15-th International Symposium MTNS 2002*, University of Notre Dame, U.S.A.

[284] Y. Yamamoto and A. Imamoglu, *Mesoscopic Quantum Optics*, Wiley-Interscience Publication, New York, 1999.

[285] X. Wang, D. Burgarth and S. G. Schirmer, Subspace controllability of spin $\frac{1}{2}$ chains with symmetries, *Phys. Rev. A,* 94, 052319, (2016).

[286] X. Wang, P. Pemberton-Ross and S. G. Schirmer, Symmetry and controllability for spin networks with a single node control, *IEEE T. Automat. Contr.*, 57, No. 8, 1945–1956, (2012).

[287] X. Wang and S. G. Schirmer, Analysis of effectiveness of Lyapunov control for non-generic quantum states, *IEEE T. Automat. Contr.*, 55, No. 6, 1406–1411, (2010).

[288] X. Wang and S. G. Schirmer, Analysis of Lyapunov for control of quantum states, *IEEE T. Automat. Contr.*, 55, No. 10, 2259–2270, (2010).

[289] Y. Wang, D. Dong, B. Qi, J. Zhang, I. R. Petersen and H. Yonezawa, A quantum Hamiltonian identification algorithm: computational complexity and error analysis, *IEEE T. Automat. Contr.*, 63, No. 5, 1388–1403, (May 2018).

[290] F. W. Warner, *Foundations of Differentiable Manifolds and Lie Groups*, Springer, New York, 1983.

[291] N. Weaver, On the universality of almost every quantum logic gate, *J. Math. Phys*, 41, No. 1, 240–243, (2000).

[292] A. M. Weiner, D. E. Leaird, J. S. Patel and J. R. Wullert II, Programmable shaping of femtosecond optical pulses by use of 128-element liquid crystal phase modulator, *IEEE J. Quantum Elect.*, 28, No. 4, 908–920, (1992).

[293] H. Weyl, *The Classical Groups*, Princeton University Press, Princeton, NJ, 1939.

[294] A. M. Weiner, Ultrafast optical pulse shaping: a tutorial review, *Opt. Communi.*, 284, 3669–3692, (2011).

[295] H. M. Wiseman, *Quantum Trajectories and Feedback*, Ph.D. Thesis, Physics Department, University of Queensland, 1994.

[296] H. M. Wiseman, *Quantum Measurement and Control*, Cambridge University Press, New York, 2010 (paperback edition 2014).

[297] P. Woit, *Quantum Theory, Groups and Representations; An Introduction*, Springer International Publishing, 2017.

[298] A. Wong and N. Christensen, A potential multipartite entanglement measure, *Phys. Rev. A*, 63, 044301, (2001).

[299] W. K. Wootters, Entanglement of formation of an arbitrary state of two qubits, *Phys. Rev. Lett.*, 80, 2245–2248, (1998).

[300] W. K. Wootters, Entanglement of formation and concurrence, *Quantum Inform. Compu.*, 1, No. 1, 27–44, (2001).

[301] R. B. Wu, H. Rabitz and M. Hsieh, Characterization of the critical submanifolds in quantum ensemble control landscapes, *J. Phys. A: Math. Theor.*, 41, 015006, (2008).

[302] K. Wüthrich, *NMR in Biological Research: Peptides and Proteins*, North-Holland, Amsterdam, 1976.

[303] R. Zare, Laser control of chemical reactions, *Science*, 279, 1875–1879, (1998).

[304] R. Zeier and T. Schulte-Herbrüggen, Symmetry principles in quantum systems theory, *J. Math. Phys.*, 52, 113510, (2011).

[305] Z. Zimborás, R. Zeier, T. Schulte-Herbrüggen and D. Burgarth, Symmetry criteria for quantum simulability of effective interactions, *Phys. Rev. A*, 92, 042309, (2015).

[306] C. J. Zhang, Y. S. Zhang, S. Zhang and G. C. Guo, Entanglement detection beyond the cross-norm or realignment criterion, *Phys. Rev. A*, 77, 060301(R), (2008).

[307] J. Zhang, J. Vala, S. Sastry and K. Whaley, Geometric theory of non-local two-qubit operations, *Phys. Rev. A*, 67, 042313, (2003).

[308] J. Zhang, Y. Liu, R. Wu, K. Jacobs and F. Nori, Quantum feedback: theory, experiments, and applications, *Phys. Rep.* 679, 1–60, (2017).

[309] W. Zhu and H. Rabitz, A rapid monotonically convergent iteration algorithm for quantum optimal control over the expectation value of a positive definite operator, *J. Chem. Phy.*, 109, No. 2, 385–401, (1998).

[310] W. Zhu and H. Rabitz, Attaining optimal controls for manipulating quantum systems, *Int. J. Quantum Chem.*, 93, 50–58, (2003).

[311] W. H. Zurek, Decoherence and the transition from quantum to classical, *Phys. Today*, 44, 10, 36, (1991)

[312] Royal Swedish Academy of Sciences, *Measuring and Manipulating Individual Quantum Systems*, Scientific Background on the Nobel Prize in Physics, 2012, October 9, 2012.

Index

398 *Index*

Printed in the United States
by Baker & Taylor Publisher Services